# Advanced Materials Structures for Sound and Vibration Damping

# Advanced Materials Structures for Sound and Vibration Damping

Guest Editor
**Martin Vašina**

Basel • Beijing • Wuhan • Barcelona • Belgrade • Novi Sad • Cluj • Manchester

*Guest Editor*
Martin Vašina
Department of Physics and
Materials Engineering
Tomas Bata University in Zlín
Zlín
Czech Republic

*Editorial Office*
MDPI AG
Grosspeteranlage 5
4052 Basel, Switzerland

This is a reprint of the Special Issue, published open access by the journal *Materials* (ISSN 1996-1944), freely accessible at: www.mdpi.com/journal/materials/special_issues/Materials_Damping.

For citation purposes, cite each article independently as indicated on the article page online and using the guide below:

Lastname, A.A.; Lastname, B.B. Article Title. *Journal Name* **Year**, *Volume Number*, Page Range.

ISBN 978-3-7258-3052-7 (Hbk)
ISBN 978-3-7258-3051-0 (PDF)
https://doi.org/10.3390/books978-3-7258-3051-0

© 2025 by the authors. Articles in this book are Open Access and distributed under the Creative Commons Attribution (CC BY) license. The book as a whole is distributed by MDPI under the terms and conditions of the Creative Commons Attribution-NonCommercial-NoDerivs (CC BY-NC-ND) license (https://creativecommons.org/licenses/by-nc-nd/4.0/).

# Contents

About the Editor ..................................................... vii

Preface ............................................................. ix

**Martin Vašina**
Advanced Materials Structures for Sound and Vibration Damping
Reprinted from: *Materials* **2022**, *15*, 1295, https://doi.org/10.3390/ma15041295 .......... 1

**Shuichi Sakamoto, Kentaro Toda, Shotaro Seino, Kohta Hoshiyama and Takamasa Satoh**
Theoretical and Experimental Analyses on the Sound Absorption Coefficient of Rice and Buckwheat Husks Based on Micro-CT Scan Data
Reprinted from: *Materials* **2023**, *16*, 5671, https://doi.org/10.3390/ma16165671 .......... 3

**Meisam Ansari, Fabiola Tartaglione and Carsten Koenke**
Experimental Validation of Dynamic Response of Small-Scale Metaconcrete Beams at Resonance Vibration
Reprinted from: *Materials* **2023**, *16*, 5029, https://doi.org/10.3390/ma16145029 .......... 21

**Jianjun Zhou, Bowen Yang, Shuaiyuan Li and Junzhou Huo**
Fretting Fatigue Life Prediction of Dovetail Structure Based on Plastic Effect and Sensitivity Analysis of Influencing Factors
Reprinted from: *Materials* **2023**, *16*, 3521, https://doi.org/10.3390/ma16093521 .......... 38

**Takamasa Satoh, Shuichi Sakamoto, Takunari Isobe, Kenta Iizuka and Kastsuhiko Tasaki**
Mathematical Model for Estimating the Sound Absorption Coefficient in Grid Network Structures
Reprinted from: *Materials* **2023**, *16*, 1124, https://doi.org/10.3390/ma16031124 .......... 61

**Shuichi Sakamoto, Kyosuke Suzuki, Kentaro Toda and Shotaro Seino**
Estimation of the Acoustic Properties of the Random Packing Structures of Granular Materials: Estimation of the Sound Absorption Coefficient Based on Micro-CT Scan Data
Reprinted from: *Materials* **2022**, *16*, 337, https://doi.org/10.3390/ma16010337 .......... 80

**Shuichi Sakamoto, Kyosuke Suzuki, Kentaro Toda and Shotaro Seino**
Mathematical Models and Experiments on the Acoustic Properties of Granular Packing Structures (Measurement of Tortuosity in Hexagonal Close-Packed and Face-Centered Cubic Lattices)
Reprinted from: *Materials* **2022**, *15*, 7393, https://doi.org/10.3390/ma15207393 .......... 93

**Fei Yang, Enshuai Wang, Xinmin Shen, Xiaonan Zhang, Qin Yin and Xinqing Wang et al.**
Optimal Design of Acoustic Metamaterial of Multiple Parallel Hexagonal Helmholtz Resonators by Combination of Finite Element Simulation and Cuckoo Search Algorithm
Reprinted from: *Materials* **2022**, *15*, 6450, https://doi.org/10.3390/ma15186450 .......... 107

**Mateusz Żurawski and Robert Zalewski**
Experimental Studies on Adaptive-Passive Symmetrical Granular Damper Operation
Reprinted from: *Materials* **2022**, *15*, 6170, https://doi.org/10.3390/ma15176170 .......... 127

**Shuichi Sakamoto, Juung Shin, Shota Abe and Kentaro Toda**
Addition of Two Substantial Side-Branch Silencers to the Interference Silencer by Incorporating a Zero-Mass Metamaterial
Reprinted from: *Materials* **2022**, *15*, 5140, https://doi.org/10.3390/ma15155140 .......... 141

**Shuichi Sakamoto, Ryo Iizuka and Takumi Nozawa**
Effect of Sheet Vibration on the Theoretical Analysis and Experimentation of Nonwoven Fabric Sheet with Back Air Space
Reprinted from: *Materials* **2022**, *15*, 3840, https://doi.org/10.3390/ma15113840 . . . . . . . . . . **155**

**Christin Zacharias, Carsten Könke and Christian Guist**
A New Efficient Approach to Simulate Material Damping in Metals by Modeling Thermoelastic Coupling
Reprinted from: *Materials* **2022**, *15*, 1706, https://doi.org/10.3390/ma15051706 . . . . . . . . . . **166**

**Nabeel Taiseer Alshabatat**
Natural Frequencies Optimization of Thin-Walled Circular Cylindrical Shells Using Axially Functionally Graded Materials
Reprinted from: *Materials* **2022**, *15*, 698, https://doi.org/10.3390/ma15030698 . . . . . . . . . . . **183**

**Lubomír Lapčík, Martin Vašina, Barbora Lapčíková and Yousef Murtaja**
Effect of Conditioning on PU Foam Matrix Materials Properties
Reprinted from: *Materials* **2021**, *15*, 195, https://doi.org/10.3390/ma15010195 . . . . . . . . . . . **197**

**Marek Pöschl and Martin Vašina**
Study of the Mechanical, Sound Absorption and Thermal Properties of Cellular Rubber Composites Filled with a Silica Nanofiller
Reprinted from: *Materials* **2021**, *14*, 7450, https://doi.org/10.3390/ma14237450 . . . . . . . . . . **207**

# About the Editor

**Martin Vašina**

Martin Vašina is an academic and researcher at Tomas Bata University in Zlín and at VŠB-Technical University of Ostrava in the Czech Republic. He deals mainly with the mechanical, vibration isolation, sound absorption, and light-related technical properties of various materials. Another part of his research is focused on fluid mechanics. He is a member of the Czech Society for Mechanics.

# Preface

This reprint provides a comprehensive overview of modern research in dynamics, acoustics, and materials engineering, focusing on vibration suppression, material optimization, and sound absorption improvement. It encompasses both theoretical and experimental studies, including the optimization of natural frequencies of thin-walled cylindrical shells, new methods for modeling material damping, and the integration of acoustic metamaterials. Practical applications cover the design of silencers, the analysis of granular material acoustic properties, and the study of the mechanical and thermal characteristics of composites. The motivation for this work stems from the need to address challenges in vibrations and acoustics using innovative approaches that combine simulations, experiments, and advanced theoretical models. The reprint is intended for researchers, engineers, and students seeking inspiration and practical solutions in these fields. This achievement was made possible through collaboration among authors from various disciplines and the support of institutions, colleagues, and experimental teams, to whom we extend our gratitude. I hope this compilation contributes to further progress and innovation.

**Martin Vašina**
*Guest Editor*

*Editorial*

# Advanced Materials Structures for Sound and Vibration Damping

Martin Vašina

Department of Physics and Materials Engineering, Tomas Bata University in Zlín, Vavrečkova 5669, 760 01 Zlín, Czech Republic; vasina@utb.cz

**Citation:** Vašina, M. Advanced Materials Structures for Sound and Vibration Damping. *Materials* **2022**, *15*, 1295. https://doi.org/10.3390/ma15041295

Received: 31 January 2022
Accepted: 7 February 2022
Published: 10 February 2022

**Publisher's Note:** MDPI stays neutral with regard to jurisdictional claims in published maps and institutional affiliations.

**Copyright:** © 2022 by the author. Licensee MDPI, Basel, Switzerland. This article is an open access article distributed under the terms and conditions of the Creative Commons Attribution (CC BY) license (https:// creativecommons.org/licenses/by/ 4.0/).

The studies of sound and vibration are closely related. Noise is an unwanted sound that is considered unpleasant, loud, or disturbing to the organs of hearing. Mechanical vibration is caused by the oscillation of a mechanical or structural system around its equilibrium position. Noise and mechanical vibration are, in many cases, among the negative environmental factors. They can have an adverse effect on human health, production accuracy, the life of processing equipment and tools, labor protection, etc. For these reasons, it is necessary to eliminate unwanted noise and mechanical vibration by appropriate means. There are various options to reduce excessive noise and mechanical vibration. In general, it is necessary to convert the excessive mechanical energy of oscillating motion and acoustic energy into other types of energy, especially heat.

This Special Issue focuses on the collection of scientific papers dealing with the development of advanced material structures for sound and vibration damping. Various measures that can help eliminate unwanted noise and mechanical vibration are explored in this Special Issue.

In [1], the authors investigated the mechanical, acoustic, and thermal properties of cellular rubber composites, which were produced with different concentrations of silica nanofillers at the same blowing agent concentration. It was found that a higher concentration of silica nanofillers generally led to an increase in the mechanical stiffness and thermal conductivity and to a decrease in the sound absorption and thermal degradation of the investigated rubber composites. It was also possible to find a correlation between the mechanical stiffness of the tested rubber composites evaluated using conventional and vibroacoustic measurement techniques. Vibration damping and sound absorption methods are relatively simple, inexpensive, fast, and non-destructive compared to the conventional methods used to determine the mechanical stiffness of solids. Therefore, the vibroacoustic methods can also be easily applied to compare the mechanical stiffness of different materials.

Monkova et al. [2] studied the sound absorption properties of 3D printed open porous acrylonitrile butadiene styrene (ABS) specimens that were produced with four different lattice structures (Cartesian, Starlit, Rhomboid, and Octagonal). In this work, various factors affecting the sound absorption properties were evaluated, namely, the type of lattice structure, the excitation frequency of acoustic waves, the specimen thickness, and the air gap size behind the sound-absorbing materials inside the acoustic impedance tube. It could be concluded that the application of 3D printed materials is promising in terms of sound absorption. Three-dimensional printing technology allows the production of lightweight materials of various shapes and structures compared to other manufacturing technologies, which saves time and energy and reduces the weight of materials.

Muhazeli et al. [3] dealt with the sound-damping properties of a magneto-induced (called magnetorheological) foam containing different concentrations of carbonyl iron particles. It was found that the addition of a magnetic field led to a peak frequency shift from the middle to higher frequency ranges. Therefore, the change in the magnetic field has a significant influence on sound absorption, and the MR foam could be applied as a noise-controllable material.

The vibration-damping properties of environmentally friendly concrete using crumb rubber recycled from waste rubber tires were investigated in [4]. It could be concluded that a mix of 180- and 400-micron crumb rubber significantly improved the concrete's damping ratio (improvement of 100%) compared to normal concrete. For this reason, the use of crumb rubber in concrete presents one of the best alternatives while dealing with rubber waste, as it will help both environmental protection and the reduction in railway sleeper costs.

The effect of conditioning on the vibration damping of polyurethane (PU) foams, which were subjected to conditioning at two different temperatures (45 and 80 °C) and relative humidity values (45 and 80%) for different time intervals, was evaluated in [5]. It was found that the conditioning process had a significant influence on the properties of the tested PU foams. In general, the thermal degradation and permanent deformation of the PU foams increased with the increasing conditioning time and temperature. The higher permanent deformation was accompanied by a decrease in Young's modulus of elasticity and subsequently confirmed by non-destructive dynamical-mechanical vibration tests, which confirmed the samples' higher vibration damping, resulting in the loss of elasticity.

The author of [6] investigated the optimization of the natural frequencies of circular cylindrical shells using axially functionally graded materials. The constituents of the functionally graded materials (FGMs) were graded in the axial direction. It could be concluded that the spatial change in material properties changed the structural stiffness and mass, which then affected the structure's natural frequencies. Therefore, axially FGMs can be a useful technique for the optimization of natural frequencies.

Martin Vašina is an academic and researcher at Tomas Bata University in Zlín and at VŠB-Technical University of Ostrava in the Czech Republic. He deals mainly with mechanical, vibration isolation, sound absorption and light-technical properties of various materials. Another part of his research is focused on fluid mechanics. He is a member of the Czech Society for Mechanics.

**Funding:** This research received no external funding.

**Acknowledgments:** The Guest Editor expresses his gratitude for the valuable contributions by all authors, referees, the editorial team of *Materials*, and especially Lili Li for helping to manage this Special Issue.

**Conflicts of Interest:** The author declares no conflict of interest.

## References

1. Pöschl, M.; Vašina, M. Study of the Mechanical, Sound Absorption and Thermal Properties of Cellular Rubber Composites Filled with a Silica Nanofiller. *Materials* **2021**, *14*, 7450. [CrossRef] [PubMed]
2. Monkova, K.; Vasina, M.; Monka, P.P.; Kozak, D.; Vanca, J. Effect of the Pore Shape and Size of 3D-Printed Open-Porous ABS Materials on Sound Absorption Performance. *Materials* **2020**, *13*, 4474. [CrossRef] [PubMed]
3. Muhazeli, N.S.; Nordin, N.A.; Ubaidillah, U.; Mazlan, S.A.; Abdul Aziz, S.A.; Nazmi, N.; Yahya, I. Magnetic and Tunable Sound Absorption Properties of an In-Situ Prepared Magnetorheological Foam. *Materials* **2020**, *13*, 5637. [CrossRef] [PubMed]
4. Kaewunruen, S.; Li, D.; Chen, Y.; Xiang, Z. Enhancement of Dynamic Damping in Eco-Friendly Railway Concrete Sleepers Using Waste-Tyre Crumb Rubber. *Materials* **2018**, *11*, 1169. [CrossRef] [PubMed]
5. Lapčík, L.; Vašina, M.; Lapčíková, B.; Murtaja, Y. Effect of Conditioning on PU Foam Matrix Materials Properties. *Materials* **2022**, *15*, 195. [CrossRef] [PubMed]
6. Alshabatat, N.T. Natural Frequencies Optimization of Thin-Walled Circular Cylindrical Shells Using Axially Functionally Graded Materials. *Materials* **2022**, *15*, 698. [CrossRef]

Article

# Theoretical and Experimental Analyses on the Sound Absorption Coefficient of Rice and Buckwheat Husks Based on Micro-CT Scan Data

Shuichi Sakamoto [1,*], Kentaro Toda [2], Shotaro Seino [2], Kohta Hoshiyama [2] and Takamasa Satoh [3]

[1] Department of Engineering, Niigata University, Ikarashi 2-no-cho 8050, Nishi-ku, Niigata 950-2181, Japan
[2] Graduate School of Science and Technology, Niigata University, Ikarashi 2-no-cho 8050, Nishi-ku, Niigata 950-2181, Japan; f21b105g@gmail.com (K.T.); t17a057k@gmail.com (S.S.); kohtahoshiyama@gmail.com (K.H.)
[3] Fukoku Co., Ltd., 6 Showa Chiyoda-machi, Oura-gun, Gunma 370-0723, Japan; takamasa_satoh@fukoku-rubber.co.jp
\* Correspondence: sakamoto@eng.niigata-u.ac.jp; Tel.: +81-25-262-7003

**Abstract:** In this study, the sound absorption coefficients of rice and buckwheat husks were estimated. Computed tomography (CT) images were processed to determine the circumference and surface area of voids in the granular material, and the normal incident sound absorption coefficients were derived. In addition, the tortuosity, which expresses the complexity of the sound wave propagation through the structure, was measured for each material. The theoretical sound absorption coefficients were then compared to the measured sound absorption coefficients with and without consideration of the tortuosity. A correction factor was used to bring the surface area of the granular material closer to the actual surface area and observed that the tortuosity obtained theoretical values that matched the trend of the measured values. These results indicate that using CT images to estimate the sound absorption coefficient is a viable approach.

**Keywords:** rice husks; buckwheat husks; micro-CT scan; tortuosity; sound absorption coefficient

**Citation:** Sakamoto, S.; Toda, K.; Seino, S.; Hoshiyama, K.; Satoh, T. Theoretical and Experimental Analyses on the Sound Absorption Coefficient of Rice and Buckwheat Husks Based on Micro-CT Scan Data. *Materials* **2023**, *16*, 5671. https://doi.org/10.3390/ma16165671

Academic Editor: Martin Vašina

Received: 13 July 2023
Revised: 11 August 2023
Accepted: 12 August 2023
Published: 17 August 2023

**Copyright:** © 2023 by the authors. Licensee MDPI, Basel, Switzerland. This article is an open access article distributed under the terms and conditions of the Creative Commons Attribution (CC BY) license (https:// creativecommons.org/licenses/by/ 4.0/).

## 1. Introduction

Plant biomass from grains, vegetables, and trees is used for various purposes, including food, fuel, and building materials. The shape and size of these materials indicate that they may also have useful sound absorption effects. For example, rice straw has a microtubular structure [1–3]. Bastos et al. made sound-absorbing panels from coconut fiber [4], Gabriel et al. did the same with corn fiber [5], Khai et al. used oil palm fiber [6], and Sezgin used discarded tea leaves [7]. Umberto et al. made sound absorbers from pulverized cane [8], and Rubén et al. used cork bark [9]. Moreover, previous reports of biomass sound-absorbing materials can be found in various instances, such as natural bamboo fibers [10], straw [11], sugarcane wasted fibers [12], and Tatami mats [13]. These studies were all based on the principle that the boundary-layer viscosity of air at the walls of granular/flake-filled structures attenuates the energy of incident sound waves. The continuous voids of such structures cause them to exhibit the same acoustic behavior as porous materials, and the acoustic properties can be adjusted according to the layer thickness, grain size, and packing structure [14,15].

Japan generates about 1.6 million tons of rice husks [16] and 20,000–30,000 tons of buckwheat husks [17] annually, of which much is incinerated and disposed of without being utilized. Rice husks are a byproduct from the production of rice, which is a staple food of Japan and is widely dispersed throughout the country, so a stable supply can be expected. Buckwheat husks are commonly used to make pillows, but this application has been decreasing, so most husks are at present disposed of as waste [17]. Recently, restrictions have been placed on burning these materials in the open due to environmental concerns, which has made their disposal a serious problem. The sound absorption of rice and buckwheat husks has previously been reported [1], but the sound absorption coefficient was only measured. A method of estimating the sound absorption coefficient has not been developed. This is because the packing structures of rice and buckwheat husks do not have a periodic arrangement and vary with the grain size, which makes constructing a mathematical model difficult. However, rice and buckwheat husks maintain a stable flake shape after threshing and are expected to exhibit acoustic properties as granular materials with stable shape, such as porous materials. This study elaborates on the sound absorption principle caused by the viscous friction of the boundary layer of the wall surface, which has been discussed in a previous study [15]. The sound wave attenuation discussed in this paper is independent of microscale molecular structures. Meanwhile, nano-fibers, such as nanocellulose fibers, act to enhance the sound absorption properties of nonwoven fabrics [18].

In this study, the sound absorption coefficients of rice and buckwheat husks were estimated. Computed tomography (CT) images were processed to determine the circumference and surface area of voids in the granular material, and the normal incident sound absorption coefficients were derived. However, micro-CT scans have only been used to observe plant structure conventionally [19]. In addition, the tortuosity, which expresses the complexity of the sound wave propagation through the structure, was measured for each material. The theoretical sound absorption coefficients were then compared to the measured sound absorption coefficients with and without consideration of the tortuosity.

## 2. Experimental Measurements

### 2.1. Sound Absorption

Two types of biomass materials were tested for their sound absorption: rice husks and buckwheat husks. Both materials were from Japan. Figure 1 shows the measurement samples, and Table 1 presents the specifications. Each sample was used to directly fill an aluminum alloy tube with an inner diameter of 29 mm and length of 20 mm.

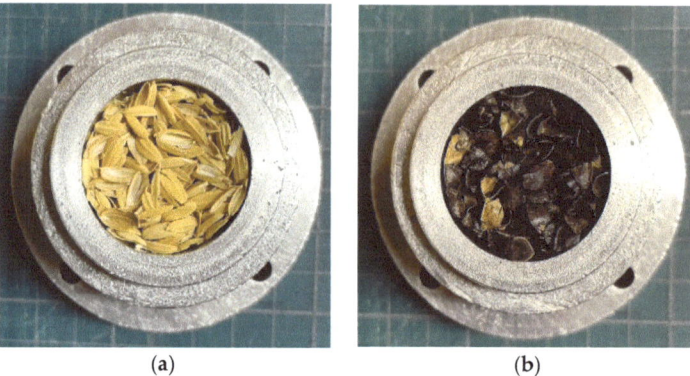

(a)      (b)

**Figure 1.** Samples: (**a**) rice husks; (**b**) buckwheat husks.

**Table 1.** Sample specifications.

| Material | Average Grain Size (mm) | Mass per Grain (mg) | Bulk Density (kg/m$^3$) | Measured Tortuosity |
|---|---|---|---|---|
| Rice husk | 7.3 × 3.6 | 2.15 | 105.36 | 1.92 |
| Buckwheat husk | 5.7 × 4.1 | 4.54 | 110.66 | 1.74 |

During threshing, rice and buckwheat husks separate on their own owing to their flake shapes (Figure 1), and stable shapes can be obtained. This study focused on rice and buckwheat husks owing to their ability to form a stable shape through this process. Therefore, this study proposes to leverage these naturally occurring and readily obtainable threshed shapes. In general, rice and buckwheat husks do not attract insect damage after being washed and dried, as they retain minimal seed powder. Note that, if the rice husks were heated to ash, they could manifest the sound absorption properties of finer-grained powders [20].

As shown in Figure 2, a Brüel and Kjær Type 4206 (Nærum, Denmark) two-microphone impedance measurement tube was used to measure the normal-incident sound absorption coefficient. A sound wave based on a sinusoidal signal from a signal generator with a built-in fast Fourier-transform (FFT) analyzer DS-3000 fabricated by Ono Sokki (Yokohama, Japan) was output into the measurement tube containing the sample. The sound pressure in the tube was measured by the two microphones, and the transfer function between the sound pressure signals was calculated by using the FFT analyzer. The measured transfer function was used to derive the normal incident sound absorption coefficient in accordance with ISO 10534-2 [21]. The critical frequency at which a plane wave forms depends on the inner diameter of the acoustic tube. Because the tube used in this study had an inner diameter of 29 mm, the upper limit of the measurement range was 6400 Hz. The voltage of the input signal to the loudspeaker was 0.2 V.

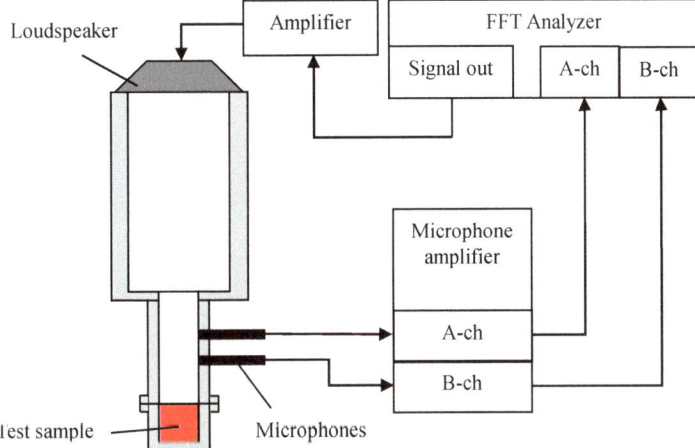

**Figure 2.** Configuration of the two-microphone impedance tube used for sound absorption measurements.

## 2.2. Tortuosity

The tortuosity is an acoustic parameter that expresses the complexity of the path of a sound wave passing through a sound-absorbing material with a complex internal geometry. In this study, ultrasonic sensors were used to measure the tortuosity of the materials so that its effect on sound absorption can be considered. The tortuosity $\alpha_\infty$ can be derived from the velocity of sound in air $c_0$ and the apparent velocity of sound in a material $c$ [22]:

$$\alpha_\infty = \left(\frac{c_0}{c}\right)^2 \tag{1}$$

If the sample material contains no obstacles and the sound wave travels in a straight path, then $\alpha_\infty = 1$ because the $c_0 = c$. If the path is complex, then $c$ decreases, and $\alpha_\infty$ becomes >1. In other words, a greater tortuosity indicates a longer path for sound waves within a structure. This effect is similar to an increase in the material thickness.

To measure the tortuosity, sound waves were transmitted through the bottom of the sample holder, as shown in Figure 3. The wire mesh bottom allowed only sound waves to penetrate the sample, and no granular material fell. The sample holder was fabricated from light-cured resin using the 3D printer Form2 manufactured by Formlabs Inc. (Somerville, MA, USA). Figure 4 shows the tortuosity measurement setup.

**Figure 3.** Samples for tortuosity measurement: (**a**) rice husks; (**b**) buckwheat husks.

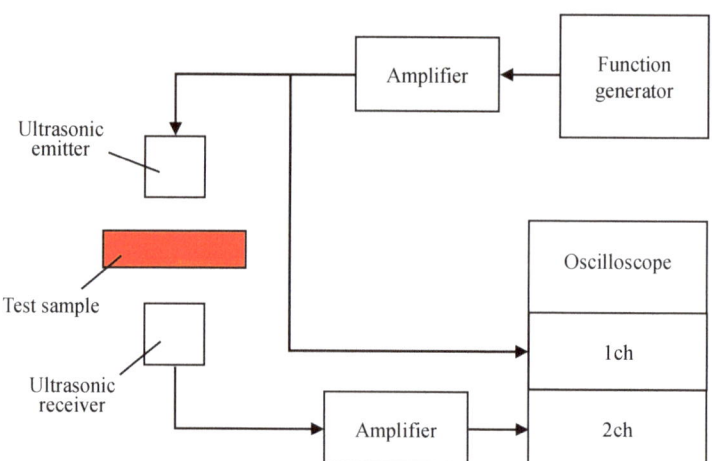

**Figure 4.** Tortuosity measurement setup.

The tortuosity was measured by the same method reported previously [14]. Ultrasonic sensors with central frequencies of 32.7, 40, 58, 110, 150, 200, and 300 kHz were used. First, the tortuosity $\alpha_\infty$ was measured for each filling structure at each frequency. Then, the inverse of the square root of the measurement frequency was taken as the value on the horizontal axis, and the tortuosity $\alpha_\infty$ at each frequency was plotted as the value on the vertical axis. The least-squares method was applied to obtain a linear approximation of these points, which yielded a straight line that increased steadily to the right. The tortuosity $\alpha_\infty$ of the material was defined as the extreme value of the tortuosity when the frequency of the approximated line was set to infinity, i.e., the $y$-intercept of the graph. To improve the signal-to-noise ratio and measurement accuracy, the results of 300 measurements were summed synchronously. The signals were measured at a resolution of 16 bits.

Figures 5 and 6 show the measured tortuosities, which were $\alpha_\infty = 1.92$ for rice husks and $\alpha_\infty = 1.74$ for buckwheat husks.

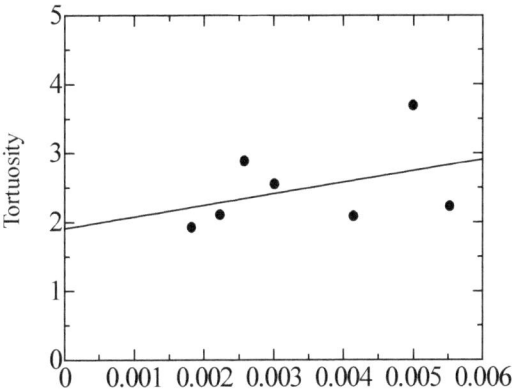

**Figure 5.** Measured tortuosity of the rice husks.

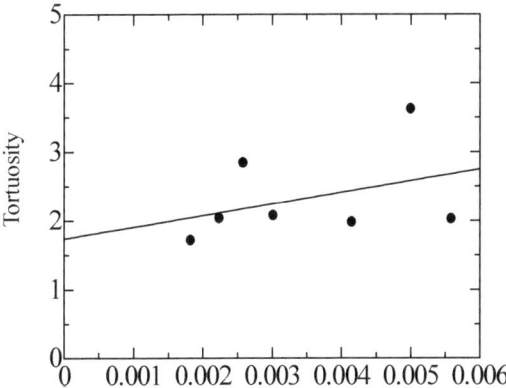

**Figure 6.** Measured tortuosity of the buckwheat husks.

## 3. Theoretical Analysis

### 3.1. Overview

Figure 7 shows a flowchart of the theoretical analysis used to derive the sound absorption coefficient. Micro-CT was used to obtain tomographic images, but these contained too much information for theoretical analysis in their original state. Thus, the images were processed by binarization and edge extraction to obtain the cross-sectional area and circumference of the rice and buckwheat husks in the tomographic plane. The cross-sectional area and circumference were approximated as the clearance between two planes, which was used to calculate the propagation constants and characteristic impedance to consider the attenuation of sound waves. The transfer matrix method was performed to obtain the transfer matrix for the entire sample, which could then be used to derive the normal incident sound absorption coefficient of the sample.

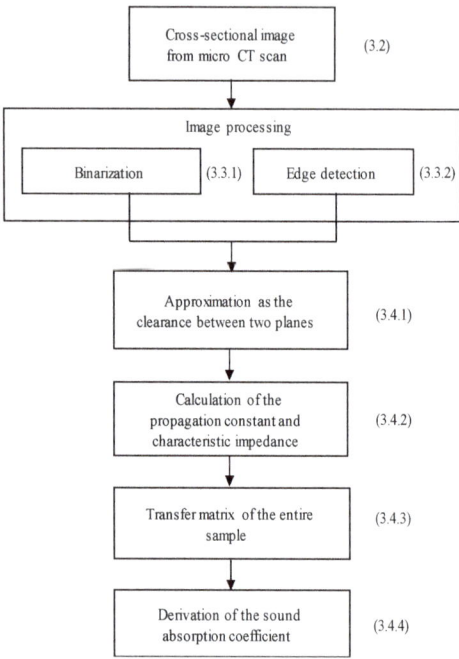

**Figure 7.** Overview of the theoretical analysis.

### 3.2. Image Acquisition

Figure 8a shows a tomographic image of a rice husk taken by a micro-CT scan (NIKON Corp. (Tokyo, Japan) MCT225 Metrology CT). As shown in Figure 8b, the image was sliced along the $y$–$z$ plane, which was perpendicular to the incident direction of the sound wave (i.e., $x$-direction). The image had dimensions of 20 mm in the $x$-direction and 25.7 mm$^2$ in the $y$–$z$ plane. The theoretical analysis used 884 images spanning 20 mm in the $x$-direction. The thickness $d$ of an element corresponded to the pitch in the $x$-direction, which was about 22.6 µm.

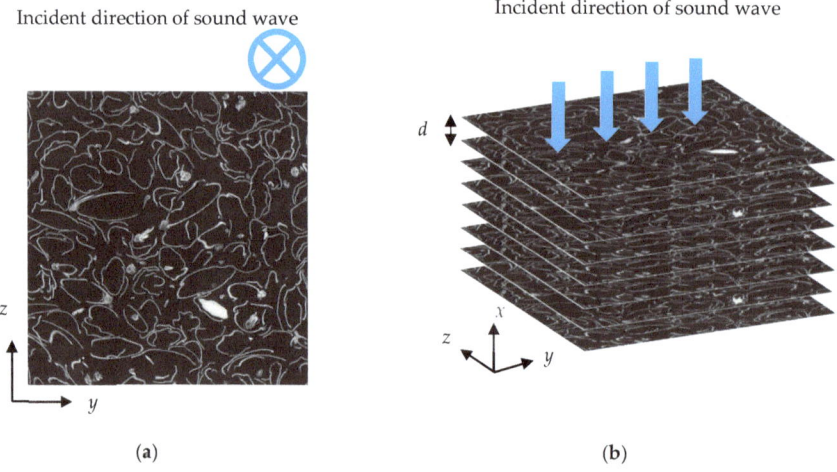

(a)      (b)

**Figure 8.** Cross-sectional image of a rice husk: (**a**) typical image with an arbitrary point in the $x$-direction; (**b**) schematic analysis.

## 3.3. Image Processing

### 3.3.1. Binarization

Binarization was performed to obtain the cross-sectional area of the clearance from a CT image. Binarization is an image-processing technique that converts an image with many shades of gray into a binary image with only two colors of black and white. A CT image is an 8-bit grayscale image in which each pixel can have a value of 0–255. It can be converted into a binary image by using a threshold. Otsu's binarization [23] was used to determine the threshold, which involved finding the threshold value at which the histogram has maximum separation. As an example, Figure 9a shows a CT image of a rice husk at an arbitrary location and Figure 9b shows the corresponding histogram. The horizontal axis is the luminance, and the vertical axis is the number of pixels. Class 1 was defined as luminance values that fall on the left side of threshold value, and class 2 was defined as those that fall on the right side. The average luminance values $m_1$ and $m_2$ for each class are expressed by:

$$m_1 = \frac{1}{n_1}\sum_{i=1}^{n_1} x_i \tag{2}$$

$$m_2 = \frac{1}{n_2}\sum_{i=1}^{n_2} x_i \tag{3}$$

where $n_1$ and $n_2$ are the number of pixels in each class and $x_i$ is the luminance value of the $i$-th pixel.

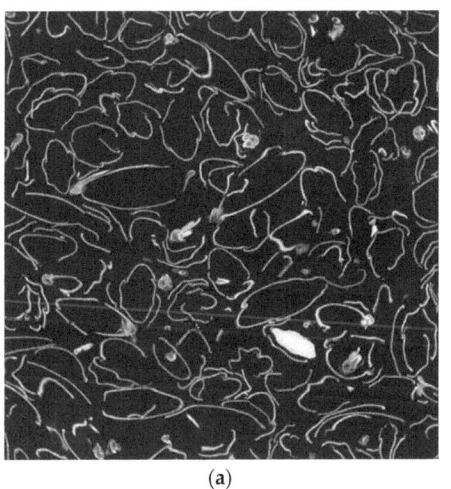

(a)

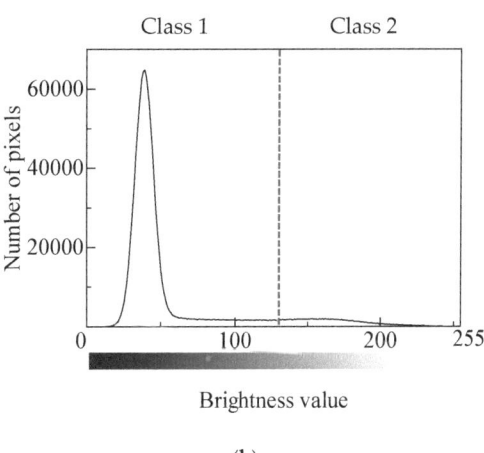

(b)

**Figure 9.** Otsu's binarization method: (**a**) typical cross-sectional image of an arbitrary point in the *x*-direction; (**b**) the corresponding histogram (red dashed line: threshold value between class 1 and 2).

Maximizing the degree of separation $\sigma^2$ is synonymous with maximizing the interclass separation $\sigma_b^2$, which can be defined as follows:

$$\sigma_b^2 = \frac{n_1 n_2 (m_1 - m_2)^2}{(n_1 + n_2)^2} \tag{4}$$

Let $t$ be the threshold value at which the degree of separation between classes is maximized. Then, the cross-sectional area of the gap $S$ is expressed as:

$$S = l_p^2 \times \sum_{i=1}^{t} n_i \quad (5)$$

where $l_p$ is the pixel size and $n_i$ is the number of pixels at the $i$-th threshold value.

Figure 10 shows a binarized image. The cross-sectional area of the gap $S$ was obtained by summing the number of pixels in the area determined to be black.

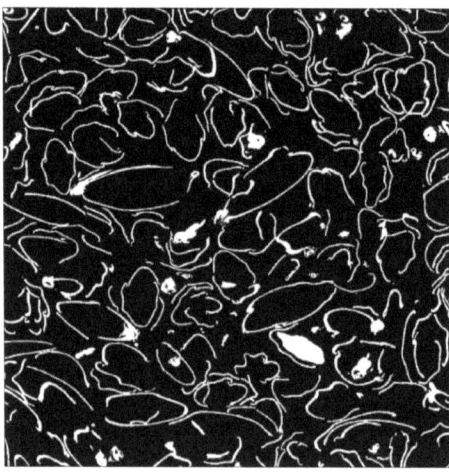

**Figure 10.** Binarized image.

3.3.2. Edge Extraction

In this study, the Canny edge detection method [24] was used for edge extraction. As a preliminary step, a Gaussian filter was applied to remove noise, which involves smoothing a pixel based on the luminance values of adjacent pixels. The weighting of the luminance values decreases according to the distance of the adjacent pixel from the target pixel. The Gaussian filter is expressed as

$$f(x,y,\varepsilon) = \frac{1}{2\pi\varepsilon^2} \exp\left(-\frac{x^2 - y^2}{2\varepsilon^2}\right) \quad (6)$$

where $\varepsilon$ is the standard deviation of the two-dimensional Gaussian distribution, which in this case was set to $\varepsilon = 5.0$. This corresponded to smoothing within a radius of about 7 pixels. The intensity $G$ of the luminance gradient and its connecting direction $\theta$ in a two-dimensional digital image can be defined from the horizontal luminance derivative $G_x$ and normal luminance derivative $G_y$ as follows:

$$|G| = \sqrt{G_x^2 + G_y^2} \quad (7)$$

$$\theta = tan^{-1}\frac{G_y}{G_x} \quad (8)$$

$\theta$ is used to select the optimal pixel among the eight pixels tangent to the target pixel as the tangential direction of the luminance gradient, and the line segment connecting these pixels is recognized as the edge (i.e., contour) of the image. Next, the three pixels in the direction normal to the calculated edge are considered. If the intensity of the center pixel is greater than that of the pixels at both ends, that pixel is considered the maximum, and the rest of the image is deleted. In other words, because areas with large luminance gradients inevitably have a certain width at the edges of an image, the information of differential-value pixels (i.e., non-maximum areas) that are not related to the direction of edge extension are suppressed to make individual edges stand out.

Then, the hysteresis thresholding process is applied to the extracted edges to distinguish real edges from fake edges. Two thresholds are set: the upper and lower thresholds of the luminance value gradient. If the target edge is always higher than the upper threshold, it is considered a real edge. If it is lower than the lower threshold, it is deleted as a fake edge. An edge is considered a real edge if it is connected to a portion of the edge extension that is above the upper threshold. This process also removes edges with a small number of pixels based on the assumption that edges are long lines.

Figure 11 shows an example of the final image after edge extraction. Let $n_v$ be the number of edge pixels that are connected vertically or parallel to adjacent edges and $n_d$ be the number of edge pixels that are connected diagonally to adjacent edges. Then, the edge length $l$ can be expressed as follows:

$$l = \left(n_v + \left(n_d \times \sqrt{2}\right)\right) \times l_p \tag{9}$$

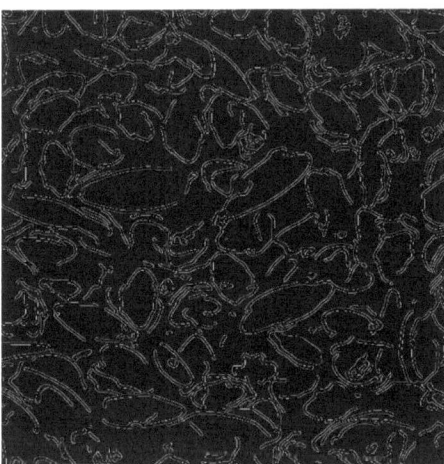

**Figure 11.** Image of rice husks with the edges extracted by the Canny edge detection method.

*3.4. Derivation of the Sound Absorption Coefficient*

3.4.1. Approximation to Clearance between Two Planes

After the image processing, the sound absorption coefficient can be derived. The first step is to approximate the voids as the clearance between two planes. Figure 12 shows the shapes before and after approximation. Multiplying the cross-sectional area of the gap $S$ obtained from image binarization (Section 3.3.1) by the pitch $d$ of the image yields the volume of the gap $V_n$, as shown in Figure 12a. Similarly, multiplying the total circumference of the cross-section obtained from edge extraction (Section 3.3.2) by $d$ yields $S_n$, as shown in Figure 12a. From this, for a single image with the pitch $d$, the gap thickness $b_n$ between the two planes shown in Figure 12b can be obtained and expressed as follows:

$$b_n = \frac{2V_n}{S_n \times F} \tag{10}$$

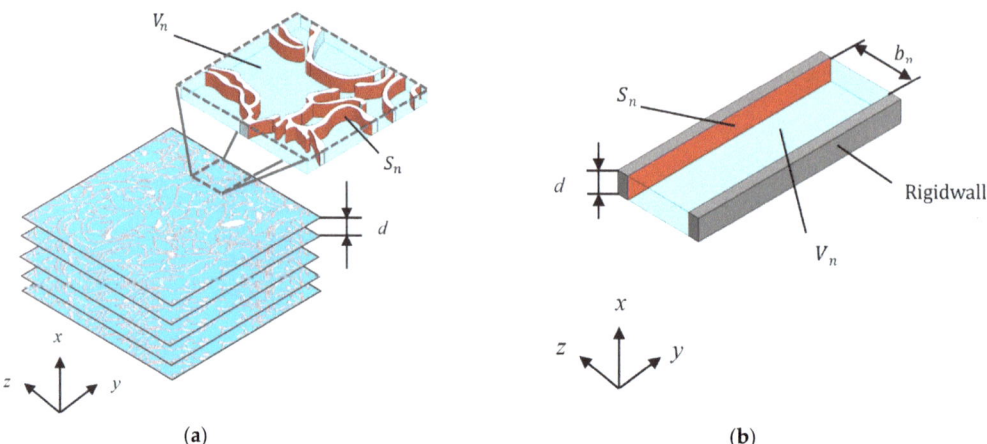

**Figure 12.** Surface area of the rice husks and volume of the clearance: (**a**) cross-sectional image of an arbitrary point in the *x*-direction; (**b**) approximation as two planes.

$F$ is a correction factor for obtaining the real surface area, and it is defined as the ratio of the real surface area to $S_n$. As shown in the left side of Figure 13a, if a flake is assumed a hemisphere, then the flake on $x$–$y$ plane is stepped as shown on the right side as a result of the CT scan, and $F = \pi/2 \cong 1.507$. As it is also shown on the left side of Figure 13b, if a flake is assumed to have a flat plate inclined 45° in the $x$-direction, then $F = \sqrt{2} \cong 1.414$.

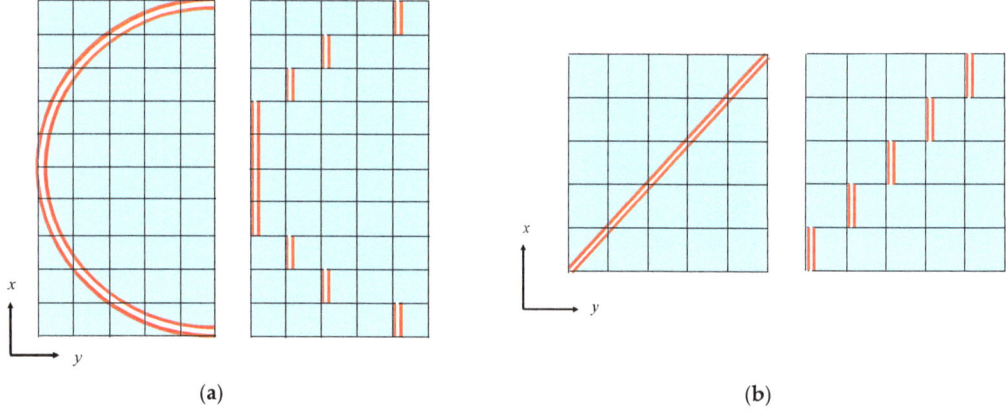

**Figure 13.** Cross-sectional view of two typical shapes: (**a**) hemispherical flake; (**b**) flat flake inclined at 45°.

To accurately calculate the surface area of samples, the Simpleware software (https://www.synopsys.com/company.html, accessed on 10 August 2023) was used to generate three-dimensional curved surfaces for the front and back of flakes. For the CT image shown in the upper side of Figure 14a, the front and back surfaces of the flakes were generated by complementing the line segments comprising the cross-section of each flake between adjacent images with a curved surface, as shown in the lower side of Figure 14a. Figure 14b shows an example image of a 3D model of flakes with the generated surfaces. The surface area of the 3D model was considered close to the surface area of the sample captured by the CT image. The ratio of the surface area $S_n$ to the surface area of the 3D model is expressed by the correction factor $F$. Based on the 3D model, the correction factors for rice and buckwheat husks were calculated as $F = 1.37$ and $F = 1.43$, respectively. In addition, the effects of varying the correction factor $F$ from 1.0 to 1.5 in steps of 0.1 were investigated for both rice and buckwheat husks.

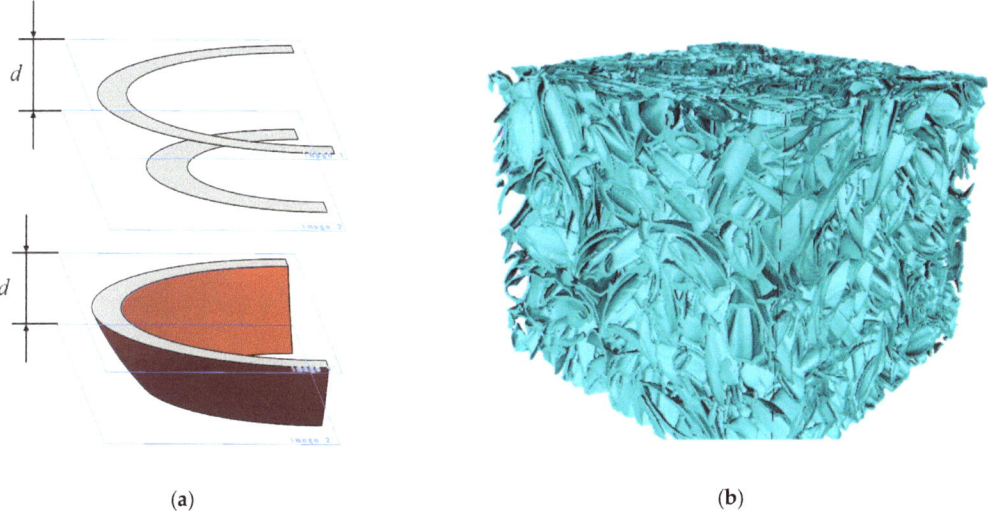

(a)          (b)

**Figure 14.** Three-dimensional model created from the CT images: (**a**) complementing the line segments with a curved surface; (**b**) 3D model of rice husks created with Simpleware.

3.4.2. Propagation Constants and Characteristic Impedance

After the clearance between two planes is approximated, the propagation constants, and characteristic impedances can be derived while considering the attenuation of sound waves. Previous studies have obtained the propagation constants and characteristic impedances considering the viscosity of air inside a tube. Tijdeman [25] and Stinson [26] considered circular tubes, Stinson and Champou [27] considered equilateral triangular tubes, and Beltman et al. [28] considered rectangular tubes. Allard [29] considered the degree of tortuosity. In this study, the methods of Stinson and Champou [27] and Allard [29] were applied.

Figure 15 shows a Cartesian coordinate system for the space between two parallel planes, for which the effective density $\rho_s$ and compressibility $C_s$ can be derived from a three-dimensional analysis using the Navier–Stokes equations, gas equation of state, continuity equation, energy equation, and the dissipative function representing heat transfer:

$$\rho_s = \rho_0 \left[ 1 - \frac{\tanh\left(\sqrt{j}\lambda_s\right)}{\sqrt{j}\lambda_s} \right]^{-1}, \quad \lambda_s = \frac{b_n}{2}\sqrt{\frac{\omega \rho_0}{\eta}} \quad (11)$$

$$C_s = \left(\frac{1}{\kappa P_0}\right)\left\{1 + (\kappa - 1)\left[\frac{\tanh\left(\sqrt{jN_{pr}}\lambda_s\right)}{\sqrt{jN_{pr}}\lambda_s}\right]\right\} \quad (12)$$

where $\rho_0$ is the density of air, $\lambda_s$ is the mediator variable, $b_n$ is the clearance between two planes, $\omega$ is the angular frequency, $\eta$ is the viscosity of air, $\kappa$ is the specific heat ratio of air, $P_0$ is the atmospheric pressure, and $N_{pr}$ is the Prandtl number.

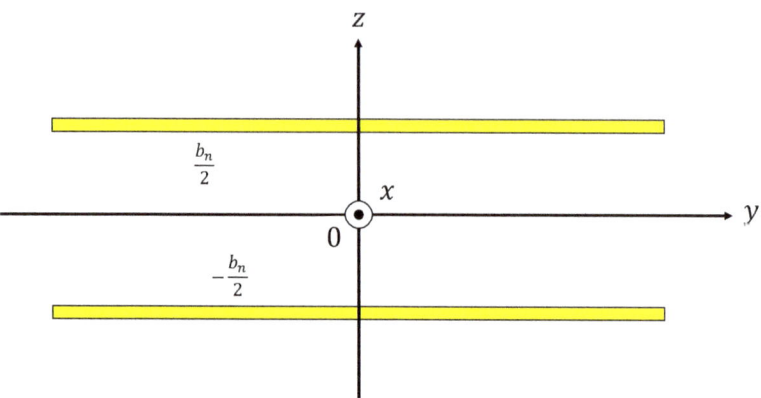

**Figure 15.** Cartesian coordinate system for the space between two parallel planes.

The propagation constant $\gamma$ and characteristic impedance $Z_c$ can be expressed by using the effective density $\rho_s$ and compressibility $C_s$:

$$\gamma = j\omega\sqrt{\rho_s C_s} \tag{13}$$

$$Z_c = \sqrt{\frac{\rho_s}{C_s}} \tag{14}$$

By using the effective density multiplied by the tortuosity, the propagation constant and characteristic impedance considering tortuosity can be obtained [29]. Therefore, the propagation constant and characteristic impedance considering the tortuosity $\alpha_\infty$ are expressed as follows:

$$\gamma = j\omega\sqrt{\alpha_\infty \rho_s C_s} \tag{15}$$

$$Z_c = \sqrt{\frac{\alpha_\infty \rho_s}{C_s}} \tag{16}$$

3.4.3. Transfer Matrix

The clearance between two planes was analyzed by using the transfer matrix method for the sound pressure and volume velocity based on the one-dimensional wave equation. Figure 16 shows a schematic diagram of the clearance between two planes shown in Figure 15, which is expressed as one element in the $x$-direction. The cross-sectional area $S$ of the clearance, the pitch $d$ per layer, the characteristic impedance $Z_c$, and the propagation constant $\gamma$ can be used to obtain the transfer matrix $T$ and four-terminal constants $A$–$D$ of the acoustic tube element:

$$T = \begin{bmatrix} \cosh(\gamma d) & \frac{Z_c}{S}\sinh(\gamma d) \\ \frac{S}{Z_c}\sinh(\gamma d) & \cosh(\gamma d) \end{bmatrix} = \begin{bmatrix} A & B \\ C & D \end{bmatrix} \tag{17}$$

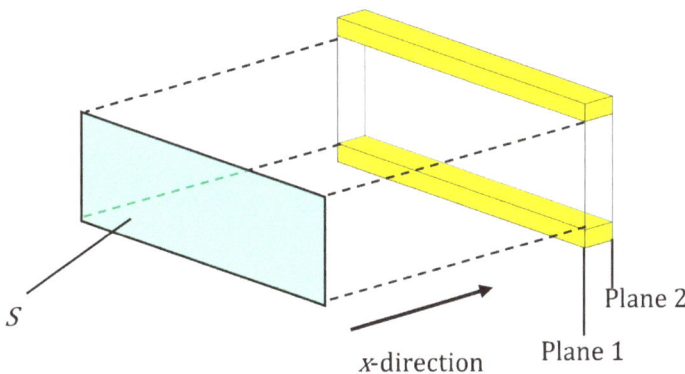

**Figure 16.** Sound incident area, incident plane, and transmission plane of the approximated clearance between two planes shown in Figure 15.

Let the sound pressure and particle velocity in plane 1 be $p_1$ and $u_1$, respectively, and the sound pressure and particle velocity in plane 2 be $p_2$ and $u_2$, respectively. Then, the transfer matrix is expressed as follows:

$$\begin{bmatrix} p_1 \\ Su_1 \end{bmatrix} = \begin{bmatrix} A & B \\ C & D \end{bmatrix} \begin{bmatrix} p_2 \\ Su_2 \end{bmatrix} \qquad (18)$$

Applying Equation (18) to the gap between the two planes obtains the transfer matrix for each divided element. Because each divided element is continuous in the $x$-direction, the transfer matrix $T_{all}$ for the entire sample can be calculated by cascading the transfer matrices of each divided element based on the equivalent circuit shown in Figure 17. Here, $n$ = 884 transfer matrices were cascaded, which is equivalent to the number of images for each sample.

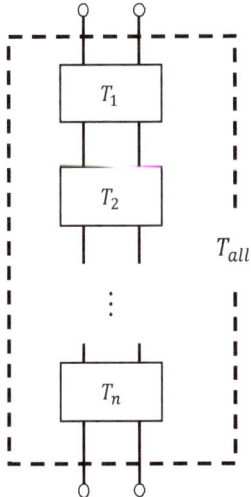

**Figure 17.** Equivalent circuit of the whole sample.

### 3.4.4. Normal Incident Sound Absorption Coefficient

The sound absorption coefficient was calculated from the transfer matrix $T_{all}$. For the acoustic tube shown in Figure 16, plane 2 can be considered a rigid wall. Therefore, because the particle velocity $u_2 = 0$, Equation (18) can be transformed as follows:

$$\begin{bmatrix} p_1 \\ Su_1 \end{bmatrix} = \begin{bmatrix} A & B \\ C & D \end{bmatrix} \begin{bmatrix} p_2 \\ 0 \end{bmatrix} \tag{19}$$

This allows us to obtain:

$$\begin{bmatrix} p_1 \\ Su_1 \end{bmatrix} = \begin{bmatrix} Ap_2 \\ Cp_2 \end{bmatrix} \tag{20}$$

Let the sound pressure and particle velocity immediately outside plane 1 be $p_0$ and $u_0$, respectively. Then, the specific acoustic impedance $Z_0$ from the sample's plane of incidence in the interior can be expressed as follows:

$$Z_0 = \frac{p_0}{u_0} \tag{21}$$

Therefore, by $p_0 = p_1$, $S_0 u_0 = S u_1$, and Equation (21), the specific acoustic impedance $Z_0$ of the sample can be expressed as follows:

$$Z_0 = \frac{p_0}{u_0} = \frac{p_0}{u_0 S_0} S_0 = \frac{p_1}{u_1 S} S_0 = \frac{A}{C} S_0 \tag{22}$$

The relationship between the specific acoustic impedance $Z_0$ and reflectance $R$ is expressed as follows:

$$R = \frac{Z_0 - \rho_0 c_0}{Z_0 + \rho_0 c_0} \tag{23}$$

The following relationship between the sound absorption coefficient and reflectance and Equation (23) can be used to obtain the theoretical value for the normal incident sound absorption coefficient $\alpha$ of the sample:

$$\alpha = 1 - |R|^2 \tag{24}$$

Thus, we present a supplemental note for the readers interested in sound insulation. Using the four-terminal constants $A$–$D$ in Equation (17), the normal incident transmission loss can be determined according to a previous report [30]. However, the high porosity of the sample used in this paper does not exhibit high sound-insulation performance.

## 4. Results and Discussion

The measured and theoretical values of the normal incident sound absorption coefficient were compared for the rice husk and buckwheat husk samples. For the theoretical values, the correction factor $F$ in Equation (10) was varied from 1.0 to 1.5 to evaluate its effect. The theoretical values using the correction factor $F$ (shown in Table 2) to obtain the real surface area derived from the 3D model were also obtained. Figure 18 shows the results for the rice husks, and Figure 19 shows the results for the buckwheat husks. Tables 3 and 4 present the measured and theoretical values, respectively, of the peak sound absorption frequency and peak sound absorption.

**Table 2.** Surface area from the CT images and 3D model.

|  | Surface Area Calculated from CT Images (mm²) | Surface Area in the 3D Model (mm²) | Correction Factor $F$ |
| --- | --- | --- | --- |
| Rice husk | 49,383 | 67,481 | 1.37 |
| Buckwheat husk | 41,461 | 59,403 | 1.43 |

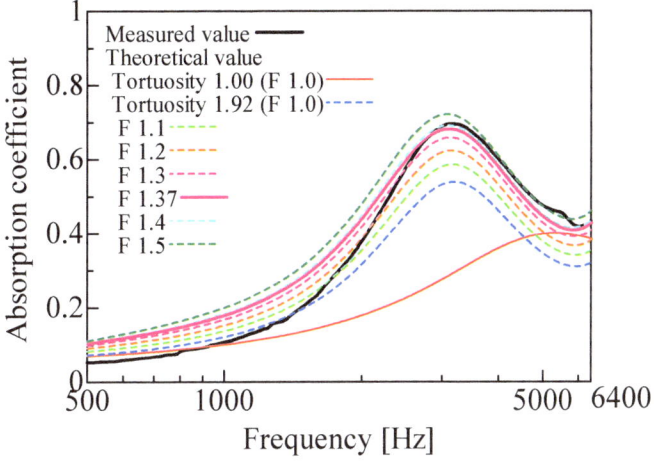

**Figure 18.** Comparison between the experiment and calculations (rice husks, $l$ = 20 mm).

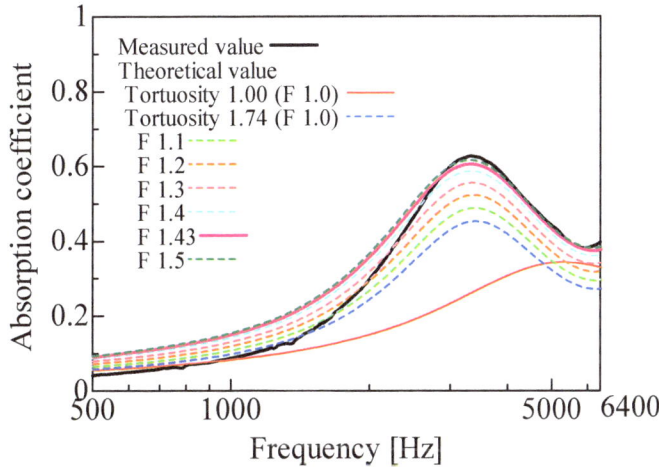

**Figure 19.** Comparison between experiment and calculations (buckwheat husks, $l$ = 20 mm).

**Table 3.** Frequency and absorption coefficient at peak (rice husks, $l$ = 20 mm).

|  | Peak Frequency (Hz) | Absorption Coefficient at Peak | Tortuosity | Correction Factor $F$ |
| --- | --- | --- | --- | --- |
| Measured value | 3150 | 0.696 | - | - |
| Theoretical value | 5288 | 0.401 | 1.00 | 1.00 |
| Theoretical value (Considering tortuosity) | 3175 | 0.540 | 1.92 | 1.00 |
| Theoretical value (Considering surface correction) | 3100 | 0.682 | 1.92 | 1.37 |

Table 4. Frequency and absorption coefficient at peak (buckwheat husks, $l$ = 20 mm).

| | Peak Frequency (Hz) | Absorption Coefficient at Peak | Tortuosity | Correction Factor $F$ |
|---|---|---|---|---|
| Measured value | 3325 | 0.627 | - | - |
| Theoretical value | 5275 | 0.343 | 1.00 | 1.00 |
| Theoretical value (Considering tortuosity) | 3400 | 0.453 | 1.74 | 1.00 |
| Theoretical value (Considering surface correction) | 3313 | 0.605 | 1.74 | 1.43 |

First, the measured values (black line) and theoretical values (red line) were compared without considering the tortuosity. For both the rice and buckwheat husks, the theoretical peak sound absorption frequency without considering tortuosity was higher than the measured frequency, and the theoretical peak sound absorption was lower than the measured value. When the tortuosity was considered, the peak sound absorption frequency shifted lower for both the rice and buckwheat husks, which increased the peak sound absorption. Thus, considering the tortuosity decreased the difference between the theoretical and measured values, a tortuosity greater than 1 means that the propagation path of sound waves in the sample increased in length, which has the same effect as an increase in the sample thickness. This explains the lower frequency. Based on the aforementioned findings, the peak frequency of sound absorption is related to the thickness of the layer of the sound-absorbing material, which must be thick enough to accommodate low frequencies.

In general, the sound absorption coefficient of a porous material is greatly affected by its thickness [31,32]. In other words, the sound absorption peak appears at a frequency where the thickness of the porous material corresponds to one-fourth of the wavelength of the sound wave. The peak sound absorption frequency can be decreased by a decrease in the apparent sound velocity of a material due to boundary-layer friction or tortuosity, which is equivalent to an increase in the apparent thickness of the material. In other words, increasing the tortuosity decreases the peak sound absorption frequency, which often also increases the sound absorption coefficient. For the same reason, the theoretical peak sound absorption increased when the tortuosity was considered. Overall, the accuracy of the theoretical values improved when the tortuosity was considered, and the measured tortuosity values for both the rice and buckwheat husks appear reasonable.

The theoretical values for both the rice and buckwheat husks with the correction factor $F$ were close to the experimental values. The peak sound absorption increased with an increasing correction factor, which is because the attenuation of sound waves due to boundary-layer viscosity increases with a greater surface area. The theoretical values with the correction factor $F$ calculated from the 3D model were close to the experimental values near the sound absorption peaks. This indicates that the theoretical surface area was closer to the actual surface area of the sample when the correction factor $F$ was included. Therefore, the theoretical values with the correction factor $F$ calculated from the 3D model were generally valid. Further, theoretical analyses reveal that the sound absorption characteristics rely on the geometric features of the voids; therefore, the sound absorption characteristics do not vary considerably until water is impregnated in the voids. However, the sound absorption properties may slightly vary owing to the variation in the gaps between the particles upon the inclusion of water droplets among the particles [33].

The differences between the measured and theoretical values are discussed in this section. A factor may be that the method used to estimate the attenuation of sound waves in gaps [27] has previously been shown [34] to be inaccurate for larger gaps and grain sizes. When the correction factor $F$ was considered, the sound absorption coefficient was calculated to be as large as 0.05 in the low-frequency range. The sound absorption coefficient is defined as 1 minus the reflectance. Therefore, the estimation error was not large because it was 0.05 at most for a reflection coefficient of 0.8–0.9 (0.2–0.1 for the sound absorption coefficient). For both the rice and buckwheat husks, the theoretical values of the peak sound absorption frequency and peak sound absorption were close to the measured values when the tortuosity and correction factor were considered. Therefore, the theoretical values calculated by the mathematical model were considered reasonable.

## 5. Conclusions

The normal incident sound absorption coefficients of rice and buckwheat husks were calculated theoretically. The tortuosity was measured for each material and was considered in the calculation. Then, the theoretical values were compared with the measured values. The following results were obtained:

1. The structures filled with rice and buckwheat husks were not periodic, which made constructing a geometric model difficult. Therefore, the sound absorption coefficient was estimated theoretically by first processing CT images.
2. The tortuosity increased the theoretical value of the peak sound absorption and lowered the frequency, which decreased the difference with the measured values. Therefore, the measured tortuosity was considered reasonable.
3. We used a correction factor to bring the surface area of the granular material closer to the actual surface area and observed that the tortuosity obtained theoretical values that matched the trend of the measured values. These results indicate that using CT images to estimate the sound absorption coefficient is a viable approach.
4. Mass production application studies based on this research are under consideration.

**Author Contributions:** Conceptualization, S.S. (Shuichi Sakamoto); Software, K.T. and T.S.; Formal analysis, K.T., S.S. (Shotaro Seino), K.H. and T.S.; Data curation, S.S. (Shotaro Seino) and K.H.; Supervision, S.S. (Shuichi Sakamoto); Project administration, S.S. (Shuichi Sakamoto). All authors have read and agreed to the published version of the manuscript.

**Funding:** This work was supported by the Japan Society for the Promotion of Science (JSPS) KAKENHI Grant No. 20K04359.

**Institutional Review Board Statement:** Not applicable.

**Informed Consent Statement:** Not applicable.

**Data Availability Statement:** Data is contained within the article.

**Conflicts of Interest:** The authors declare no conflict of interest.

## References

1. Sakamoto, S.; Takauchi, Y.; Yanagimoto, K.; Watanabe, S. Study for Sound Absorbing Materials of Biomass Tubule etc (Measured Result for Rice Straw, Rice Husks, and Buckwheat Husks). *J. Environ. Eng.* **2011**, *6*, 352–364. [CrossRef]
2. Sakamoto, S.; Tsurumaki, T.; Fujisawa, K.; Yamamiya, K. Study for sound-absorbing materials of biomass tubule (Oblique incident sound-absorption coefficient of oblique arrangement of rice straws). *Trans. JSME* **2016**, *15*, 16–00344. (In Japanese) [CrossRef]
3. Sakamoto, S.; Tanikawa, H.; Maruyama, Y.; Yamaguchi, K.; Ii, K. Estimation and experiment for sound absorption coefficient of three clearance types using a bundle of nested tubes. *J. Acoust. Soc. Am.* **2018**, *144*, 2281–2293. [CrossRef]
4. Bastos, L.P.; Melo, G.d.S.V.d.; Soeiro, N.S. Panels Manufactured from Vegetable Fibers: An Alternative Approach for Controlling Noises in Indoor Environments. *Adv. Acoust. Vib.* **2012**, *2012*, 698737. [CrossRef]
5. Buot, P.G.C.; Cueto, R.M.; Esguerra, A.A.; Pascua, R.I.C.; Magon, E.S.S.; Gumasing, M.J.J. Design and Development of Sound Absorbing Panels using Biomass Materials. In Proceedings of the 2nd African International Conference on Industrial Engineering and Operations Management Harare, Harare, Zimbabwe, 7–10 December 2020.
6. Or, K.H.; Putra, A.; Selamat, M.Z. Oil palm empty fruit bunch fibres as sustainable acoustic absorber. *Appl. Acoust.* **2017**, *119*, 9–16. [CrossRef]
7. Ersoy, S.; Küçük, H. Investigation of industrial tea-leaf-fibre waste material for its sound absorption properties. *Appl. Acoust.* **2009**, *70*, 215–220. [CrossRef]
8. Berardi, U.; Iannace, G. Acoustic characterization of natural fibers for sound absorption applications. *Build. Environ.* **2015**, *94*, 840–852. [CrossRef]

9. Maderuelo-Sanz, R.; Morillas, J.M.B.; Escobar, V.G. Acoustical performance of loose cork granulates. *Eur. J. Wood Wood Prod.* **2014**, *72*, 321–330. [CrossRef]
10. Koizumi, T.; Tsujiuchi, N.; Adachi, A. The development of sound absorbing materials using natural bamboo fibers. *High Perform. Struct. Mater.* **2002**, *4*, 157–166. [CrossRef]
11. McGinnes, C.; Kleiner, M.; Xiang, N. An environmental and economical solution to sound absorption using straw. *J. Acoust. Soc. Am.* **2005**, *118*, 1869. [CrossRef]
12. Putra, A.; Abdullah, Y.; Efendy, H.; Farid, W.M.; Ayob, R.M.; Py, M.S. Utilizing sugarcane wasted fibers as a sustainable acoustic absorber. *Procedia Eng.* **2013**, *53*, 632–638. [CrossRef]
13. Tsuchiya, Y.; Kobayashi, M. Examination of practical sound absorption coefficient by material and the composition. *Toda Tech. Res. Rep.* **2005**, *31*, 6. (In Japanese)
14. Sakamoto, S.; Suzuki, K.; Toda, K.; Seino, S. Mathematical Models and Experiments on the Acoustic Properties of Granular Packing Structures (Measurement of tortuosity in hexagonal close-packed and face-centered cubic lattices). *Materials* **2022**, *15*, 7393. [CrossRef]
15. Sakamoto, S.; Suzuki, K.; Toda, K.; Seino, S. Estimation of the Acoustic Properties of the Random Packing Structures of Granular Materials: Estimation of the Sound Ab-sorption Coefficient Based on Micro-CT Scan Data. *Materials* **2023**, *16*, 337. [CrossRef] [PubMed]
16. Furubayashi, T. O-13 Analysis of Rice Husk Energy Utilization System Considering Monthly Generation Amount. In Proceedings of the Conference on Biomass Science, Online, 20–21 January 2021. (In Japanese) [CrossRef]
17. Kojima, Y. Studies on Utilization of Husks from Buckwheat. In Proceedings of the Annual Conference of the Japan Institute of Energy, Hokkaido, Japan, 30–31 July 2003; pp. 3–41. (In Japanese) [CrossRef]
18. Sakamoto, S.; Shintani, T.; Hasegawa, T. Simplified Limp Frame Model for Application to Nanofiber Nonwovens (Selection of Dominant Biot Parameters). *Nanomaterials* **2022**, *12*, 3050. [CrossRef]
19. Maeda, E.; Miyake, H. A non-destructive tracing with an X-ray micro CT scanner of vascular bundles in the ear axes at the base of the lower level rachis-branches in Japonica type rice (*oryza sativa*). *Jpn. J. Crop Sci.* **2009**, *78*, 382–386. [CrossRef]
20. Sakamoto, S.; Takakura, R.; Suzuki, R.; Katayama, I.; Saito, R.; Suzuki, K. Theoretical and Experimental Analyses of Acoustic Characteristics of Fine-grain Powder Considering Longitudinal Vibration and Boundary Layer Viscosity. *J. Acoust. Soc. Am.* **2021**, *149*, 1030–1040. [CrossRef] [PubMed]
21. ISO 10534-2; Acoustics–Determination of Sound Absorption Coefficient and Impedance in Impedance Tubes–Part 2: Transfer-Function Method. International Organization for Standardization: Geneva, Switzerland, 2002.
22. Allard, J.F.; Castagnede, B.; Henry, M.; Lauriks, W. Evaluation of tortuosity in acoustic porous materials saturated by air. *Rev. Sci. Instrum.* **1994**, *65*, 754–755. [CrossRef]
23. Otsu, N. A Threshold Selection Method from Gray-Level Histograms. *IEEE Trans. Syst. Man Cybern.* **1979**, *9*, 62–66. [CrossRef]
24. Canny, J. A Computational Approach to Edge Detection. *IEEE Trans. Pattern Anal. Mach. Intell.* **1986**, *8*, 679–698. [CrossRef]
25. Tijdeman, H. On the propagation of sound waves in cylindrical tubes. *J. Sound Vib.* **1975**, *39*, 1–33. [CrossRef]
26. Stinson, M.R. The propagation of plane sound waves in narrow and wide circular tubes and generalization to uniform tubes of arbitrary cross-sectional shape. *J. Acoust. Soc. Am.* **1991**, *89*, 550–558. [CrossRef]
27. Stinson, M.R.; Champoux, Y. Propagation of sound and the assignment of shape factors in model porous materials having simple pore geometries. *J. Acoust. Soc. Am.* **1992**, *91*, 685–695. [CrossRef]
28. Beltman, W.; van der Hoogt, P.; Spiering, R.; Tijdeman, H. Implementation and experimental validation of a new viscothermal acoustic finite element for acousto-elastic problems. *J. Sound Vib.* **1998**, *216*, 159–185. [CrossRef]
29. Allard, J.-F.; Daigle, G. Propagation of sound in Porous Media Modeling Sound Absorbing Materials. *J. Acoust. Soc. Am.* **1994**, *95*, 2785. [CrossRef]
30. Sakamoto, S.; Shin, J.; Abe, S.; Toda, K. Addition of Two Substantial Side-Branch Silencers to the Interference Silencer by Incorporating a Zero-Mass Metamaterial. *Materials* **2022**, *15*, 5140. [CrossRef]
31. Taban, E.; Khavanin, A.; Jafari, A.J.; Faridan, M.; Tabrizi, A.K. Experimental and mathematical survey of sound absorption performance of date palm fibers. *Heliyon* **2019**, *5*, e01977. [CrossRef]
32. Cuiyun, D.; Guang, C.; Xinbang, X.; Peisheng, L. Sound absorption characteristics of a high-temperature sintering porous ceramic material. *Appl. Acoust.* **2012**, *73*, 865–871. [CrossRef]
33. Sakamoto, S.; Tsutsumi, Y.; Yanagimoto, K.; Watanabe, S. Study for Acoustic Characteristics Variation of Granular Material by Water Content. *Trans. Jpn. Soc. Mech. Eng. (Part C)* **2009**, *75*, 2515–2520. [CrossRef]
34. Sakamoto, S.; Higuchi, K.; Saito, K.; Koseki, S. Theoretical analysis for sound-absorbing materials using layered narrow clearances between two planes. *J. Adv. Mech. Des. Syst. Manuf.* **2014**, *8*, JAMDSM0036. [CrossRef]

**Disclaimer/Publisher's Note:** The statements, opinions and data contained in all publications are solely those of the individual author(s) and contributor(s) and not of MDPI and/or the editor(s). MDPI and/or the editor(s) disclaim responsibility for any injury to people or property resulting from any ideas, methods, instructions or products referred to in the content.

*Article*

# Experimental Validation of Dynamic Response of Small-Scale Metaconcrete Beams at Resonance Vibration

Meisam Ansari *, Fabiola Tartaglione and Carsten Koenke

Institute of Structural Mechanics, Bauhaus-Universität Weimar, 99423 Weimar, Germany
* Correspondence: meisam.ansari@uni-weimar.de

**Abstract:** Structures and their components experience substantially large vibration amplitudes at resonance, which can cause their failure. The scope of this study is the utilization of silicone-coated steel balls in concrete as damping aggregates to suppress the resonance vibration. The heavy steel cores oscillate with a frequency close to the resonance frequency of the structure. Due to the phase difference between the vibrations of the cores and the structure, the cores counteract the vibration of the structure. The core-coating inclusions are randomly distributed in concrete similar to standard aggregates. This mixture is referred to as metaconcrete. The main goal of this work is to validate the ability of the inclusions to suppress mechanical vibration through laboratory experiments. For this purpose, two small-scale metaconcrete beams were cast and tested. In a free vibration test, the metaconcrete beams exhibited a larger damping ratio compared to a similar beam cast from conventional concrete. The vibration amplitudes of the metaconcrete beams at resonance were measured with a frequency sweep test. In comparison with the conventional concrete beam, both metaconcrete beams demonstrated smaller vibration amplitudes. Both experiments verified an improvement in the dynamic response of the metaconcrete beams at resonance vibration.

**Keywords:** metaconcrete; damping aggregate; vibration absorber; free vibration test; frequency sweep test

**Citation:** Ansari, M.; Tartaglione, F.; Koenke, C. Experimental Validation of Dynamic Response of Small-Scale Metaconcrete Beams at Resonance Vibration. *Materials* 2023, *16*, 5029. https://doi.org/10.3390/ma16145029

Academic Editor: Martin Vašina

Received: 24 May 2023
Revised: 13 July 2023
Accepted: 14 July 2023
Published: 16 July 2023

**Copyright:** © 2023 by the authors. Licensee MDPI, Basel, Switzerland. This article is an open access article distributed under the terms and conditions of the Creative Commons Attribution (CC BY) license (https://creativecommons.org/licenses/by/4.0/).

## 1. Introduction

The resonance vibration is an unpleasant incident that occurs when the exciting frequency is close to the natural frequency of a structure. The vibration amplitude of the structure at resonance increases substantially, which can lead to the collapse of the structure. The most iconic example of such a failure is the collapse of the Tacoma Narrows Bridge in 1940 when the right wind condition drove the bridge to its resonance frequency [1]. Since the structures are subjected to a variety of dynamic loads throughout their lifespan, it is crucially important to avoid resonance vibration by taking necessary measures.

Using a damper tuned to the natural frequency of a structure is the classical solution for reducing the vibration amplitude at resonance [2]. The damper is an auxiliary mass that is secured to the structure, and it starts to vibrate by the vibration of the structure with a frequency close to the resonance frequency of the structure. Due to the phase difference, the impact of the damper's vibration is in the opposite direction of the vibration of the structure. Figure 1 illustrates the damper's mechanism. Figure 1a shows the structure and the damper in their neutral positions. When the structure is experiencing a downward displacement, the damper is displacing upwards and vice versa (Figure 1b,c). Therefore, the damper's vibration counteracts the vibration of the structure. The vibration absorber of Herman Frahm [3] is known to be one of the earliest devices that were configured to suppress vibration. He mainly used his device on ships to reduce their vibration during sailing.

The efficiency of a damper depends significantly on its tuning. The study by Den Hartog in [4] derived the formulation for optimum tuning parameters of a damper. The optimum frequency and the self-damping of the damper were determined with the ratio of the damper's mass to the structure's mass. The studies in [5,6] extended Den

Hartog's work to problems including structures with a light damping ratio, as well as Multi-Degree-Of-Freedom systems (MDOF) that can be represented with the single mode without the damper.

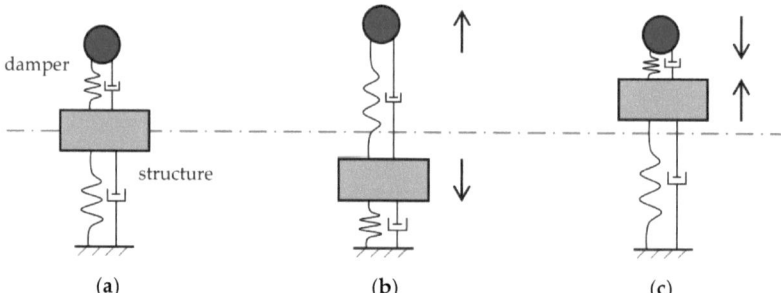

**Figure 1.** Illustrative presentation of a damper's mechanism: (**a**) neutral position; (**b**) expansion phase; (**c**) contraction phase.

In practical problems, it is challenging to satisfy the optimized tuning parameters of a damper. Several works, such as [2,7,8], reviewed and discussed the sensitivity of a Tuned Mass Damper (TMD) to deviation from the optimum tuning parameters. Using multiple TMDs was introduced in [8] as a solution to the sensitivity of a TMD. The frequencies of the TMDs were distributed over a frequency range around the natural frequency of the structure. Such a configuration was called a multi-TMD setup. Ref. [9] demonstrated the application of multiple TMDs to passive vibration control. Comprehensive studies were conducted in [10,11] to solve the complicated eigenvalue problem of a multi-TMD setup composed of a structure with multiple connected subsystems.

Our study in [12] regarded the randomly distributed damping aggregate in concrete as a multi-TMD setup. The damping aggregates were silicone-coated steel balls that were randomly distributed in concrete. We showed in [12] that the core-coating inclusions had a mechanism similar to a TMD. The soft silicone coating provided a suspension for the heavy steel core to oscillate. When the inclusions were randomly distributed in concrete, they were TMDs incorporated in the structure forming a multi-TMD setup. The lower sensitivity of the multi-TMD setup to the tuning error [8] is the main advantage of the damping aggregate compared to the single TMD. In addition, distributed damping aggregates in concrete are implemented TMDs in structural components. Therefore, they eliminate the need for an external TMD unit. They are preventive measures for the resonance vibration of the structural components.

In the study in [12], we modeled the metaconcrete beams with randomly distributed core-coating inclusions and verified their ability to suppress the resonance vibration. The work in this paper aimed to validate the findings in [12] by conducting experiments with metaconcrete beams.

*1.1. Metaconcrete*

The term "metamaterial" was brought to attention by [13] to emphasize the purpose of the engineered materials: " ... to achieve material performance beyond the limits of conventional composites". The author of [14] defined a metamaterial as an artificial composite with properties that mainly depend on the engineered microstructure. In our study, metaconcrete specifically refers to the mixture of concrete with the damping aggregates. The damping aggregates are used to improve the dynamic response of structural members made from concrete. The utilization of engineered aggregates in concrete has been the subject of other studies too. A summary of some of the major contributions to the topic is provided in this section.

In an extensive study in [14], the standard aggregates of concrete were replaced with silicone-coated steel balls to achieve a higher attenuation property than the conventional concrete demonstrates. The new composite was named "metaconcrete". The engineered inclusions created frequency band gaps and greatly reduced the energy transmission of the applied wave motion.

The authors of [15] studied the mitigation properties of metaconcrete under blast loading. Cylindrical specimens of metaconcrete with randomly distributed core-coating inclusions were cast. The specimens showed a 2-order reduction in the magnitude of the transmitted signal.

The study in [16] investigated the use of a metaconcrete thin plate in passive vibration control. The rubber-coated steel inclusions were embedded periodically in the concrete plate. The steel inclusions with circular and square geometries were used. The dispersion diagrams of the thin plate with the inclusions revealed the appearance of the frequency band gaps.

A frequency sweep test with a linear sweep rate was conducted in other studies to investigate the attenuation properties of metaconcrete. Ref. [17] carried out the test with cubic specimens of metaconcrete. The core-coating inclusions were placed inside the specimen in lattice-like patterns. The number of inclusions and the spacing between them were varied to study the impact of the inclusion's arrangement on the attenuation of the applied wave. Ref. [18] conducted the test similarly with cylindrical specimens at low sonic frequencies. Ref. [19] employed the cylindrical specimens in the test too. The inclusions were arranged in a semi-regular lattice-like pattern. The pattern was rotated in every specimen to obtain a pseudo-random arrangement. In the studies in [17–19], the specimen of concrete with standard aggregate was used as a reference. Comparing the results of the metaconcrete specimen with the reference specimen revealed a remarkable attenuation of the transmitted signal in the proximity of the inclusion's natural frequencies.

The reviewed studies in this section investigated the attenuation characteristics of the core-coating inclusions in wave propagation. In contrast, the work in this paper investigated whether the inclusions can suppress the resonance vibration by functioning as dampers. Furthermore, the current work disregarded the uniform arrangement of the inclusions in the specimens because the work intended to utilize the inclusions in concrete similar to the standard aggregates with a random distribution.

*1.2. Scope and Outline of the Work*

The main goal of this study was to validate the findings of our previous study [12] with laboratory experiments. In our previous study [12], we first investigated the ability of the individual core-coating elements to function similarly to a TMD. We used the free vibration test to determine the damping ratio of small-scale beams cast from conventional concrete. After securing individual core-coating elements to the beams with adhesive tape, we observed an increase in the damping ratio. The higher damping ratio of the beams with the external core-coating elements indicated that the individual core-coating elements functioned like a TMD. In the second stage of the study in [12], we investigated the ability of the randomly distributed core-coating inclusions in concrete to function similarly to a multi-TMD setup. We employed numerical simulation to model and analyze small-scale metaconcrete beams. The Frequency Response Analysis (FRA) of the models showed the ability of the randomly distributed inclusions to suppress the resonance vibration.

This article is a complementary work to [12]. The goal of this work was to validate the improved dynamic response of small-scale metaconcrete beams through laboratory experiments. In contrast to our previous work in [12], the core-coating inclusions were not utilized here as external elements but rather as randomly distributed aggregates in fresh concrete. Two small-scale beams were cast from the metaconcrete for the experimental investigation. For the validation of the improved dynamic response, two objectives were defined. Firstly, an increase in the damping ratio of the metaconcrete beams was investigated through the free vibration test. Secondly, a reduction in the vibration amplitude of the beams was verified with the frequency sweep test.

Section 2 of this manuscript describes the materials and the specimens used in this study.

Section 3 mainly addresses the first objective highlighted earlier. It provides the description and results of the free vibration test, which was conducted to determine the damping ratio of the specimens. The test setup, procedure, and conditions were similar to the free vibration test conducted in [12]. The description is provided here for the completeness of this paper and for easy reference.

Section 4 is dedicated to the frequency sweep test for the evaluation of the vibration amplitude. The second objective is addressed in this section.

In the final section, the conclusive findings are summarized.

## 2. Materials and Specimens

There were two types of experiments employed in this study: the free vibration test and the frequency sweep test. The experiments were conducted to evaluate the improved dynamic response of the metaconcrete specimens. In this regard, one specimen was cast from conventional concrete to serve as the reference specimen. The two other specimens were cast from metaconcrete. The tests were carried out with all of the specimens. The results of the tests with the metaconcrete specimens and the conventional concrete specimen were compared to evaluate the improvement in the dynamic response. This section provides an overview of the materials and specimens used in the experiments. The detailed procedure of the tests and results will be discussed in Sections 3 and 4.

### 2.1. Concrete Mixture

The cementitious mixture for casting the specimens was designed per conventional concrete C35/45. Cement type-I was used in the mixture. Due to the small sectional dimensions of the specimens, the maximum size of the coarse aggregate in the mixture was limited to 8 mm. The mix design of concrete is provided in Table 1.

Table 1. Concrete mix design for casting the specimens.

| Component | Volume Fraction (%) | Volume (dm$^3$/m$^3$) | Unit Mass (kg/dm$^3$) | Mass (kg/m$^3$) |
|---|---|---|---|---|
| Aggregates | | | | |
| 0~2 mm | 35 | 229 | 2.65 | 608 |
| 2~8 mm | 65 | 426 | 2.60 | 1108 |
| Sum | 100 | 655 | | 1716 |
| Chosen W/C: 0.45 | | | | |
| Water | | | | 189 |
| Cement | | | | 420 |
| Sum | | | | 609 |
| Total sum | | | | 2325 |

### 2.2. Damping Aggregates

The engineered inclusions in this study are silicone-coated stainless-steel balls, which are utilized in concrete as damping aggregates (Figure 2). We reviewed the mechanism and properties of these inclusions in [12] (Section 2). Only two core-coating sizes were chosen for the experiments in this paper. Table 2 introduces their configuration. The mechanical properties of the core and coating material are provided in Table 3.

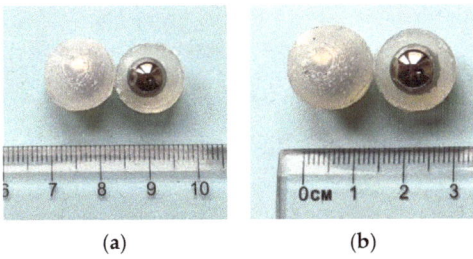

**Figure 2.** The silicone-coated steel balls used in the experiments: (a) K7 size: diameter = 14 mm; (b) K9 size: diameter = 16 mm.

**Table 2.** The core-coating sizes used in the experiments.

| Size | Stainless-Steel Core Diameter (mm) | Mass (gr) | Silicone Coating Thickness (mm) | Overall Diameter (mm) | Overall Volume (mm³) |
|---|---|---|---|---|---|
| K7 | 8 | 2.14 | 3 | 14 | 1437 |
| K9 | 10 | 4.19 | 3 | 16 | 2145 |

**Table 3.** Mechanical properties of the core and coating materials.

| Component | Material | Density (kg/m³) | E (MPa) |
|---|---|---|---|
| Core | Stainless-Steel | 8000 | 200,000 |
| Coating | Silicone | 1040 | 0.024 |

### 2.3. Specimens

A total of three specimens were cast for this study. The geometry and dimensions of all specimens were identical. Figure 3 provides a schematic illustration of the specimens.

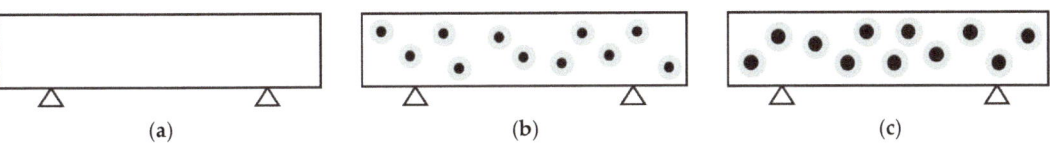

**Figure 3.** Schematic illustration of the specimens: (a) P720: specimen without damping aggregates; (b) P720-K7: specimen with K7 inclusions; (c) P720-K9: specimen with K9 inclusions. The distribution of the inclusions in the specimens is illustrative.

The first specimen, P720, represents the conventional concrete throughout this paper. It contains no damping aggregates and serves as the reference specimen (Figure 3a). The other two specimens contain damping aggregates. They represent metaconcrete in this study (Figure 3b,c). The second specimen, P720-K7, was cast from metaconcrete with the K7-size, and the third specimen, P720-K9, with the K9-size. Table 4 summarizes the specifications of the specimens.

**Table 4.** The specifications of the specimens used in the experiments.

| Specimen | Core-Coating in Concrete Size | $n_{req.}$ | Length (mm) | Section (mm) (Width × Depth) | Mass (gr) |
|---|---|---|---|---|---|
| P720 | - | - | 720 | 60 × 40 | 3980 |
| P720-K7 | K7 | 60 | 720 | 60 × 40 | 4064 |
| P720-K9 | K9 | 40 | 720 | 60 × 40 | 4178 |

The total volume fraction of the damping aggregates in a cubic meter of concrete was chosen to be 5%. This means, the total required volume of the inclusions in every beam was

$$V_{req.} = (720 \times 60 \times 40) \times 0.05 = 8.64 \times 10^4 \quad \text{mm}^3 \tag{1}$$

and by computing the volume of one inclusion with the following equation

$$V_{Ki} = \left(\frac{4}{3}\right) \pi \left(\frac{D}{2}\right)^3 \tag{2}$$

where $D$ is the overall diameter of the core-coating inclusion, the required number of the inclusions in one specimen was obtained by

$$n_{req.} = \frac{V_{req.}}{V_{Ki}} \tag{3}$$

that is also provided in Table 4. The inclusions were gently mixed in the fresh concrete before pouring the mixture into the mold.

## 3. Determination of the Damping Ratio with the Free Vibration Test

The free vibration test is the experimental procedure of measuring the free oscillation of an object. The test object is usually excited by an impact hammer or a shaker. The vibration is then allowed to decay freely. Different types of sensors can be used to capture and record the vibration, such as accelerometers, laser-vibrometers, etc. The recorded vibration can be used to obtain the dynamic properties of the test object, such as its natural frequencies. The test has a wide range of applications in structural dynamics. For example, it can be conducted to determine the damping ratio [20,21] or to investigate the frequency-dependent modulus of materials [22,23]. The test was mainly employed in this study to investigate the damping ratio of the specimens.

With the free vibration tests in [12] (Section 3), we investigated the ability of the individual core-coating elements to function similarly to a TMD. In our study in [12], we secured the silicone-coated steel balls externally to the small-scale beams cast from conventional concrete. The external core-coating elements were tuned to the first vibratory mode of the beams similar to a classical TMD. An increase in the damping ratio of the beams demonstrated the ability of the core-coating elements to suppress the vibration.

In contrast to our study in [12], we randomly mixed the core-coating inclusions in fresh concrete for casting small-scale metaconcrete beams for the experiments in this study. The free vibration test with the same procedure was conducted with the metaconcrete beams. The goal was to investigate whether the metaconcrete beams with randomly distributed core-coating inclusions demonstrate a higher damping ratio compared to conventional concrete beams. In [12], we explained the procedure and setup of the test. However, for the sake of completeness and ease of reference, we provide a more detailed description of the test in this paper.

### 3.1. Description of the Test Setup and Procedure

Similar to our study in [12], the vibratory mode of interest for evaluating the damping ratio was the first mode of the beams. The first mode shape of a beam with free ends is pictured in Figure 4. At a distance ratio of 0.244 from each end, the deflection of the mode shape is zero. These points are referred to as "nodal points" [24] (p. 176). ISO 7626-2 specifies that the supporting points should have the least possible effect on the intended measurements. The standard recommends supporting near the nodal points of the test object. By point-supporting the metaconcrete beams at the nodal points in the test, the interaction of the supports with the beam was minimized. Furthermore, this support alignment facilitated the activation of the first vibratory mode in the test.

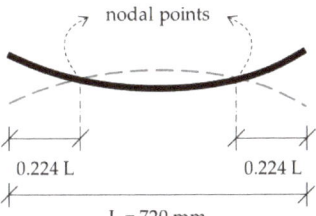

**Figure 4.** The first mode shape of the beam with free ends and its nodal points.

We used the same setup from our study in [12] to point-support the beams. At first, the nodal points were marked on the beams. Then, stainless steel short rods with pointy tips were secured in wooden blocks as shown in Figure 5b. The rods had an approximate length of 5 cm and a diameter of 8 mm. Supporting the beam with only one rod at each nodal point was not possible due to a lack of stability. Therefore, two rods supported the beam at the left and one rod at the right nodal point (Figure 5a,c).

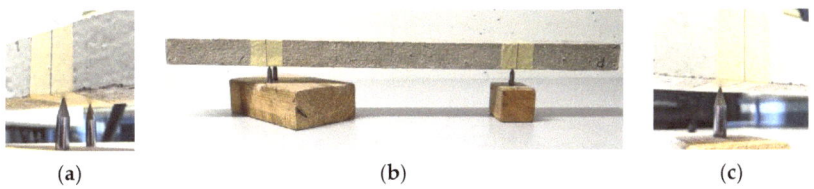

**Figure 5.** Point-supporting the beam in the laboratory: (**a**) left support with two rods; (**b**) overall view; (**c**) right support with one rod; (similar to the support alignment in [12]).

We used the same equipment and test setup from our study in [12] for conducting the test in this study. The beams were manually excited with an impact hammer (Brand: Sigmatest, Model: IH02). The excitation point was approximately at the mid-span of the beam. A single-point laser vibrometer measured the velocity of the vibration (Brand: Polytec, Model: PDV-100). The monitoring point was at the mid-span of the beam. Figure 6 shows an overall view of the test setup.

**Figure 6.** An overall view of the test setup showing the laser vibrometer pointing at the mid-span of the beam (similar to the test setup in [12]).

The decaying free vibration of the beam in the time domain was obtained from the test. The damping ratio was determined by approximating the upper decay (Figure 7) with the following exponential function

$$V(t) = a \cdot e^{-\zeta \omega t} \qquad (4)$$

where $a$ is the initial amplitude, $\zeta$ is the unitless damping ratio, $\omega$ is the circular natural frequency of the specimen, and $t$ is the time [12]. The natural frequency was verified with the Fourier transform of the recorded signal.

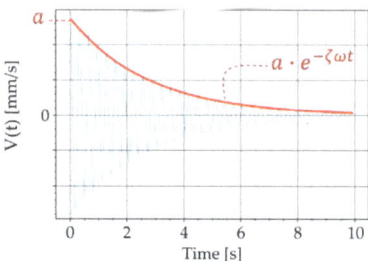

**Figure 7.** Approximating the upper decay with an exponential function.

The damping ratio in Equation (4) is a unitless value between 0 to 1.0. However, it is common in civil engineering and structural design codes to describe the damping ratio as a percentage. Therefore, we present the damping ratio throughout this manuscript as a percentage. To convert the damping ratio to a percentage, the unitless value is multiplied by 100.

*3.2. Results and Discussion*

The free vibration test was conducted in two steps. In the first step, the test was carried out with the conventional concrete beam, P720. Figure 8a shows the free vibration diagram obtained from the test. The Butterworth filter with a bandpass of 200 to 500 Hz was used to filter the recorded signal. The upper and lower limit of the bandpass was adjusted to have the natural frequency of the specimen approximately in the middle of the bandpass. The Fourier transform of the free vibration is provided in Figure 8b. This diagram is also known as the Fourier Spectrum. The horizontal axis shows the frequency components of the signal and the vertical axis represents the amplitude of the frequency components. The amplitude describes how dominant a particular frequency component is [25] (pages 51 to 70). The Fourier transform of the signal was obtained in this study to determine the natural frequency of the beam. The peak of the diagram at 304 Hz corresponds to the first natural frequency of the beam. By applying Equation (4) to the free vibration, the exponential curve approximating the upper decay was obtained, which resulted in a value of 0.36% for the damping ratio (Figure 8a). The signal processing was programmed in Python with the SciPy library [26].

In the second step, the test was repeated with the metaconcrete beams P720-K7 and P720-K9. The diagrams of the free vibration are provided in Figures 9a and 10a, respectively. The corresponding Fourier transforms are shown in Figures 9b and 10b. For both metaconcrete beams, the damping ratio was estimated with Equation (4). The first metaconcrete beam, P720-K7, exhibited a damping ratio of 1.83%. For the second metaconcrete beam, P720-K9, the damping ratio was estimated at 2.72%. Both metaconcrete beams demonstrated a larger damping ratio than that of the conventional concrete beam. A larger damping ratio indicates a faster decay of the free vibration. This is apparent from a comparison between the diagrams of the free vibration of the beams. The diagram in Figure 8a shows that the vibration of the conventional concrete beam decreased to zero in approximately 0.5 s. The vibration of the metaconcrete beam P720-K7 took about 0.1 s to decay

(Figure 9a). The decay time for the vibration of the metaconcrete beam P720-K9 was around 0.7 s (Figure 10a).

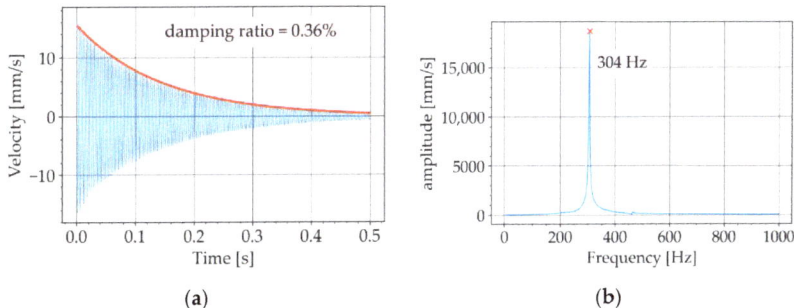

**Figure 8.** Experiment with P720: (**a**) decaying free vibration; (**b**) Fourier transform of vibration.

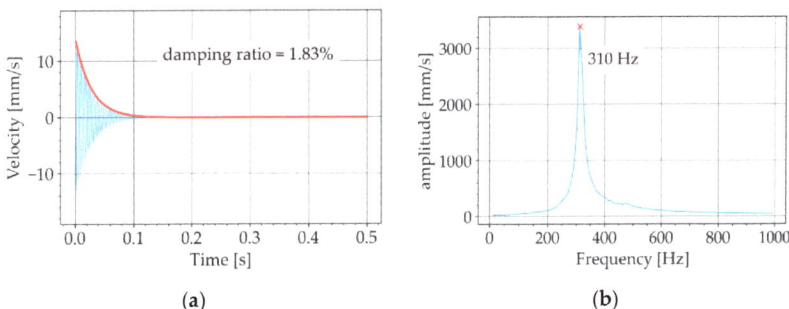

**Figure 9.** Experiment with P720-K7: (**a**) decaying free vibration; (**b**) Fourier transform of vibration.

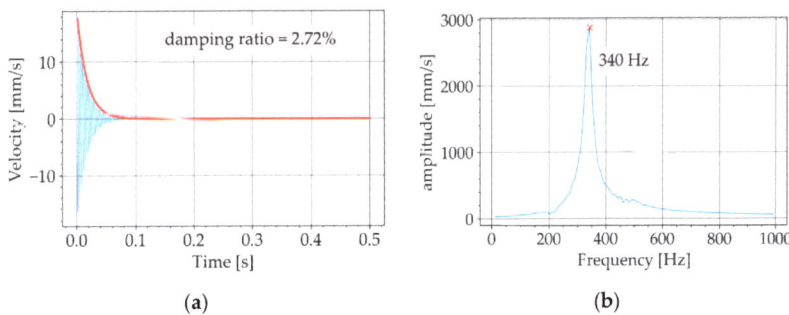

**Figure 10.** Experiment with P720-K9: (**a**) decaying free vibration; (**b**) Fourier transform of vibration.

The free vibration test was repeated with every beam a total number of four times. This was done to ensure that the results are reproducible. The tests were conducted under the same conditions each time. The test setup including the manual excitation with the impact hammer remained unchanged in all tests. The damping ratio was estimated in every test. For every beam, the mean, maximum, and minimum values of the damping ratio were verified. The diagrams in Figure 11 illustrate these values.

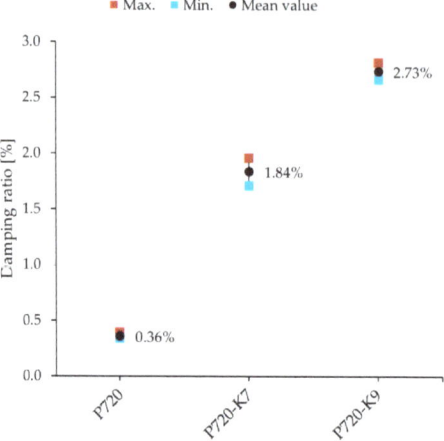

**Figure 11.** Mean values of the damping ratio of the conventional concrete beam, P720, and the metaconcrete beams P720-K7 and P720-K9.

As shown in Figure 11, the conventional concrete beam, P720, exhibited a mean damping ratio of 0.36%. Both metaconcrete beams, P720-K7 and P720-K9, demonstrated a significantly higher damping ratio. For the metaconcrete beam with the K7 size, the mean value of the damping ratio was measured at 1.84% (Figure 11). This is more than five times larger than the measured value with the conventional concrete beam. The metaconcrete beam with the K9 size demonstrated an even higher damping ratio with a mean value of 2.73% (Figure 11). This is more than seven times larger than the damping ratio of the conventional concrete beam.

The mean values of the damping ratio of the metaconcrete beams were greater than that of the conventional concrete beam in all of the repeated tests. Therefore, the first objective of this study was fulfilled. The increase in the damping ratio shows suppression of the beam's vibration. This is because a larger damping ratio signifies that the free vibration of the beam decays faster.

Figure 11 shows that the beam with K9-inclusions demonstrated a higher damping ratio than the beam with K7-inclusions did. The higher damping ratio of P720-K9 was caused by the larger mass ratio in the multi-TMD setup that the beam and the inclusions built. To elaborate more on this, the mass ratio is computed here for every beam. The mass ratio is one of the parameters for tuning a damper that Den Hartog introduced in [4]. We used Den Hartog's tuning procedure in our experimental study in [12] (Sections 3.3 and 4.4). We showed how the core-coating inclusions were tuned to the vibratory modes of the small-scale beam under investigation. Similarly, the mass ratio in this study was computed with

$$\mu = \frac{m_d}{m_m} \qquad (5)$$

where $m_d$ was the mass of the damper and $m_m$ was the kinetic equivalent mass of the main system in the mode under consideration. According to [4], the equivalent mass of a single-span beam in the first vibratory mode is a quarter of the total mass. As mentioned earlier, the volume fraction of the inclusions in a cubic meter of concrete was 5%. Therefore, the equivalent mass, $m_m$, was computed from the mass of P720 given in Table 4 as follows

$$m_m = (3980 \times (1 - 0.05)) \times \frac{1}{4} = 945.25 \quad \text{gr} \qquad (6)$$

The damper's mass was the sum of the masses of the steel cores. By obtaining the mass of the steel cores from Table 2 and the number of them from Table 4, the damper's mass in every metaconcrete beam was

$$m_{d;K7} = 2.14 \times 60 = 128.4 \quad \text{gr} \tag{7}$$

$$m_{d;K9} = 4.19 \times 40 = 167.6 \quad \text{gr} \tag{8}$$

Finally, the mass ratio was computed for every metaconcrete beam with Equation (5)

$$\mu_{P720-K7} = \frac{128.4}{945.25} \times 100 = 13.6\% \tag{9}$$

$$\mu_{P720-K9} = \frac{167.6}{945.25} \times 100 = 17.7\%. \tag{10}$$

Petersen in [27] conducted a parametric study for Den Hartog's tuning parameters to investigate the impact of the frequency, damping ratio, and mass of the damper on the response of the main system. An increase in the mass of the damper resulted in greater suppression of the system's dynamic response [27] (Chapter 18). In this study, the greater mass ratio of P720-K9 caused a higher damping ratio compared to P720-K7, which was in line with Petersen's finding.

## 4. Determination of the Vibration Amplitude with the Frequency Sweep Test

The frequency sweep test is the experimental procedure of measuring the quasi-steady-state response of the test object to a harmonic excitation with a varied frequency. The excitation in the test is a sinusoidal signal with a constant amplitude. The frequency of the signal continuously sweeps from the lower to the upper limit of the frequency range of interest. The test has a wide range of applications in structural dynamics. The purpose of the test in this study was to investigate whether the metaconcrete beams with the randomly distributed damping aggregates demonstrate a lower vibration amplitude compared to the conventional concrete beam. The test procedure followed the given requirements in ISO 7626-2: 1990.

*4.1. Description of the Test Setup and Procedure*

Similar to the free vibration test, the first vibratory mode of the beam was the mode of interest in the frequency sweep test. This means the amplitude of the resonance vibration at the first mode was to be evaluated. In addition, the requirements of ISO 7626-2 for the supports were applicable here too. Therefore, the supports of the beams were aligned similarly to the free vibration test in Section 3.1.

A vibration speaker was employed to generate the swept sinusoidal excitation (Figure 12). The speaker was connected to a digital amplifier to boost the signal (Brand: Sigmatest, Model: DPA4-700). The signal frequency swept linearly from 100 to 600 Hz. The sweep rate was chosen carefully to fulfill the requirements of ISO 7626-2. The standard specifies that the sweep rate shall be slow enough to achieve a quasi-steady-state response of the structure. For linearly swept excitation, the upper limit of the sweep rate, $(df/dt)_{max}$, in Hz/min, was computed with

$$\left(\frac{df}{dt}\right)_{max} \leq 54 \frac{f_n^2}{Q^2} \tag{11}$$

where $f_n$ was the estimated natural frequency and $Q$ was the dynamic amplification factor [28] (Section 3.2.3), which was determined with

$$Q = \frac{1}{\sqrt{(1-\eta^2)^2 + (2\zeta\eta)^2}} \tag{12}$$

where $\eta$ was the ratio of the exciting frequency and the natural frequency. Knowing that at resonance the exciting frequency and natural frequency are equal, we substituted in Equation (12) to obtain

$$\eta = 1 \tag{13}$$

$$Q = \frac{1}{2\zeta} \tag{14}$$

where $\zeta$ was the unitless damping ratio of the test object.

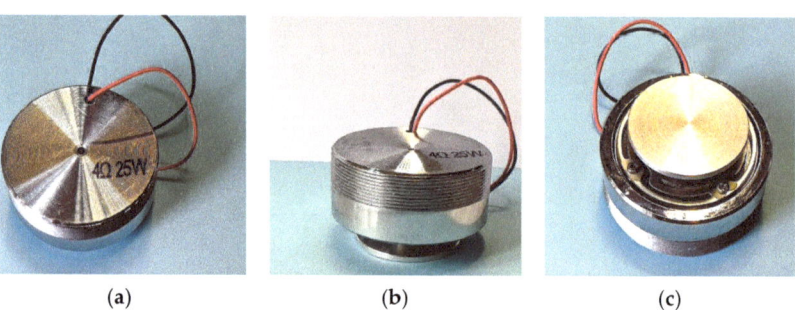

**Figure 12.** Vibration speaker: (**a**) top view; (**b**) side view; (**c**) bottom view.

Table 5 summarizes the first natural frequency, $f_n$, and the mean damping ratio $\zeta_{mean}$ of the beams, which were obtained from the free vibration test in Section 3. The dynamic amplification factor, $Q$, was determined with Equation (14) with the unitless damping ratio, and the corresponding maximum sweep rate was computed with Equation (11). The maximum allowable sweep rate for P720 at 258 Hz/min was the dominating value because it was the smallest. Therefore, the sweep rate in the test with all of the beams was set to 258 Hz/min.

**Table 5.** Determination of the maximum sweep rate in accordance with ISO 7626-2.

| Beam | $f_n$ [Hz] | $\zeta_{mean}$ [%] | Q [-] | $(df/dt)_{max}$ [Hz/min] |
|---|---|---|---|---|
| P720 | 304 | 0.36 | 139 | 258 |
| P720-K7 | 310 | 1.84 | 27 | 7119 |
| P720-K9 | 340 | 2.73 | 18 | 192,667 |

Similar to the test in Section 3, the mid-span of the beams was monitored with the laser vibrometer. Figure 13a shows an overall view of the test setup. The vibration speaker was placed approximately at the mid-span of the beams (Figure 13b).

*4.2. Results and Discussion*

The frequency sweep test was conducted in two steps. In the first step of the experiment, the conventional concrete beam, P720, was tested. Figure 14a shows the response of the beam to the swept sinusoidal excitation in the time domain. Since the laser vibrometer recorded the velocity of the vibration, the unit of the response amplitude is mm/s. The Butterworth filter with a bandpass of 100 to 600 Hz was used to filter the recorded response. The upper and lower limit of the bandpass matched the frequency range of the swept excitation. The response diagram in Figure 14a shows a peak with an amplitude of 24.7 mm/s at around 45 s. The exciting frequency reached 300 Hz at around 45 s because it started at 100 Hz with a sweep rate of 258 Hz/min. Therefore, the peak at this time is the amplitude of the resonance vibration at the first vibratory mode of the beam. The Fourier transform of the response is provided in Figure 14b. The peak of the Fourier transform at 302 Hz corresponds to the first natural frequency of the beam.

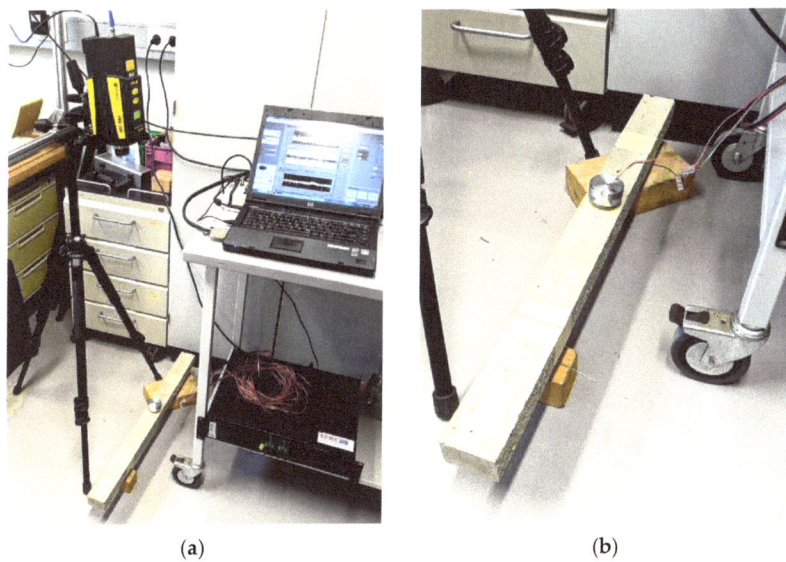

**Figure 13.** Frequency sweep test setup: (**a**) an overall view; (**b**) a close-up of the specimen with the vibration speaker placed on it.

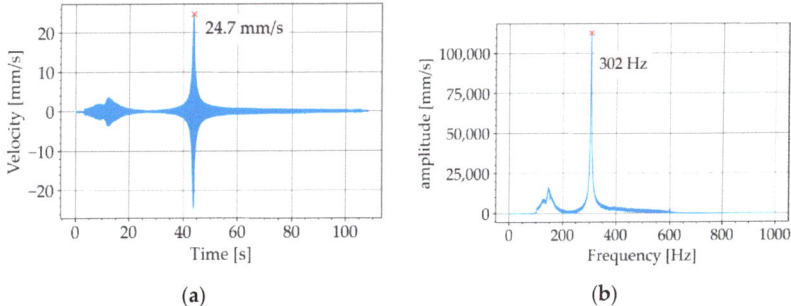

**Figure 14.** Frequency sweep test with P720: (**a**) time history of response; (**b**) Fourier transform of response.

In the second step, the frequency sweep test was repeated with both metaconcrete beams. The response of the first metaconcrete beam, P720-K7, in the time domain, is provided in Figure 15a. The Fourier transform of the response is shown in Figure 15b. The peak of the response appeared after 45 s with a much smaller amplitude of 6.2 mm/s, which is the amplitude of the resonance vibration of the beam.

Similarly, the response diagram of the second metaconcrete beam, P720-K9, is provided in Figure 16a, and the Fourier transform in Figure 16b. The response diagram shows the peak of the resonance just below 50 s with an amplitude of 4.4 mm/s, which corresponds to the first natural frequency of the beam.

The response diagrams of all three beams show peaks between 0 to 20 s (Figures 14a, 15a and 16a). These peaks correspond to frequencies around 150 Hz on the FFT diagrams (Figures 14b, 15b and 16b). These frequencies are generally related to the resonance of the suspensions in the test setup. ISO 7626-2 specifies that the frequencies of the suspension resonances shall be well away from the modal frequencies of the test object. The standard requires the resonance frequencies of the suspension to be less than half of the lowest frequency of interest. This requirement was met by the test setup in this study since the frequency of the first vibratory mode of all three beams was equal to or higher than 300 Hz.

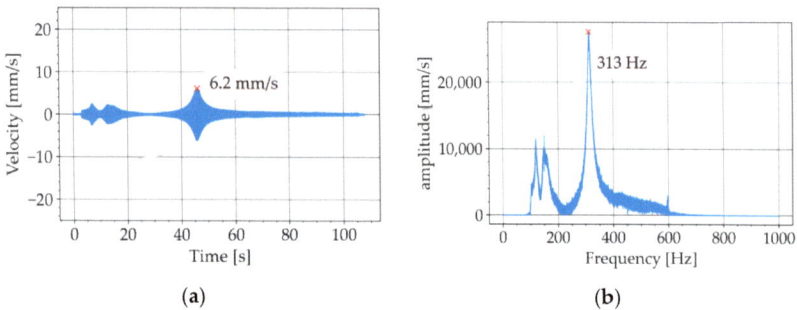

**Figure 15.** Frequency sweep test with P720-K7: (**a**) time history of response; (**b**) Fourier transform of response.

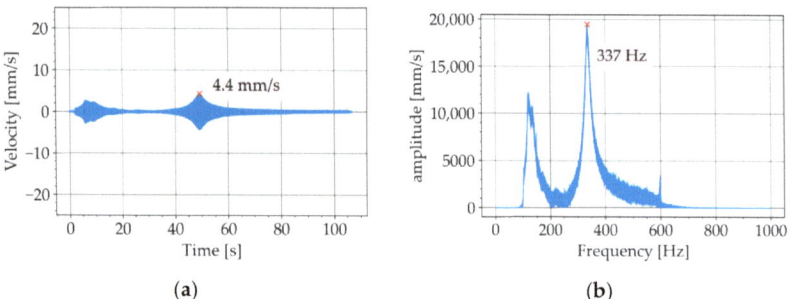

**Figure 16.** Frequency sweep test with P720-K9: (**a**) time history of response; (**b**) Fourier transform of response.

Similar to the experiments in Section 3, the frequency sweep test with every beam was repeated a total number of four times. For every test, the amplitude of the resonance vibration was obtained from the response diagram. The diagram in Figure 17 compares the response of the three beams at resonance vibration. The maximum and minimum measurements of the response together with the mean values are shown in the diagrams. In the following, the mean values are compared and reviewed.

As presented in Figure 17, the vibration of the conventional concrete beam, P720, had a mean amplitude as high as 24.8 mm/s. The response of both metaconcrete beams, P720-K7 and P720-K9, showed a significantly smaller vibration amplitude (Figure 17). The metaconcrete beam with the K7 damping aggregate, P720-K7, demonstrated a mean amplitude of 6.1 mm/s. This is one-fourth of the vibration amplitude of the beam P720. The mean amplitude of the response of the second metaconcrete beam, P720-K9, was measured at 4.7 mm/s, which is smaller than one-fifth of the mean amplitude of P720.

The frequency sweep test demonstrated that the amplitude of the resonance vibration of the metaconcrete beams was significantly smaller than that of the conventional concrete beam. This means that the core-coating inclusions in concrete suppressed the vibration of the beam. Therefore, the outcome of the frequency sweep test fulfilled the second objective of this study.

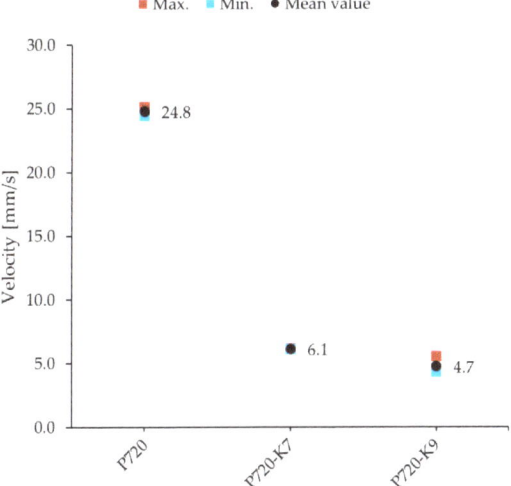

**Figure 17.** Mean values of the vibration amplitude of the conventional concrete beam, P720, and the metaconcrete beams, P720-K7 and P720-K9, at resonance.

## 5. Conclusions

Structures and their components exhibit a substantially large vibration amplitude at resonance, which can lead to their failure. The classical solution for suppressing resonance vibration is a damper, which is tuned to the resonance frequency of the structure. As an alternative solution, this study investigated the utilization of silicone-coated steel balls in concrete as damping aggregates. This mixture was referred to as metaconcrete. The ability of the silicone-coated steel balls to function similarly to a damper was verified in our study in [12]. As a complementary work, this study validated the improved dynamic response of two small-scale metaconcrete beams through laboratory experiments.

The authors demonstrated the improvement in the dynamic response of two small-scale metaconcrete beams by measuring their damping ratio and vibration amplitude at resonance. A similar beam made from conventional concrete was used in both measurements as the reference specimen. To verify the improved dynamic response of the metaconcrete beams, their measurements were compared to those of the reference beam.

Firstly, the authors conducted the free vibration test to determine the damping ratio of the beams. The test with the first metaconcrete beam, P720-K7, measured a mean damping ratio of five times larger than the damping ratio of the conventional concrete beam. The test with the second metaconcrete beam, P720-K9, estimated a mean damping ratio of more than seven times larger than that of the reference beam. A higher damping ratio means a faster decay of the free vibration. The test showed that the free vibration of the metaconcrete beams decayed to zero in a shorter time compared to the reference beam. The higher damping ratio and shorter decay time proved the ability of the distributed core-coating inclusions to suppress the vibration of the beams.

Secondly, the authors conducted the frequency sweep test to measure the vibration amplitude of the beams at resonance. The test determined the response of the beams to a linearly swept sinusoidal excitation with constant amplitude. The vibration amplitude of the first metaconcrete beam was measured at one-fourth of the vibration amplitude of the reference beam. The measurement with the second metaconcrete beam showed an amplitude of one-fifth of the vibration amplitude of the reference beam. The smaller vibration amplitudes of the metaconcrete beams at resonance proved the ability of the distributed core-coating inclusions to suppress resonance vibration.

The larger damping ratio and the smaller vibration amplitude of the metaconcrete beams verified their improved dynamic response compared to the conventional concrete beam with similar characteristics. Therefore, this study validated that the utilization of the randomly distributed core-coating inclusions in concrete is an alternative solution for suppressing resonance vibration. However, this work has some limitations. The natural frequencies of the structural components in practice are significantly lower than the natural frequencies of the small-scale beams. Although the small-scale beams were intentionally chosen as the test object for this study, similar experiments with larger test objects and lower frequencies are to be done in future work.

We conducted the experiments with only one specimen for every core-coating size. To understand how the random distribution of inclusions affects the outcome of the experiments, it is recommended that a variety of specimens be tested, each with a different random distribution of the inclusions. This approach should provide valuable insights into the impact of the distribution on experimental results. This task will be addressed in future work.

Utilizing engineered aggregates in concrete should not have a negative impact on the strength and mechanical properties of concrete. Despite the fact that the volume fraction of the inclusions in this work had a relatively small value of 5% in a cubic meter of concrete, it is necessary to conduct the compressive and tensile test with metaconcrete specimens to examine its strength. The limited scope of this paper did not allow for such an investigation. However, the scope of future work will include it.

**Author Contributions:** Conceptualization, M.A. and C.K.; methodology, M.A. and C.K.; validation, M.A. and C.K.; formal analysis, M.A.; investigation, M.A.; resources, M.A. and C.K.; writing—original draft preparation, M.A.; writing—review and editing, M.A., F.T. and C.K.; visualization, M.A.; supervision, C.K.; project administration, C.K.; funding acquisition, C.K. All authors have read and agreed to the published version of the manuscript.

**Funding:** This manuscript was prepared as part of the project "Functionalization of smart materials for multi-feld requirements of transport infrastructure" funded by the Carl Zeiss Foundation (www.carl-zeiss-stiftung.de) as part of "Förderlinie Durchbrüche 2019—Intelligente Werkstoffe".

**Institutional Review Board Statement:** Not applicable.

**Informed Consent Statement:** Not applicable.

**Data Availability Statement:** Not applicable.

**Acknowledgments:** The authors would like to thank the Material Research and Testing Institute (Materialforschungs- und -prüfanstalt, www.mfpa.de) for providing the laboratory and instruments used in this study.

**Conflicts of Interest:** The authors declare no conflict of interest.

## References

1. Olson, D.W.; Wolf, S.F.; Hook, J.M. The tacoma narrows bridge collapse. *Phys. Today* **2015**, *68*, 64–65. [CrossRef]
2. Petersen, C. *Schwingungsdämpfer im Ingenieurbau*; Maurer Soehne GmbH & Co. KG: Munich, Germany, 2001.
3. Frahm, H. Device for Damping Vibration of Bodies. U.S. Patent 989,958, 30 October 1909.
4. Den Hartog, J. *Mechanical Vibrations*, 3rd ed.; McGraw-Hill: New York, NY, USA, 1947.
5. Warburton, G.; Ayorinde, E. Optimum absorber parameters for simple systems. *Earthq. Eng. Struct. Dyn.* **1980**, *8*, 197–217. [CrossRef]
6. Warburton, G. Optimum absorber parameters for various combinations of response and excitation parameters. *Earthq. Eng. Struct. Dyn.* **1982**, *10*, 381–401. [CrossRef]
7. Abe, M.; Fujino, Y. Dynamic characterization of multiple tuned mass dampers and some design formulas. *Earthq. Eng. Struct. Dyn.* **1994**, *23*, 813–835. [CrossRef]
8. Yamaguchi, H.; Harnpornchai, N. Fundamental characteristics of Multiple Tuned Mass Dampers for suppressing harmonically forced oscillations. *Earthq. Eng. Struct. Dyn.* **1993**, *22*, 51–62. [CrossRef]
9. Igusa, T.; Xu, K. Dynamic characteristics of multiple substructures with closely spaced frequencies. *Earthq. Eng. Struct. Dyn.* **1992**, *21*, 1059–1070.

10. Igusa, T.; Achenbach, J.D.; Min, K.-W. Resonance Characteristics of Connected Subsystems: Theory and Simple Configurations. *J. Sound Vib.* **1991**, *146*, 407–421. [CrossRef]
11. Igusa, T.; Achenbach, J.D.; Min, K.-W. Resonance Characteristics of Connected Subsystems: General Configurations. *J. Sound Vib.* **1991**, *146*, 423–437. [CrossRef]
12. Ansari, M.; Zacharias, C.; Koenke, C. Metaconcrete: An Experimental Study on the Impact of the Core-Coating Inclusions on Mechanical Vibration. *Materials* **2023**, *16*, 1836. [CrossRef] [PubMed]
13. Wasler, R.M. Complex Mediums II: Beyond Linear Isotropic Dielectrics. *Electromagn. Metamater.* **2001**, *4467*, 1–15.
14. Mitchell, S.J. Metaconcrete: Engineered Aggregates for Enhanced Dynamic Performance. Ph.D. Thesis, California Institute of Technology, Pasadena, CA, USA, 2016.
15. Briccola, D.; Ortiz, M.; Pandolfi, A. Experimental Validation of Metaconcrete Blast Mitigation Properties. *J. Appl. Mech.* **2017**, *84*, 031001. [CrossRef]
16. Miranda, E.J.P., Jr.; Angelin, A.F.; Silva, F.M.; Dos Santos, J.M.C. Passive vibration control using a metaconcrete thin plate. *Cerâmica* **2019**, *65*, 27–33. [CrossRef]
17. Briccola, D.; Tomasin, M.; Netti, T.; Pandolfi, A. The Influence of a Lattice-Like Pattern of Inclusions on the Attenuation Properties of Metaconcrete. *Front. Mater.* **2019**, *6*, 35. [CrossRef]
18. Briccola, D.; Cuni, M.; Juli, A.D.; Ortiz, M.; Pandolfi, A. Experimental Validation of the Attenuation Properties in the Sonic Range of Metaconcrete Containing Two Types of Resonant Inclusions. *Exp. Mech.* **2021**, *61*, 515–532. [CrossRef]
19. Briccola, D.; Pandolfi, A. Analysis on the Dynamic Wave Attenuation Properties of Metaconcrete Considering a Quasi-Random Arrangement of Inclusions. *Front. Mater.* **2021**, *7*, 615189. [CrossRef]
20. Kaewunruen, S.; Li, D.; Chen, Y.; Xian, Z. Enhancement of Dynamic Damping in Eco-Friendly Railway Concrete Sleepers Using Waste-Tyre Crumb Rubber. *Materials* **2018**, *11*, 1169. [CrossRef] [PubMed]
21. Razak, H.; Choi, F.C. The effect of corrosion on the natural frequency and modal damping of reinforced concrete beams. *Eng. Struct.* **2001**, *23*, 1126–1133. [CrossRef]
22. Gaul, L.; Schmidt, A. Experimental Determination and Modeling of Material Damping. In *Schwingungsdämpfung: Modellbildung, Numerische Umsetzung, Experimentelle Verfahren, Praxisrelevante Passive und Adaptive Anwendungen*; VDI Verlag GmbH: Wiesloch, Germany, 2007.
23. Botelho, E.; Campos, A.; de Barros, E.; Pardini, L.; Rezende, M. Damping behavior of continuous fiber/metal composite materials by the free vibration method. *Compos. Part B Eng.* **2005**, *37*, 255–263. [CrossRef]
24. Rogers, G.L. *Dynamics of Framed Structures*; John Wiley & Sons, Inc.: Hoboken, NJ, USA, 1959.
25. Gasquet, C.; Witomski, P. *Fourier Analysis and Applications*; Springer: New York, NY, USA, 1999.
26. SciPy documentation-Fourier Transforms, The SciPy Community, 28 June 2023. Available online: https://docs.scipy.org/doc/scipy/tutorial/fft.html (accessed on 12 July 2023).
27. Petersen, C.; Werkle, H. *Dynamik der Baukonstruktionen*, 2nd ed.; Springer Vieweg: Berlin/Heidelberg, Germany, 2017.
28. Chopra, A.K. *Dynamics of Structures*, 5th ed.; Pearson Education: London, UK, 2020.

**Disclaimer/Publisher's Note:** The statements, opinions and data contained in all publications are solely those of the individual author(s) and contributor(s) and not of MDPI and/or the editor(s). MDPI and/or the editor(s) disclaim responsibility for any injury to people or property resulting from any ideas, methods, instructions or products referred to in the content.

Article

# Fretting Fatigue Life Prediction of Dovetail Structure Based on Plastic Effect and Sensitivity Analysis of Influencing Factors

Jianjun Zhou [1], Bowen Yang [2,*], Shuaiyuan Li [1] and Junzhou Huo [2]

[1] State Key Laboratory of Shield Machine and Boring Technology, Zhengzhou 450001, China
[2] School of Mechanical Engineering, Dalian University of Technology, Dalian 116024, China
* Correspondence: ybw19920218@163.com

**Abstract:** Micro relative sliding exists on the contact surface of the main primary equipment's surface structures, resulting in serious fretting fatigue. The plastic effect causes serious fatigue to the structure under alternating loads. Existing fatigue life prediction models fail to fully consider the shortcomings of fretting and plastic effects, which causes the prediction results to be significantly different to real-lifeworld in engineering situations. Therefore, it is urgent to establish a fretting damage fatigue life prediction model of contact structures which considers plastic effects. In this study, a plastic fretting fatigue life prediction model was established according to the standard structural contact theory. The location of dangerous points was evaluated according to a finite element simulation. The cyclic load maximum stress value was compared with the fretting fatigue test data to confirm the error value, and the error between the proposed fretting fatigue life model and the test value was within 15%. Concurrently, we combined this with mass data analysis and research, as it is known that the contact zone parameters have an impact on fretting fatigue and affect the structural lifespan. With the help of ABAQUS, the fretting numerical calculation of the dovetail tenon model was carried out to analyze the sensitive factors affecting the fretting fatigue life of the dovetail tenon structure. By keeping the fretting load unchanged, the contact area parameters such as contact surface form, contact area width and friction coefficient were changed in order to calculate the fretting stress value, $\sigma_{fretting}$ and the dovetail structure was improved to extend its fretting fatigue life. Finally, it was concluded that fretting fatigue was most sensitive to the width and contact form of the contact area. In actual engineering design, multiple factors should be considered comprehensively to determine a more accurate and suitable width and form of the contact area. For the selection of friction coefficient, on the premise of saving costs and meeting the structural strength requirements, the friction coefficient should be as small as possible, and the problem can also be solved through lubrication during processing.

**Keywords:** dovetail structure; fretting fatigue; life prediction; sensitivity factors

Citation: Zhou, J.; Yang, B.; Li, S.; Huo, J. Fretting Fatigue Life Prediction of Dovetail Structure Based on Plastic Effect and Sensitivity Analysis of Influencing Factors. *Materials* 2023, 16, 3521. https://doi.org/10.3390/ma16093521

Academic Editor: Davide Palumbo

Received: 14 March 2023
Revised: 13 April 2023
Accepted: 24 April 2023
Published: 4 May 2023

**Copyright:** © 2023 by the authors. Licensee MDPI, Basel, Switzerland. This article is an open access article distributed under the terms and conditions of the Creative Commons Attribution (CC BY) license (https://creativecommons.org/licenses/by/4.0/).

## 1. Introduction

The extreme environmental complexity of major equipment requires the dynamic load of the main load-bearing structure to have strong mutation characteristics, which cause fatigue fretting wear. During actual service, serious damage and failure of key load-bearing components and other engineering problems occur, as shown in Figure 1. These dangerous parts can very easily produce micro-sliding, which leads to fretting wear, fretting fatigue, and other phenomena, resulting in a significant impact on fatigue life. Under the cyclic load, fretting fatigue refers to fretting wear, fretting fatigue, and other phenomena caused by the slight slip of structural members. Therefore, it is necessary to accurately predict the fretting fatigue life of structural members and understand the fretting damage mechanism on an in-depth level. The biggest difference between fretting fatigue and normal fatigue is the contact area of two components. Affected by the complex multiaxial load state in the contact area and the stress gradient at the edge of the contact area, the need to obtain accurate load values and the change trend of the contact area has become an important factor for scholars across the world when studying fretting fatigue. Studying fretting fatigue mainly involves the stress calculation under a multi-axial load to accurately predict the location of fretting crack initiation and fretting fatigue life under different external factors [1,2]. As shown in Figure 1, high and low circumferential cyclic fatigue due to stress concentration can be caused by the leaf root, leaf body, and other structural fatigue damage. The centrifugal tensile stress high-temperature pneumatic load causes a transient thermal strain long-time high temperature state, resulting in fatigue damage to the turbine curved blade body. Fretting fatigue causes fatigue damage at the drive shaft and turbine disk seat connection, the flange connection between the drive shaft and turbine disk or helical gear, and the dovetail-type connection of the blade wheel disk.

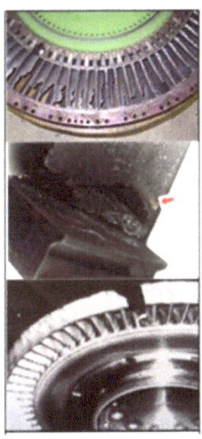

**Figure 1.** Fatigue damage diagram.

Fretting fatigue research covers the multi-axial load and severe stress distribution. Due to the limited harsh conditions of actual service conditions, it is difficult to predict the location of fretting crack initiation and fretting fatigue life. Therefore, many scholars have proposed multi-angle research methods and theories to solve the above problems, for example, based on empirical theory and experimental verification, the critical plane method [3,4], as well as the damage mechanics method [5,6], fretting specific parameter method [7], crack initiation and growth rate [8], etc. The critical plane method is the most commonly used stress–strain system with important damage parameters [9], followed by the KBM parameter method [10] and the FS parameter method [11]. The FP parameter method [12], SSR parameter method [13], and MSSR parameter method [14] are based on stress, while the L-parameter method is based on energy [15], as is the S-W-T parameter method [16]. The main classical continuous damage mechanical model is the Lemaitre damage model [17]. Zeng et al. [18] investigated the susceptibility of the steady-state properties of frictional interaction in preliminary conditions through digital analysis but without experimental verification. Mario et al. [19] proposed the computation of the maximum contact stress at the edge of the dovetail structure, based on the contact theory model. Lemoine et al. [20] analyzed the stress pattern in the contact zone of an aircraft blade under cyclic loading. The fatigue life theory based on stress–strain gradient cycles was established. Macdonald et al. [21] predicted the critical interface orientation and fretting fatigue life based on the stress–strain guidelines. However, normal stress and strain can only be used to predict primary life. The high stresses induced by fretting lead to the fatigue damage of the structure, which results in a plastic zone at the crack tip and rapid growth. Wahab et al. [22] predicted fatigue life by performing fretting fatigue tests. However, the model did not involve the effect of plasticity effects on fatigue. Han et al. [23] considered the effect of plasticity on estimating the free fatigue life, but only the onset of free fatigue was considered and the complete life expectation analysis, such as crack growth, was not investigated. Bhatti et al. [3] analyzed in depth the plastic effect of dovetails under frictional loading and the effects of different stress ratios but did not perform any experiments to verify the model accuracy. Pereira et al. [4] considered the combined influence of frictional and fatigue stresses on the structural parts but neglected to examine the effect of residual stress build-up on the lifespan of the structural parts. Sun et al. [24] investigated the effect of plastic transformation on the evolution of fretting fatigue damage, but the study did not consider the effects of plasticity on the lifespan of the material. Han et al. [25] assessed the effects of different conditions on crack clamping. Although crack tensioning could be analyzed, an in-depth study of fatigue life was lacking. Wu et al. [26] explored the plastic wear behavior of titanium alloys during friction and further analyzed the plasticity effect by monitoring the friction coefficient. Wang et al. [27] analyzed the effect of plastic deformation on the wear process using the FEM method, developed a computational model of cumulative plastic strain, and performed experimental validation. However, this model did not include the effects of plasticity on the fretting fatigue life. Therefore, it has a large margin of error.

Fretting fatigue depends on environmental factors. Relative humidity, corrosive media, vibration, contact pressure, friction, etc. may cause structural damage and cracking [28]. Sharma [29] measured the in-situ wear depth by micro-wear experiments under a normal load of 400 N. It was found that friction had a serious effect on the structure, with the least damage occurring when the friction coefficient was reduced by 50%. Sun [30] analyzed the effect of high temperature on the micro-action fatigue life of a dovetail structure, and established a micro-action fatigue life prediction model considering the criterion of micro-action fatigue crack sprouting failure based on circular plane contact theory. Liu [31] proposed an artificial neural network (HMNN) prediction method based on a fractal mechanism. The layers are proportional multi-axial fatigue, non-proportional multi-axial fatigue, notched fatigue and micro-motion fatigue. Based on the step-by-step construction of fatigue complexity, each layer can be used to evaluate the fatigue life of the previous layer. Mirsayar [32] investigated the effect of material anisotropy on the fatigue crack propagation (FCG) behavior of additively manufactured materials by establishing an energy-based criterion. It was found that material anisotropy, caused by both tectonic orientation and T-shaped stresses, plays a significant role in the FCG behavior of 3D printed components.

In summary, because fretting wear is not easy to observe during crack initiation and fretting damage generally occurs in the contact zone or the edge of the contact zone, which is also difficult to observe, many fatigue damage studies focus on the fracture mechanics method of predicting crack growth life, which is not applicable to predicting fretting fatigue lifespan. The prediction of free fatigue life using the continuous cumulative damage theory needs to consider many factors. Due to the complexity of the structure, the prediction of fretting fatigue initiation and propagation still lacks in-depth theoretical research. Therefore, in order to accurately control the service life impact of fretting fatigue, it is a key research necessity to establish a fretting fatigue life prediction model according to theoretical analysis and the fretting fatigue test.

## 2. Materials and Methods

Due to the special characteristics of the aero-engine dovetail structure and the complexity of multi-axial loading, there is no valid theoretical model to estimate micro-motion fatigue life (initial and total life). Based on the fretting fatigue contact theory and the progressive damage model, the effect of forming effects on the dovetail structure is considered. In order to provide accurate estimates, the research objectives of this paper are: (i) to develop a face contact fretting fatigue life estimation model using the aero-engine dovetail structure as the research object; (ii) to evaluate the validity and applicability of the estimated model by combining theoretical analysis, finite element simulation, and fretting fatigue experiments; and (iii) to analyze the sensitivity of dovetail structure parameters based on the micro-action fatigue mechanism with reference to the stress distribution law and the objective of minimizing micro-action stress.

### 2.1. Establishment of the Fretting Fatigue Life Prediction Model

Fatigue damage is caused by the severe expansion of tiny fine lines and initial defects. The internal damage variables are introduced by considering the damage variables as internal. Additionally, in order to target different material structural damage, the geometry state variable $D$ is introduced [33].

The damage variable $D$, represented as the presence of tiny defects in the unit volume of the material, is shown in Equation (1).

$$D = \frac{A - \tilde{A}}{A} \tag{1}$$

The accumulated damage $D$ is not directly accessible by existing means, so it needs to be calculated using the measured effective area $A$. Therefore, the link between the elastic modulus $E$ and the damage $D$ is established by transformation, and finally, the elastic modulus variation law is obtained experimentally.

Using Hooke's law ($\sigma = F/A$) and cumulative damage theory, the equation is established using the indirect variable of area $A$. The effective stress $\tilde{\sigma}$ is:

$$\tilde{\sigma} = \frac{F}{\tilde{A}} = \frac{\sigma}{1-D} \tag{2}$$

We proposed the concept of fretting stress $\sigma_{fretting}$ based on the structure:

$$\sigma_{fretting} = \sigma_0 + 2p_h \sqrt{\frac{\mu F_T}{F_N}} \tag{3}$$

$p_h$ is the maximum load in the contact zone, $\mu$ is the fretting coefficient of the contact surface, $F_T$ is the tangential load amplitude of standard sample and contact pad, $F_N$ is the normal load, and $\sigma_0$ is the peak value of axial stress.

In 2014, Aditya A. Walvekar of Purdue University introduced fretting stress based on theoretical research and experimental verification, $\sigma_{fretting}$, and proposed a damage evolutive formula of fretting fatigue, in which the damage parameter $\sigma_R$ is a material parameter, which is the fatigue damage parameter of materials under cyclic loading. $\sigma_R$ is the material under fully reversed loading of the resistive stress, with a stress ratio of $-1$, and is a function of the average stress. In the non-damaged state, $D = 0$ and $N = 0$, and under the completely damaged state, $N = N_f$ and $D = D_{crit}$; the derivation is as follows.

$$\frac{dD}{dN} = \left[\frac{\sigma_{fretting}}{\sigma_R(1-D)}\right]^m (\Delta \varepsilon)_p^w \tag{4}$$

In order to obtain a more accurate fretting fatigue life model of dovetail joints, it is necessary to consider the significant plastic strain deformation in the contact zone caused by the combined action of fretting and fatigue stress. The plastic effect not only affects crack generation and expansion but also leads to contact area expansion for fretting fatigue. However, the basic contact zone expansion is mainly attributed to increasing frictional work generated by the relative sliding of the contact surfaces over time, due to the constant action of the alternating load, the plastic deformation of the contact surface, and the deep penetration of the active surface into the driven surface. Thus, in this paper, the influence of the plasticity effect on fretting fatigue is characterized by the maximum plastic strain at the edge of the contact zone during the process, from the undamaged state to the fully damaged state of the specimen. This assumes that the cumulative prediction model of fretting fatigue damage considers plasticity, as shown in Equation (5):

$$N_f = \left(\frac{\sigma_R}{\sigma_{fretting}}\right)^m \left[\frac{1}{m+1} - \frac{(1-D_{crit})^{m+1}}{m+1}\right](\Delta\varepsilon)_p^w \quad (5)$$

where $D_{crit}$ is the critical damage parameter. $(\Delta\varepsilon)_p$ is obtained from experimental measurements and is expressed as the maximum plastic strain value that reaches the critical damage in the test. Based on the above formula derivation, the fretting fatigue damage accumulation model considering plastic effect is finally obtained by combining the contact area load solution, fretting fatigue test data, fitting parameters, etc. The biggest advantage of this model is that the influence of fretting stress, fatigue stress and plastic effect is comprehensively considered, so that the prediction results are more accurate.

We improve the progressive damage method in order to solve the problem of incomplete contact. A generalized stress intensity factor was introduced to specifically analyze the stress state at the edges of the asymptotic contact, and the asymptotic technique was successfully applied to predict fatigue life. Thus, the contact theory can determine the maximum contact stress $P_h$ of the structure, and the remaining unknown parameters need to be calculated using experimental data.

The contact load $P(t)$ is formulated:

$$p(t) = \begin{cases} \frac{K_N}{\sqrt{t}} & \text{if } t/a_{round} \gg 1 \\ \frac{3K_N\sqrt{t}}{a_{round}} & \text{if } 0 < t/a_{round} \ll 1 \end{cases} \quad (6)$$

$$K_N = \frac{-2E^*\sqrt{a_{round}^3}}{3\pi R}. \quad (7)$$

$$p(x) = \frac{3K_N}{4\sqrt{a_{round}^3}}\left[2\sqrt{xa_{round}} + (x - a_{round})\ln\left|\frac{\sqrt{a_{round}} - \sqrt{x}}{\sqrt{a_{round}} + \sqrt{x}}\right|\right] \quad (8)$$

where $a_{around}$ is the radius of the load contact zone, $t$ is the contact edge, $R$ is the edge radius, and $E^*$ is the modulus of elasticity.

*2.2. Fretting Fatigue Simulation Analysis and Dangerous Point Evaluation*

The ABAQUS is used to simulate and analyze the standard specimen. The analysis was performed in ABAQUS Explicit. The plastic model was used for finite element analysis. The Johnson–Cook model was used for damage. The Johnson–Cook model is shown in Equation (1), and the specific parameters are shown in Table 1.

$$\sigma = \left[A + B(\varepsilon_p)^n\right]\left[1 + C\ln\left(\frac{\dot{\varepsilon}_p}{\dot{\varepsilon}_0}\right)\right]\left[1 - \left(\frac{\theta - \theta_{room}}{\theta_{melt} - \theta_{room}}\right)^m\right] \quad (9)$$

**Table 1.** The instanton equations parameters of nickel-based alloys.

| Material Properties | A | B | C | n | m |
|---|---|---|---|---|---|
| DZ125 | 637 | 573.2 | 0.033 | 0.45 | 0.92 |

In the equation, $A$ is the yield strength under quasi-static conditions. $B$ is the strain hardening parameter. $C$ is the strain rate hardening parameter. $m$ is the thermal softening parameter. $n$ is the is the process hardening parameter. $\dot{\varepsilon}_0$ is the quasi-static strain rate. $\theta_{melt}$ is the material melting point. $\theta_{room}$ is the room temperature.

Considering the symmetry of the model, take one half of the model and apply the load under the following conditions: the left end of the standard specimen is subject to a fixed constraint and the lower end is subject to a symmetrical constraint. The left and right ends of the inching pad limit displacement in the x direction. Apply the sinusoidal cyclic axial load to the right end of the standard test piece, and apply the normal load to the upper end of the fretting pad. The contact area finite element meshing element is 100. The augmented Lagrange method was used for the finite element analysis of the contact. The finite element model is shown in Figure 2.

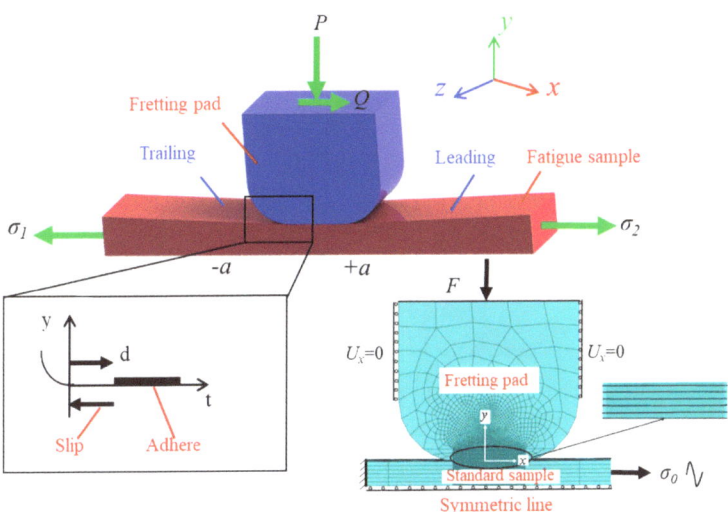

**Figure 2.** The finite element model of fretting.

Keeping the cyclic axial load at 300 MPa constant and taking the normal load as 45–150 MPa, the numerical analysis of fretting was performed on the standard specimen and the fretting pad to obtain the contact stress $P_0$ corresponding to different normal loads. The calculated theoretical result, $P_1$, was compared to the finite element simulation and the theoretical calculated analysis; the error was within 13%, as shown in Table 2, and the fitted curve is shown in Figure 3. The accuracy of the numerical analysis of the fretting force of the standard specimen with ABAQUS is proven.

The deviation between the theoretical and simulated loads may be mainly due to the friction of the contact area, the deviation of the actual size modeling, the mesh division as required in the simulation, the elastic–plastic damage theory, the force area of the contact area for different load forces, and the rounding in the calculation. We made an accurate comparison to ensure the accuracy of the simulation and that the error is within the acceptable range.

**Table 2.** Comparison of simulation and theoretical results of contact stress.

| Normal Load/MPa | a/mm | Simulation/MPa | Theoretical/MPa | Error |
|---|---|---|---|---|
| 45 | 2.538 | 824.836 | 750.504 | 9.01% |
| 50 | 2.543 | 916.573 | 830.618 | 9.38% |
| 60 | 2.548 | 1100.1 | 992.833 | 9.75% |
| 70 | 2.554 | 1283.69 | 1161.95 | 9.48% |
| 80 | 2.558 | 1467.36 | 1313.45 | 10.49% |
| 90 | 2.563 | 1651.09 | 1471.87 | 10.85% |
| 100 | 2.567 | 1834.9 | 1630.317 | 11.15% |
| 110 | 2.572 | 2018.77 | 1786.38 | 11.51% |
| 120 | 2.576 | 2202.72 | 1942.73 | 11.80% |
| 130 | 2.58 | 2386.73 | 2098.11 | 12.09% |
| 140 | 2.584 | 2570.81 | 2252.51 | 12.38% |

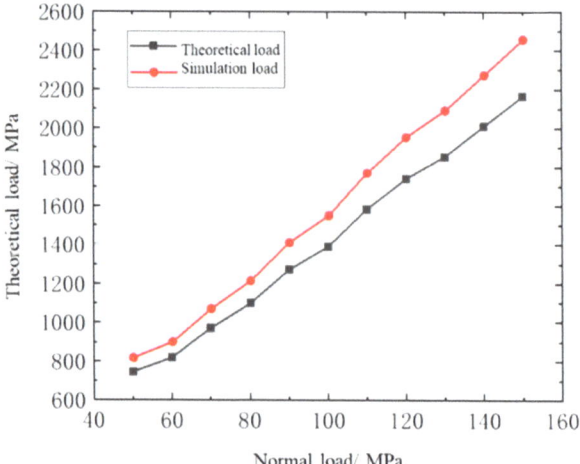

**Figure 3.** Comparison of simulation and theoretical results of contact stress.

In Figure 4, the stress cloud diagram corresponding to the normal load of 45 MPa is shown. It can be observed that different normal loads will not change the weak position of the structure. Therefore, it is possible to determine that the danger point for fretting fatigue occurs at the resulting joint.

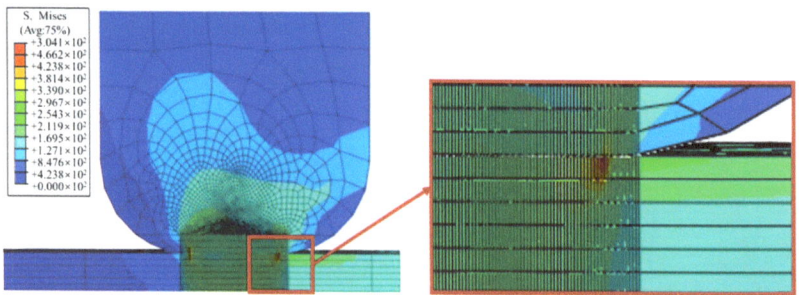

**Figure 4.** Stress nephogram (normal load is 45 MPa) and partial enlarged drawing of the trailing edge of the contact area.

## 3. Results

### 3.1. Fretting Fatigue Test and Analysis

In the fretting fatigue damage accumulation model considering the plastic effect and the contact area load solution, the unknown parameters in the assumed model are obtained from the fretting fatigue life data under different loads. Therefore, the fretting fatigue test is performed on the standard sample.

#### 3.1.1. Test Materials and Samples

The test material is a commonly used nickel base alloy, DZ125. DZ125 is a nickel base precipitation-hardening and directionally solidified columnar alloy that contains elements. It has good overall performance and outstanding fatigue properties. Using the data findings from uniaxial tensile tests, it was calibrated according to the mechanical principles. The property test is shown in Table 3, and the standard samples and fretting pads of the fretting tests are shown in Figure 5.

Table 3. Material parameters of DZ125.

| Material | Density $\rho$ (g/cm$^3$) | Elasticity Modulus $E_1$ (GPa) | Poisson's Ratio $\nu_1$ |
|---|---|---|---|
| DZ125 | 8.59 | 183 | 0.41 |

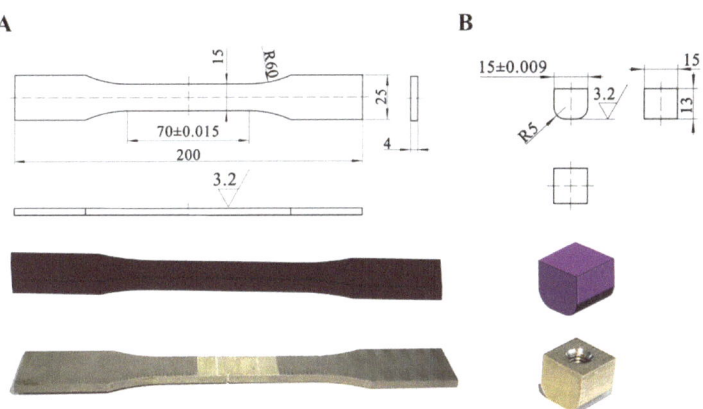

**Figure 5.** Schematic diagram of fretting test sample (**A**) standard sample; (**B**) fretting pad.

3.1.2. Test Equipment

According to the test scheme and requirements, the appropriate test equipment was selected. The fretting fatigue test system independently designed by the team of the Dalian University of Technology is used in this test. The equipment and instruments required for the fretting fatigue test include: one SDS100 electro-hydraulic servo fatigue test machine; one control system; one computer; one gateway node; and one normal loading device (consisting of four parts: platform, guide rail, sliding module, and tension sensor). As shown in Figure 6, the fretting fatigue test system of a standard specimen is adopted.

3.1.3. Test Plan

According to the overall technical scheme, the fretting fatigue test of the standard sample is divided into two parts. The test selects six standard samples with different load points, and each load point is subject to two groups of tests, which are divided into displacement control conditions and load control conditions, according to the load. The first part requires the measurement of the critical damage parameter $D_{crit}$ of the sample by changing the elastic modulus and obtaining the maximum plastic strain $(\Delta\varepsilon)_p$. Displacement control is adopted for test loading control. In the second part, fretting fatigue tests under different cyclic loads are carried out and their fatigue lives are recorded. Combined with the theoretical analysis results, a fretting fatigue life prediction model is established. Load control is adopted for testing loading control.

The fatigue life under different loading conditions was obtained, and a fatigue life model was established through analytical data. Strain plates were pasted on one side of the contact area between the standard samples and the micro pad, and the strain at the edge of the contact area was monitored in real-time. A constant normal load of 100 N was applied with a stress ratio of 0.1 and peak levels of $1.2 \times 10^4$ N, $1.5 \times 10^4$ N, $1.8 \times 10^4$ N, $2.1 \times 10^4$ N, $2.4 \times 10^4$ N, and $2.7 \times 10^4$ N cyclic sinusoidal axial loads. The load parameters are shown in Table 3. The lives of the standard specimens under different loads were monitored. The test was terminated when a strain mutation occurred at the contact area of the standard specimen and the fretting pad under the fretting load. The fretting fatigue test resulted from the standard specimen.

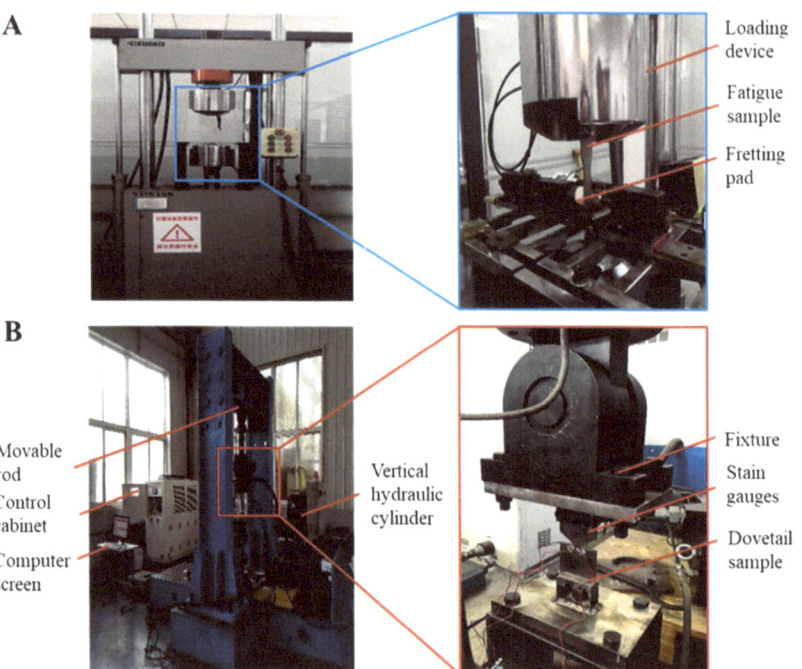

**Figure 6.** Fretting fatigue test system. (**A**) Standard sample test system and (**B**) Dovetail fatigue test system.

### 3.1.4. Fretting Test Results and Analysis

In this paper, a method of monitoring the strain at the lower edge of the contact zone in real time was used to determine the generation of cracks. This was achieved by attaching strain gauges to the side of the lower edge of the contact zone where the standard sample was in contact with the fretting plate and connecting strain nodes to monitor the strain changes at the lower edges of the contact zone during the fretting fatigue test. As the test time increases, its strain slowly increased until the strain grid was sheared at a certain point and the strain suddenly increased to infinity. At this point, cracks appeared on the sample and fractured rapidly. This moment is considered to be the fatigue life of the sample, as shown in Figure 7.

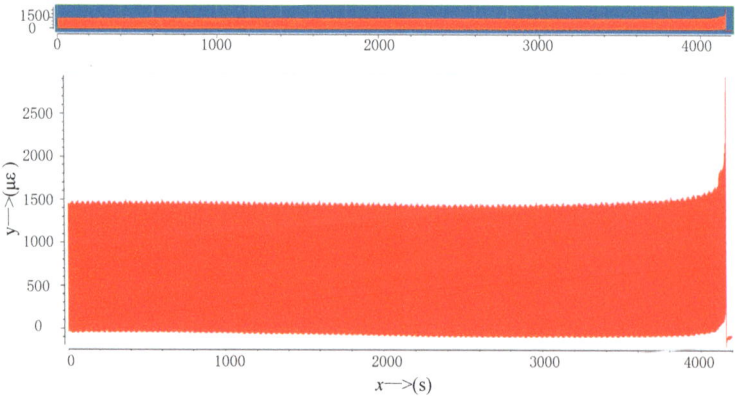

**Figure 7.** Diagram of monitoring strain.

Although many researchers use the mechanics of damage predict lifespan, they usually assume that in the preliminary state, the material has no deficiency, and so $N = 0$ and $D = 0$. In the final state, the crack occupies the entire surface of the representative volume element at the time of fracture $N = N_f$ and $D = 1$. However, according to the long-term fretting test and theoretical experience, before the fracture of the sample, due to the atomic debonding in the residual resistance area, the crack will occupy a certain area of the representative volume unit and suddenly break. Therefore, in the end state, $D$ is defined as $D_{crit}$ instead of 1. The critical damage $D_{crit}$ at the time of fracture can be calculated by measuring the change of the elastic modulus of the material during the test. This method is the elastic modulus change method proposed by Lemaitre. Accurate strain measurement is very important for obtaining $D_{crit}$, so strain gauges are installed in the constant cross-sectional area of the specimen. The test is carried out under the conditions of displacement control. In each cycle, the displacement gradually increases until plastic deformation occurs. The displacement is then reduced until the force (the stress within the specimen) returns to zero. In the next cycle, the displacement further increases, which gradually produces more plastic deformation, and then returns to the state where the force is zero. This process is repeated until the specimen breaks. Figure 8 depicts the stress–strain diagram of each cycle of the test and calculates the elastic modulus according to the slope of the unloading curve. In the expected case, the elastic modulus decreases with the increase in plastic deformation (that is, the damage to the specimen is increasing). Table 4 shows the decrease in the elastic modulus with the increase in damage. The critical damage value $D_{crit}$ at fracture is 0.382.

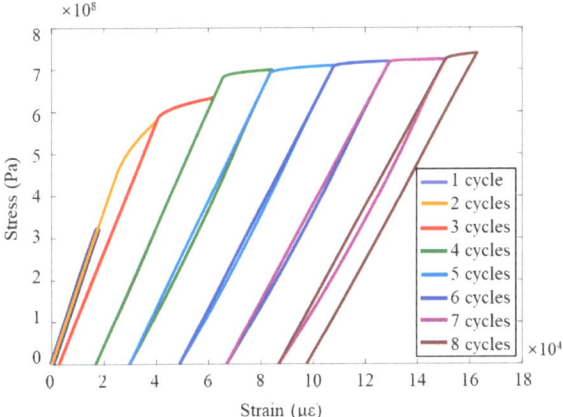

**Figure 8.** Stress–strain curve of elastic modulus change test.

**Table 4.** Variation of elasticity modulus during test.

| Cycles | E(1 − D)/GPa | D |
| --- | --- | --- |
| 1 | 184 | 0 |
| 2 | 155 | 0.156 |
| 3 | 140 | 0.238 |
| 4 | 128 | 0.303 |
| 5 | 120 | 0.347 |
| 6 | 115 | 0.374 |
| 7 | 113 | 0.382 |
| 8 | 113 | 0.382 |

As shown in Figure 8, it can be seen that the number of cycles increases and the strain value increases slowly, indicating the accumulation of plastic strain. Assuming that the plastic strain of the same material before fracture is basically constant, the maximum plastic strain during cyclic dislocation loading can be utilized to obtain the maximum plastic strain generated during the whole process from the undamaged state to the fracture of the test sample, where $(\Delta \varepsilon)_p$ is $9.73 \times 10^{-2}$. The fretting fatigue test of the standard sample was carried out through the test system. The crack fracture diagram of the sample after the test obtained the sample life and fretting stress under different axial loads; the σ fretting to obtain the fretting stress $\sigma_{fretting}$ relationship between fretting and fatigue life, N, is shown in Table 5.

**Table 5.** Summary of fretting fatigue test data of standard specimens.

| Test | Load/N | Stress Ratio ν | Frequencies/Hz | Fretting Stress/MPa | Life N/Cycles |
|---|---|---|---|---|---|
| Test-1 | $1.2 \times 10^4$ | 0.1 | 10 | 534.72 | 936,237 |
| Test-2 | $1.8 \times 10^4$ | 0.1 | 10 | 709.94 | 455,875 |
| Test-3 | $2.4 \times 10^4$ | 0.1 | 10 | 873.36 | 52,990 |
| Test-4 | $2.7 \times 10^4$ | 0.1 | 10 | 952.07 | 25,160 |

Synthesizing all the data and fitting according to the hypothetical theoretical model to determine the value of the unknown parameter, the final result is $m = 3.8$, $\sigma_R = 6774.58$, $w = -2.49$. The fitting curve of test results is shown in Figure 9 below:

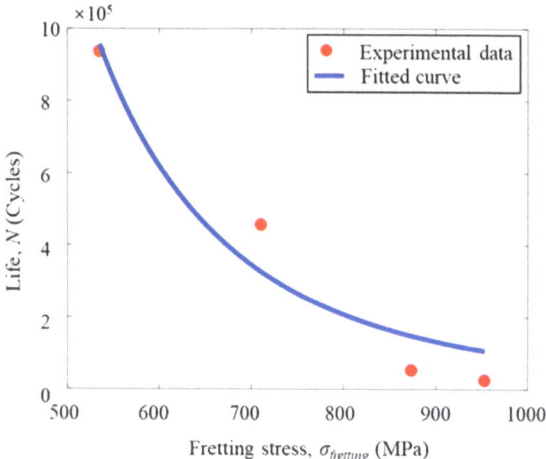

**Figure 9.** Relationship between fretting stress to life.

The schematic diagram of the fractured sample is shown in Figure 10. It can be seen that the strain value at the trailing edge of the contact area slowly increases with time as the number of fretting load cycles grow, signifying the accumulation of plastic strain. At a certain point, the strain suddenly changed in value and the sample fractured rapidly. It is obvious that the fatigue occurs near the contact zone and not at the change in the size of the standard sample, indicating that the fatigue is caused by the high stress gradient in the contact zone, which is consistent with the basic characteristics of a flying edge and proves that the fatigue is due to the flying edge and not otherwise. This proves the accuracy of the fretting fatigue test and ensures the validity of the test data, as well as the validity of the cumulative fretting fatigue life prediction model of successive damages considering the plastic effect established from the test data.

To verify the accuracy of the model based on the data of Test 2 and Test 5, as shown in Table 6, the theoretical prediction value of fretting fatigue life was compared with the test value, and the error was within 15% of the reasonable error. Therefore, the accuracy of the surface contact fatigue life prediction model based on the plastic effect is proven.

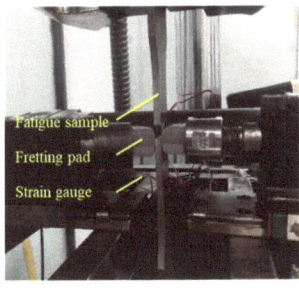
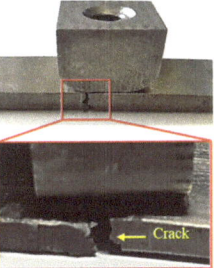

**Figure 10.** Fracture diagram of standard sample.

**Table 6.** Comparison of fretting fatigue test results of standard specimens and theoretical prediction results.

| Test | Axial Load $F_x$/N | Fretting Stress $\sigma_{fretting}$/MPa | Experimental Life N/Cycles | Predicted Life N/Cycles | Error |
|---|---|---|---|---|---|
| S2_1 | $1.5 \times 10^4$ | 624.22 | 627,663 | 534,601 | 14.8% |
| S2_2 | $1.5 \times 10^4$ | 655.37 | 590,021 | 515,102 | 12.7% |
| S4_1 | $2.1 \times 10^4$ | 792.79 | 253,563 | 215,532 | 15.0% |
| S4_2 | $2.1 \times 10^4$ | 780.34 | 149,639 | 127,983 | 14.5% |

*3.2. Fatigue Test and Analysis of Dovetail Structure*

The fretting fatigue life prediction model considering the plastic effect is also suitable for dovetail structures.

The dovetail structure test sample is designed and the fretting fatigue test is conducted according to the standard sample materials and test equipment.

3.2.1. Determining the Dangerous Location of a Dovetail Structure

Peak values are applied to the dovetail model tenon $8.58 \times 10^3$ N, $1.18 \times 10^4$ N, and $1.35 \times 10^4$ N cyclic axial load, and the mean contact stress CPRESS and the Mises equivalent stress S in the contact zone of a dovetail structure are analyzed in order to identify the structural danger points.

As shown in Figure 11, the contact stress CPRESS and the equivalent Sand stress distribution clouds are present in the contact zone at the 50th cycle. The red part is the highest concentration point in the tenon–tenon contact zone, and the maximum value of Mises equivalent stress S is 709.2 MPa. The maximum value of the contact stress CPRESS reached 486 MPa and generally showed a monotonic increasing trend from the front edges of the contact area to the rear edge of the contact zone. Under cyclic axial load, the amplitude of relative displacement increases near the contact zone rear edge. The work generated by the fretting between the mortise and tenon contact area increases, causing an extension of the stress gradient at the trailing edge of the contact area (Horizontal position → 1).

3.2.2. Fretting Fatigue Test of Dovetail Specimen

T The fretting fatigue life of dovetail test with various loading methods was found. The model based on the standard specimen was analyzed and verified to be applicable to the dovetail specimen. The dovetail samples were combined, and strain gauges were attached around the bottom edge of the dovetail contact zone of the dovetail samples. The strain change trend in the dovetail contact area was monitored in real-time in the experiment. The model dimensions and the physical drawing of the dovetail are shown in Figure 12. The sinusoidal cyclic axial load is applied with a stress ratio of 0.1 and a frequency of 10 Hz to the dovetail, and the peak values are $8.58 \times 10^3$ N and $1.18 \times 10^4$ N. We then undertook the real time monitoring and recording of the strain change trend, actuator displacement, and dovetail sample life. The experiment ended when the strain in the touch zone of the dovetail specimen changed abruptly under the cyclic load. Finally, the fretting fatigue test results of the dovetail specimens were obtained.

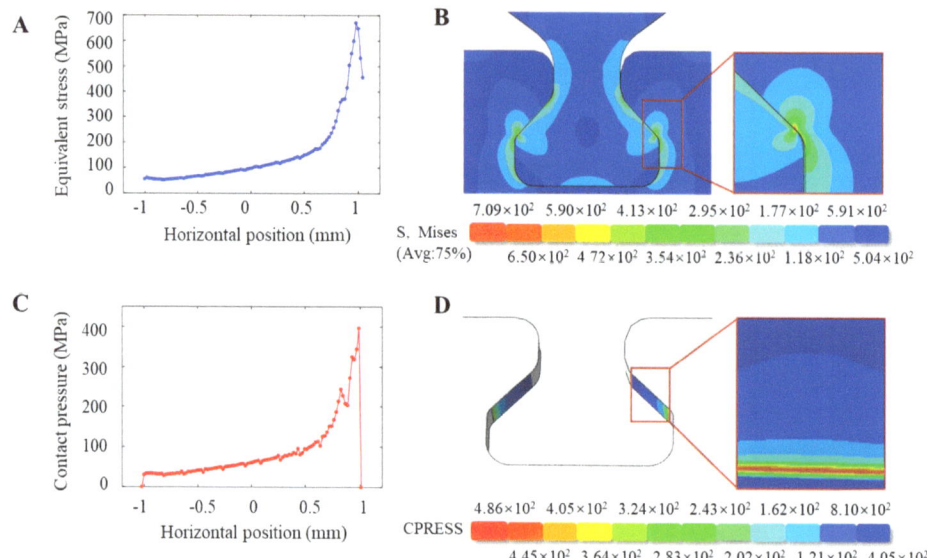

**Figure 11.** Stress nephogram distribution of dovetail model contact zone; (**A**) Mises equivalent stress force analysis of the dovetail model contact zone; (**B**) the amplification of Mises equivalent stress S nephogram and the contact zone; (**C**) the force analysis of contact stress in the contact area of dovetail; (**D**) the contact stress nephogram and contact area enlargement.

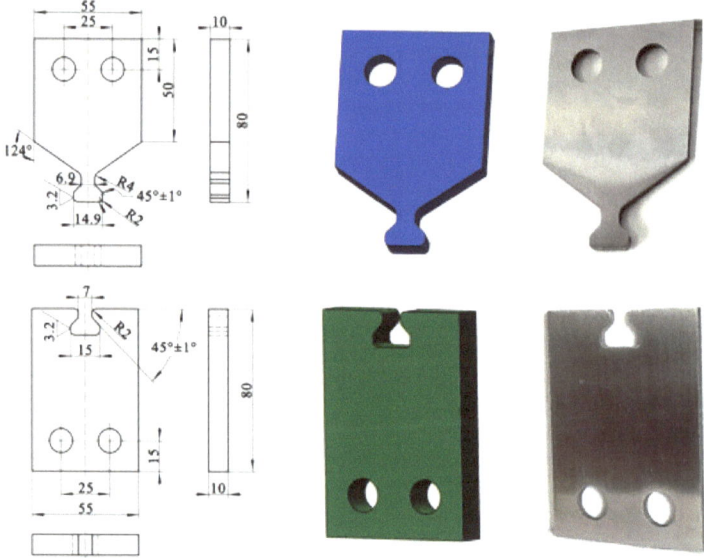

**Figure 12.** Model dimensions and physical drawing of the dovetail.

Under working conditions, there are mainly two kinds of cyclic loads borne by aero-engine turbine blades and discs: one is a centrifugal load of blades, the other is a vibration load of blades. The tenon slides up and down in the mortise under the action of centrifugal force. Under the vibration load action, the tenon swings left and right in the mortise. However, the largest swing amplitude caused by this vibration load is at the upper end of the blade, and the mortise and tenon fit. If the vibration load increases to the moving tenon, the blade will break due to ordinary fatigue, and fretting fatigue is not the main problem. Therefore, only the fretting fatigue between the dovetail and the mortise under centrifugal load is considered here.

### 3.2.3. Fretting Fatigue Test Results of Dovetail Specimens

Table 6 shows the fretting fatigue test results of dovetail with different load conditions, from which the life of the specimens at axial load is calculated. The peak axial load of the real object after the fracture of the sample (Figure 13) is $1.18 \times 10^4$ N. It can be seen that fatigue failure occurs around the contact zone, suggesting that fatigue is caused by the high-stress gradients in the contact zone. The fatigue of the dovetail structure is caused by fretting fatigue due to the relative sliding between dovetail tenon and mortise. Therefore, the test results can be used to confirm the suitability of the model.

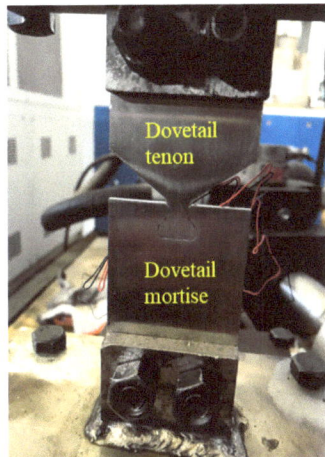
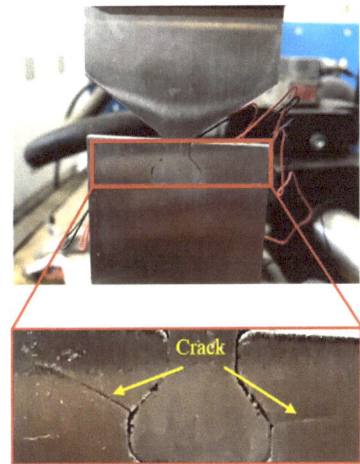

**Figure 13.** Dovetail sample crack (the peak value of cyclic axial load is $1.18 \times 10^4$ N).

The damage cracks produced by the dovetail specimens are analyzed in Figure 13. According to Equation (3), the micro-motion stress in the micro-motion test of the dovetail specimen was calculated, and then the micro-motion fatigue life prediction model of Equation (5) was used to predict the life and the results were compared with the test life, as shown in Table 7. It can be concluded that the theoretical calculation structure is relatively conservative, and the theoretical lives are all smaller than the experimental lives, but the errors are within the range of 12%. Therefore, the life prediction model is based on the fatigue life prediction of the dovetail structure specimens.

**Table 7.** Comparison of fretting fatigue test results and theoretical prediction results of dovetail specimens.

| Test | Axial Load $F_x$/N | Fretting Stress $\sigma_{fretting}$/MPa | Stress Ratio R | Frequency $f$/Hz | Experimental Life N/Cycles | Predicted Life N/Cycles | Error |
|---|---|---|---|---|---|---|---|
| T1-1 | $8.58 \times 10^3$ | 508.5 | 0.1 | 10 | 96,645 | 102,815 | 6.38% |
| T1-2 | | | | | 100,027 | 108,672 | 8.64% |
| T 2-1 | $1.18 \times 10^4$ | 658.5 | | | 34,197 | 37,998 | 11.12% |
| T 2-2 | | | | | 33,521 | 36,960 | 10.26% |
| T 3-1 | $1.35 \times 10^4$ | 684.1 | | | 21,545 | 23,653 | 9.78% |
| T 3-2 | | | | | 20,969 | 22,846 | 8.95% |

The contact area of the trailing edge of the contact area, the amount of wear of the tiny slip, the error of the equipment in the actual test process, etc., will cause uncertainty in the load loading, and the tiny error of the micro-motion stress will have some influence on the fatigue life prediction process. In addition, our actual engineering applications tend to use more conservative design ideas in order to ensure the safe and reliable performance of the structure.

## 4. Sensitivity Analysis of Structural Parameters

### 4.1. Sensitivity Analysis of Factors Affecting Fretting Fatigue Life of the Dovetail Structure

Dovetail structure connection and assembly conditions are complex. Contact geometry, loading conditions, surface treatment, etc., will affect the failure mode and damage levels of the blade disk interface. By selecting three main characteristic parameters as optimization variables, taking the stress distribution law as a reference, and taking the fretting stress, $\sigma_{fretting}$, the goal of minimizing fretting is to improve the dovetail mortise structure by using the finite element analysis software ABAQUS (CAE2018, DASSAULT SIMULIA, Provision, RI, USA).

Due to the specificity of the contact area load calculation for the dovetail structure, it is important to obtain the exact contact area load through the direct theoretical model. Thus, the finite-element simulation tool ABAQUS (CAE2018, DASSAULT SIMULIA, Provision, RI, USA) was used to solve the contact area load calculation. Due to the specificity of the contact area load calculation for the dovetail structure, it is important to obtain the exact contact area load through the direct theoretical model. Thus, the finite-element simulation tool ABAQUS (CAE2018, DASSAULT SIMULIA, Provision, RI, USA) was used to solve the contact area load. The eight-node linear element C3D8R was used to conduct numerical analysis in combination with the "master–slave" interface algorithm of the ABAQUS finite element program. The dovetail structure finite element model was established according to the processing sample, as shown in Figure 14A. The lower end was fixed, and the maximum amplitude of upper end loading was a $1.35 \times 10^4$ N, $1.18 \times 10^4$ N, and $8.6 \times 10^3$ N cyclic axial load. The dovetail structure was divided into meshes, with the non-significant regions of the model roughly divided and the contact regions were subdivided. During the mesh aggregation and analysis, the number of grids in the contact area were 50, 60, 70, 80, 90, 100, 110, 120, 130, 140, and 150, and the contact stress CPRESS and Mises equivalent stress S were obtained through numerical analysis. The contact stress CPRESS and the equivalent stress S of Mises under different grid sizes can be used for comparative analysis. The contact area element meshing element is 100. The analysis was performed in the ABAQUS Explicit. The contact surface simulation sets the friction force to 0.8. The finite element simulation is loaded with a sinusoidal fatigue cyclic load. The finite element model of the dovetail structure is shown in Figure 14.

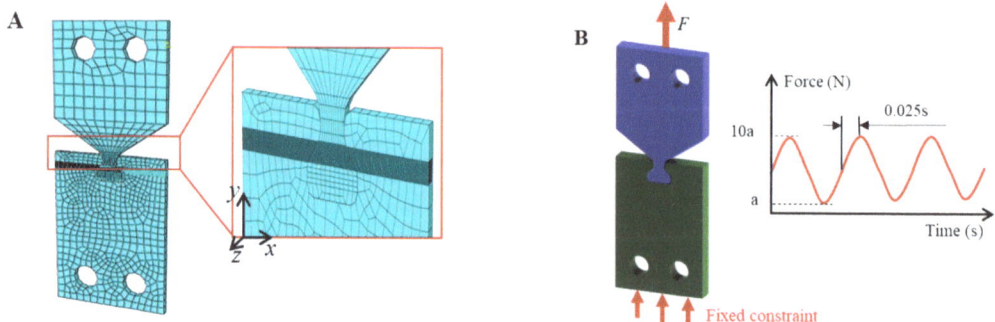

**Figure 14.** Dovetail structure finite element model (**A**) Dovetail meshing; (**B**) Boundary conditions and loading.

*4.2. Form of Contact Surface*

For the study of contact surface forms, the main purpose is to compare the contact forms of a circular tenon surface and a flat tenon surface with the tenon. The initial contact form between the circular tenon surface and the mortise plane is a line surface contact, while the initial contact form between the planar tenon surface and the mortise plane is surface contact. The size of other parts should be kept unchanged, and only the form of the tenon surface should be changed, as shown in Figure 15, which is a comparison of two different contact forms.

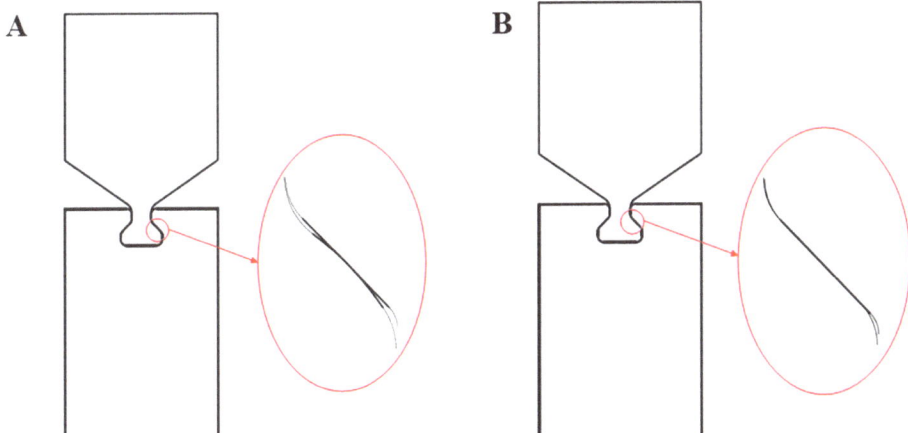

**Figure 15.** Dovetail models with two different contact forms: (**A**) arc dovetail profile (line to surface contact); (**B**) plane dovetail profile (surface to surface contact).

The different contact forms will lead to different states of the contact zone subjected to fretting load and affect the distribution of stress–strain. By means of ABAQUS, parameters are set, applying the same constraint and loading to the two models with different contact forms; the same contact area parameters are set, numerical analysis is conducted, and the Mises equivalent stress S is analyzed, along with contact stress and fretting stress $\sigma_{fretting}$ contact surface forms of fretting fatigue between dovetail models with different contact surface forms.

Holding the cyclic axial load at a constant peak value of $1.13 \times 10^4$ N, a comparison of the Mises equivalent stress S and contact stress distribution diagrams for the dovetail contact areas of the two different contact forms subjected to the tearing load is made, and thus the tearing stress $\sigma_{fretting}$, was carried out.

As shown in Figure 16, qualitatively, although the contact forms are different, under the action of fretting load, the maximum values of the equivalent stress S and the contact stress of the mortise Mises appear at the edge of the contact area. The comparison diagram of the Mises equivalent stress S and contact stress distribution of two different types of tenon profile is shown in Figure 16 (the size of the contact area between the planar tenon profile and the mortise is taken as the abscissa in the figure). From a quantitative perspective, under the same fretting load, the maximum Mises equivalent stress generated by the contact between the arc tenon profile and the mortise is 57.3 MPa lower than the maximum Mises equivalent stress generated by the contact between the plane tenon profile and the mortise. The maximum contact stress was 361.1 MPa lower than the maximum contact stress generated by the contact between the flat tenon face and the mortise, and the final fretting stress value obtained, $\sigma_{fretting}$, was 493.563 MPa lower than the fretting stress value, $\sigma_{fretting}$, generated by the contact between the flat tenon face and the mortise.

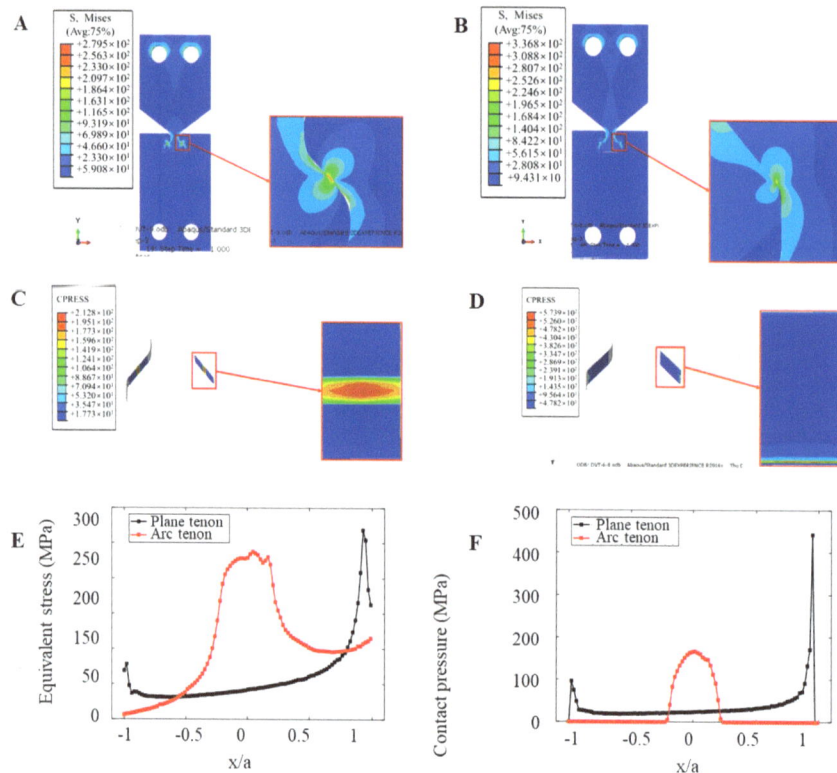

Figure 16. Comparison diagram of equivalent stress S and contact stress of two different forms of tenon profiles: (**A**) Mises equivalent stress S distribution of arc tenon profile; (**B**) Mises equivalent stress S distribution of plane tenon profile; (**C**) contact stress distribution of arc tenon profile; (**D**) contact stress distribution of plane tenon profile (**E**) Mises equivalent stress value of two different forms of tenon profiles; (**F**) contact stress value of two different forms of tenon profile.

It can be concluded that the variation of the contact zone form of the dovetail model does not affect the stress distribution and the location of the danger points of the structure under the fretting load. However, the fretting stress value of the linear surface contact form produced by the circular tenon to tenon contact is smaller than that of the surface contact form produced by the planar tenon to tenon contact, and the fretting fatigue life of the model is longer. Therefore, under the premise of satisfying the assembly requirements and structural strength requirements, the two matched structural members should adopt the form of line-to-surface contact as much as possible. Since only the axial fretting fatigue behavior of a pair of reduced diameter dovetail models is considered here, and no other factors are taken into account, a more suitable contact surface form should be selected according to various conditions in engineering practice.

### 4.3. Width of Contact Area

In addition to the contact form, the width of the contact area also affects the distribution and gradient of the load in the contact area. During the research on the influence of the width of the contact surface on fretting fatigue, the dovetail model will keep the size of other parts unchanged and only change the size of the fillets near the contact area and the width of the contact area. As shown in Figure 17, the model diagram is shown. Table 8 details the width dimensions of the corrected fillet and contact zone for each model.

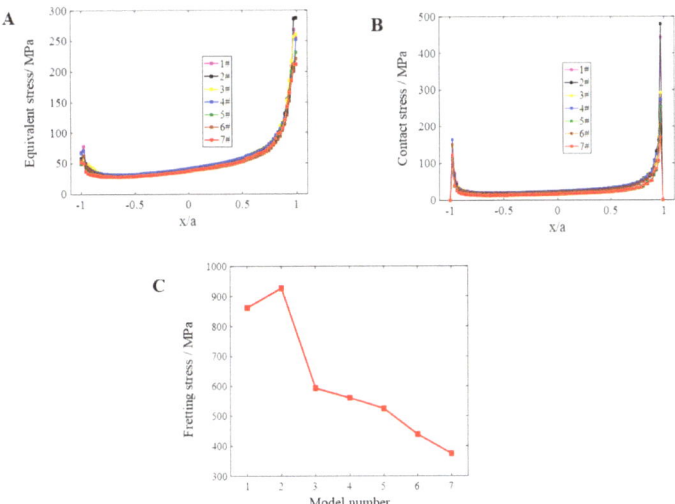

**Figure 17.** Fretting stress of models 1–7 under the same load $\sigma_{fretting}$ comparison: (**A**) Mises equivalent stress S value of models 1–7; (**B**) contact stress value of models 1–7; (**C**) fretting stress comparison under the same load conditions $\sigma_{fretting}$.

**Table 8.** Dovetail models with different width of contact area.

| Number | Tenon | | Mortise | | Contact Area Width/mm |
|---|---|---|---|---|---|
| | $R_1$/mm | $R_2$/mm | $R_5$/mm | $R_6$/mm | |
| 1# | 1.8 | 1 | 2 | 1 | 4.343 |
| 2# | 1.8 | 1.1 | 2 | 1.1 | 4.302 |
| 3# | 1.8 | 1.2 | 2 | 1.2 | 4.261 |
| 4# | 1.8 | 1.3 | 2 | 1.3 | 4.219 |
| 5# | 1.8 | 1.4 | 2 | 1.4 | 4.178 |
| 6# | 1.8 | 1.5 | 2 | 1.5 | 4.136 |
| 7# | 1.8 | 1.6 | 2 | 1.6 | 4.095 |

The different contact zone widths lead to different states of the contact zone subject to frictional loads, which in turn affect the distribution of stresses and strains. By setting parameters with the help of numerical analysis software ABAQUS, parameters are set, the same constraints and loads are applied to each model with different contact area widths, the same contact area parameters are set, numerical analysis is conducted, and the contact stress is analyzed, followed by the Mises effect force S and fretting stress between the dovetail models under different contact area widths and the $\sigma_{fretting}$ influence of the contact zone width on fretting fatigue.

Holding the peak cyclic axial load of a $1.13 \times 10^4$ N constant, the Mises effect stress S, the comparison of contact stress profiles, and the comparison of free stress $\sigma_{fretting}$ in the contact zone of the dovetail groove under different contact zone widths were carried out.

Figure 17 only lists the Mises equivalent of the stress S nephogram and contact stress nephogram of models 1 and 2, but not all models. This is because qualitatively, as long as the contact is plane contact, there is not much difference in its stress distribution under the action of fretting load. The maximum values of the equivalent force S and contact stress S in the tenon occur at the edges of the contact zone and do not vary. Therefore, only the stress nephograms 1 and 2 are shown here as representatives. The quantitative analysis shows that when the width of the contact zone decreases from 4.343 mm to 4.302 mm, the maximum Mises equivalent stress S and the maximum contact stress and fretting stress $\sigma_{fretting}$ values increase slightly, but on the whole, with the decrease in the contact zone width, the maximum Mises equivalent stress S, the maximum contact stress and the fretting stress $\sigma_{fretting}$ values tend to decrease. In general, keeping other parameters unchanged, the width of contact zone is reduced from 4.343 mm to 4.095 mm, and the reduction amplitude is only 248 µm. In the case of fretting load, the maximum Mises equivalent stress S decreases by 75.286 MPa, the maximum contact stress decreases by 308.994 MPa, and the $\sigma_{fretting}$ value decreases by 552.745 MPa; moreover, the fretting stress directly affects the fretting fatigue life, so it can be seen that fretting fatigue is very sensitive to the change of contact zone width. Separately, as the width of contact zone gradually decreases, the maximum Mises equivalent stress S, maximum contact stress, and fretting stress $\sigma_{fretting}$ is shown in Table 9. It can be seen that when the width of contact zone decreases from 4.261 mm to 4.095 mm from models 3 to 7, and the corresponding maximum contact stress and fretting stress $\sigma_{fretting}$ reduction in fretting value has little difference. In models 2 and 3, when the width of contact area is reduced from 4.302 mm to 4.261 mm, the corresponding maximum contact stress and fretting stress $\sigma_{fretting}$ value decreased abruptly. In conclusion, when the minimum width of the initial contact zone is 0 (line to surface contact), the corresponding fretting stress $\sigma_{fretting}$ value is 398.3 MPa, which is 23 MPa larger than the corresponding fretting stress value when the width of the contact area is 4.095 mm.

**Table 9.** Variation of parameters caused by the change of width of contact area.

| Model | Reduction of Maximum Mises Equivalent Stress S/MPa | Reduction of Maximum Contact Stress/MPa | Reduction of Fretting Stress/MPa |
| --- | --- | --- | --- |
| 2#–3# | 26.08 | 186.386 | 333.417 |
| 3#–4# | 8.827 | 18.176 | 32.514 |
| 4#–5# | 20.771 | 20.358 | 36.42 |
| 5#–6# | 10.464 | 48.092 | 86.03 |
| 6#–7# | 9.144 | 35.982 | 64.367 |

It appears from the results of the analysis that the change of contact zone width of dovetail model will not affect the stress distribution and the position of dangerous points of the structure under the fretting load. However, with the decrease in contact zone width, the contact stress value will decrease, the fretting stress value will also decrease accordingly, and the fretting life of the sample will be longer. It can also be seen from the basic Hertz theory that the contact stress value will decrease as the fillet angle increases and the contact zone decreases. To sum up, on the premise of meeting the assembly requirements and saving costs, the processing of structural parts in actual projects should focus on the width of the contact zone, which should be less than 4.3 mm. Therefore, only a pair of reduced dovetail models are taken as the research object, and more conditions, such as applicable location, environment, structural strength requirements, etc., need to be comprehensively considered to determine the accurate range of the selected contact zone width, so as to narrow the optimal range of the contact zone width.

### 4.4. Friction Coefficient of Contact Surface

At present, for the fretting fatigue problem of standard parts, all theoretical models for solving contact stress are established without considering the coupling effect of axial load and normal load. However, with the application of cyclic load, the maximum contact stress between the fretting pad and the standard sample changes with the effect of friction, not a fixed value. Under a certain load, the difference of friction coefficient between the fretting pad and the standard sample will affect the magnitude of friction work and the magnitude of contact stress. This paper uses ABAQUS to study and analyses the effect of friction factors on fretting fatigue.

The peak cyclic axial load of $1.13 \times 10^4$ N was kept constant and the friction coefficients were set to 0.3, 0.4, 0.5, 0.6, 0.7, and 0.8. The dovetail model loading was numerically analyzed in order to analyze the effects of contact stress, Mises equivalent force S, and the friction stress of the $\sigma_{fretting}$ friction coefficient on the friction fatigue between the dovetail models with different friction coefficients.

Figure 18 only lists the Mises equivalent stress nephogram and contact stress nephogram of dovetail model when the friction coefficient is 0.8 and 0.7, but not all models. This is because qualitatively speaking, as long as it involves plane contact, there is not much difference in its stress distribution subjected to fretting load. The contact stress and maximum Mises equivalent stress S of the mortise appear at the edge of the contact area and do not change, so only two models are shown here. Quantitatively, as the friction coefficient increases, the maximum Mises equivalent stress S, maximum contact stress, and fretting stress $\sigma_{fretting}$ values increase correspondingly, because under the same fretting load, the friction coefficient increases. With the increase in load cycles, the work carried out by the friction force increases correspondingly and the stress value also increases. This requires us to ensure the processing accuracy and assembly accuracy of the sample insofar as possible to ensure the low friction coefficient of the sample contact area, which is also true in engineering practice. In general, when keeping other parameters unchanged, the friction coefficient increases from 0.3 to 0.8, subject to fretting load; the maximum Mises equivalent stress S increases by 249.425 MPa; the maximum contact stress increases by 83.054 MPa; and the fretting stress $\sigma_{fretting}$ value increased by 366.258 MPa. Separately, with the increase in friction coefficient, the values of each parameter also increase. See Table 10 for details. It can be seen that the maximum contact stress and fretting stress increase from 0.3 to 0.7, there is little difference when the friction coefficient increases the $\sigma_{fretting}$ value of friction, but when the friction coefficient increases from 0.7 to 0.8, the maximum contact stress and fretting stress $\sigma_{fretting}$ value are increased abruptly.

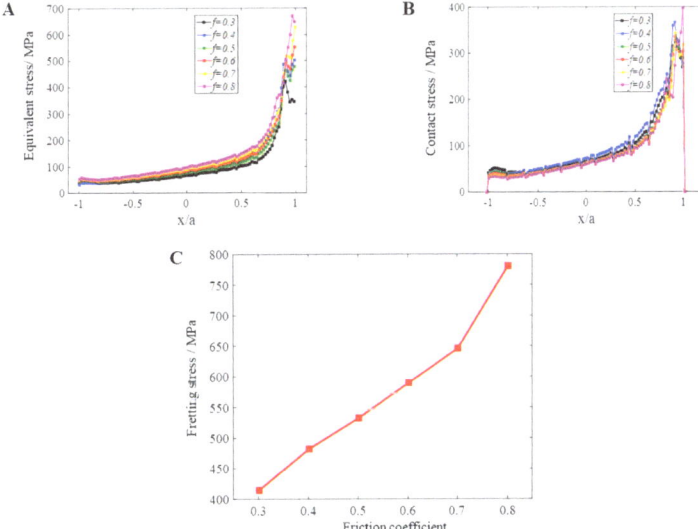

**Figure 18.** Comparison of fretting stress of the different friction coefficient models under the same load conditions: (**A**) Mises equivalent stress S value with different friction coefficients; (**B**) contact stress value with different friction coefficients; (**C**) comparison of fretting stresses of different friction coefficient models under the same load conditions.

From the analysis, it can be seen that the variation of the friction coefficient of the dovetail model does not affect the stress distribution of the structure and the location of the danger points under the fretting load. However, with the decrease in friction coefficient, the fretting stress value will decrease accordingly, and the fretting life of the sample will be longer. Therefore, in actual projects, the friction coefficient of the contact surface should be reduced as much as possible under the premise of saving costs and meeting work requirements and the friction coefficient shall be less than 0.8 insofar as possible.

Table 10. Variation of parameters caused by change of friction coefficients.

| Friction Coefficient | Reduction of Maximum Mises Equivalent Stress S/MPa | Reduction of Maximum Contact Stress/MPa | Reduction of Fretting Stress/MPa |
|---|---|---|---|
| 0.3–0.4 | 37.665 | 6.798 | 61.8 |
| 0.4–0.5 | 6.048 | 6.048 | 56.440 |
| 0.5–0.6 | 75.449 | 8.521 | 57.310 |
| 0.6–0.7 | 75.449 | 8.521 | 55.879 |
| 0.7–0.8 | 42.041 | 53.166 | 134.830 |

Based on the above fretting simulation analysis of dovetail model, it can be seen that in the fretting load action, changing the parameters of contact zone will not change the overall distribution trend of the Mises equivalent stress S and contact stress, nor will it change the location of structural dangerous points (still appearing at the edge of contact area), meeting the basic characteristics of fretting fatigue, However, changing the parameters of the contact zone will seriously affect the size of the contact stress and fretting stress, thereby changing the fatigue structure strength and life. In order to prolong the fretting fatigue life of the dovetail structure, it is necessary to conduct a further comprehensive analysis to determine the optimal solution when selecting contact zone parameters.

## 5. Conclusions

In this study, a friction fatigue life prediction theoretical based on elasticity theory was established to study the influence of the plastic effect on fretting fatigue, and the maximum plastic strain range was introduced ($\Delta\varepsilon$)$_p$. A prediction model for fretting the fatigue life of the surface-to-surface contact considering the plasticity effect was proposed. The dangerous parts of the structure were evaluated according to a finite element simulation, and the fretting fatigue test was evaluated through the standard sample and dovetail sample to verify the accuracy of the model. At the same time, an ABAQUS simulation was used to carry out fretting fatigue numerical analysis on the dovetail structures with different contact forms, different contact zone widths, and different contact surface friction coefficients. With the difference in the size and distribution of contact stress, Mises equivalent stress, and displacements as a reference, the influence degree and trend of contact forms, contact zone widths, and contact surface friction coefficients on fretting fatigue were analyzed and studied and the sensitivity of factors affecting fretting fatigue was analyzed. We found that:

(1) By monitoring the strain at the lower edge of the contact area in real time, the initiation and propagation of cracks can be judged. The introduction of the maximum plastic strain improves the accuracy of the prediction of fretting fatigue life of surface-to-surface contact structures. The error between the theoretical prediction value and the test value is within 12%. The model is applicable to predicting the fretting fatigue life of structures and provides theoretical support for the design of major equipment structures.

(2) With the goal of minimizing fretting stress, a more reasonable contact form, contact zone width, and contact surface friction coefficient should be uncovered in order to reduce fretting fatigue and extend the fretting fatigue life of the model. Under the same fretting load, changing the contact zone parameters will not change the Mises equivalent force S and the contact stress profile, nor will it change the location of structural dangerous points (which still appear at the contact zone edge); the basic characteristics of fretting fatigue are thus met.

(3) Changing the parameters of the contact zone will change the magnitude of the stress, and fretting fatigue is most sensitive to the width of the contact zone and the contact form. In the actual engineering design, the size and form should be determined by combining multiple factors. For the selection of the friction coefficient, the friction coefficient shall be as small as possible in order to save costs and meet requirements.

**Author Contributions:** Conceptualization, J.Z.; Methodology, B.Y.; Validation, S.L.; Resources, J.H. All authors have read and agreed to the published version of the manuscript.

**Funding:** This work was supported by National Key Research and Development Program (2020YFB2006803); National Key Research and Development Program of China (2020YFB2006804); Natural Science Foundation of Henan Province (212300410325); Henan Provincial Science and Technology Tackling (232102220093); Science and Technology Development Plan of China Railway Corporation (2021-Special-05).

**Institutional Review Board Statement:** Not applicable.

**Informed Consent Statement:** Not applicable.

**Data Availability Statement:** Not applicable.

**Conflicts of Interest:** The authors declare that they have no known competing financial interest or personal relationships that could have appeared to influence the work reported in this paper.

# References

1. Yang, B.; Huo, J.; Zhang, W.; Zhang, Z. Research on real-time prediction method of advanced load of service structure. *J. Northeast. Univ. Nat. Sci. Ed.* **2022**, *43*, 541–550.
2. Yan, X.; Zhang, H.; Xing, Z. Optimal design of turbine tenon connection structure against fretting fatigue damage. *Gas Turbine Test Res.* **2022**, *35*, 36–40.
3. Bhatti, N.; Wahab, M.A. Fretting fatigue crack nucleation: A review. *Tribol. Int.* **2018**, *121*, 121–138. [CrossRef]
4. Pereira, K.; Bhatti, N.; Wahab, M.A. Prediction of fretting fatigue crack initiation location and direction using cohesive zone model. *Tribol. Int.* **2018**, *127*, 245–254. [CrossRef]
5. Shen, F.; Hu, W.; Meng, Q. A damage mechanics approach to fretting fatigue life prediction with consideration of elastic–plastic damage model and wear. *Tribol. Int.* **2015**, *82*, 176–190. [CrossRef]
6. Peng, J.; Jin, X.; Xu, Z.; Zhang, J.; Cai, Z.; Luo, Z.; Zhu, M. Study on the damage evolution of torsional fretting fatigue in a 7075-aluminum alloy. *Wear* **2018**, *402–403*, 160–168. [CrossRef]
7. Leung, J.F.; Voothaluru, R.; Neu, R.W. Predicting white etching matter formation in bearing steels using a fretting damage parameter. *Tribol. Int.* **2021**, *159*, 106966. [CrossRef]
8. Shen, F.; Ke, L.-L.; Zhou, K. A debris layer evolution-based model for predicting both fretting wear and fretting fatigue lifetime. *Int. J. Fatigue* **2021**, *142*, 105928. [CrossRef]
9. Leung, J.F.; Voothaluru, R.; Neu, R.W. Experimental and numerical simulation study on fatigue life of coiled tubing with typical defects. *J. Pet. Sci. Eng.* **2021**, *198*, 108212.
10. Li, J.; Wang, X.; Li, K.; Qiu, Y. A modification of Matake criterion for considering the effect of mean shear stress under high cycle fatigue loading. *Fatigue Fract. Eng. Mater. Struct.* **2021**, *44*, 1760–1782. [CrossRef]
11. Albinmousa, J.; Al Hussain, M. Polar Damage Sum Concept for Constant Amplitude Proportional and Nonproportional Multiaxial Fatigue Analysis. *Forces Mech.* **2021**, *4*, 100025. [CrossRef]
12. Dantas, R.; Correia, J.; Lesiuk, G.; Rozumek, D.; Zhu, S.-P.; de Jesus, A.; Susmel, L.; Berto, F. Evaluation of multiaxial high-cycle fatigue criteria under proportional loading for S355 steel. *Eng. Fail. Anal.* **2021**, *120*, 105037. [CrossRef]
13. Almeida, G.M.J.; Pessoa, G.C.V.; Cardoso, R.A.; Castro, F.C.; Araújo, J.A. Investigation of crack initiation path in AA7050-T7451 under fretting conditions. *Tribol. Int.* **2021**, *144*, 106103. [CrossRef]
14. Floros, D.; Ekberg, A.; Larsson, F. Evaluation of mixed-mode crack growth direction criteria under rolling contact conditions—ScienceDirect. *Wear* **2021**, *448–449*, 203184.
15. Sunde, S.L.; Haugen, B.; Berto, F. Experimental and numerical fretting fatigue using a new test fixture. *Int. J. Fatigue* **2021**, *143*, 106011. [CrossRef]
16. Pinto, A.L.; Araújo, J.A.; Talemi, R. Effects of fretting wear process on fatigue crack propagation and life assessment. *Tribol. Int.* **2021**, *156*, 106787. [CrossRef]
17. Pandey, V.B.; Singh, I.V.; Mishra, B.K. A Strain-based Continuum Damage Model for Low Cycle Fatigue under Different Strain Ratios. *Eng. Fract. Mech.* **2020**, *242*, 107479. [CrossRef]
18. Zeng, D.; Zhang, Y.; Lu, L.; Zou, L.; Zhu, S. Fretting wear and fatigue in press-fitted railway axle: A simulation study of the influence of stress relief groove. *Int. J. Fatigue* **2018**, *18*, 225–236. [CrossRef]
19. Mario, L.; Daniele, B. Fretting Fatigue Analysis of Additively Manufactured Blade Root Made of Intermetallic Ti-48Al-2Cr-2Nb Alloy at High Temperature. *Materials* **2018**, *11*, 1052.
20. Lemoine, E.; Nélias, D.; Thouverez, F.; Vincent, C. Influence of fretting wear on bladed disks dynamic analysis. *Tribol. Int.* **2020**, *145*, 106148. [CrossRef]
21. Macdonald, B.E.; Fu, Z.; Wang, X.; Li, Z. Influence of phase decomposition on mechanical behavior of an equiatomic CoCuFeMnNi high entropy alloy. *Acta Mater.* **2019**, *181*, 25–35. [CrossRef]

22. Wahab, M.A.; Bhatti, N.A. A numerical investigation on critical plane orientation and initiation lifetimes in fretting fatigue under out of phase loading conditions. *Tribol. Int.* **2018**, *115*, 307–318.
23. Han, Q.; Rui, S.; Qiu, W.; Su, Y.; Ma, X.; He, Z.; Cui, H.; Zhang, H.; Shi, H.J. Subsurface crack formation and propagation of fretting fatigue in Ni-based single-crystal superalloys. *Fatigue Fract. Eng. Mater. Struct.* **2019**, *42*, 2520–2532. [CrossRef]
24. Sunde, S.L.; Berto, F.; Haugen, B. Predicting fretting fatigue in engineering design. *Int. J. Fatigue* **2018**, *117*, 314–326. [CrossRef]
25. Han, Q.; Lei, X.; Yang, H.; Yang, X.; Su, Z.; Rui, S.-S.; Wang, N.; Ma, X.; Cui, H.; Shi, H. Effects of temperature and load on fretting fatigue induced geometrically necessary dislocation distribution in titanium alloy. *Mater. Sci. Eng. A* **2021**, *800*, 140308. [CrossRef]
26. Wu, B. Model and Experimental Research on Fretting Fatigue Life of TC11 Titanium Alloy at Different Temperatures. Master's Thesis, Nanjing University of Aeronautics and Astronautics, Nanjing, China, 2019.
27. Wang, N.; Cui, H.; Zhang, H. Low cycle fretting fatigue tests of dovetail structure at elevated temperature. *Hangkong Dongli Xuebao J. Aerosp. Power* **2018**, *33*, 3007–3012.
28. Kumar, K.; Goyal, D.; Banwait, S. Effect of Key Parameters on Fretting Behaviour of Wire Rope: A Review. *Arch. Comput. Methods Eng.* **2019**, *27*, 549–561. [CrossRef]
29. Hills, D.A.; Thaitirarot, A.; Barber, J.R.; Dini, D. Correlation of fretting fatigue experimental results using an asymptotic approach. *Int. J. Fatigue* **2012**, *43*, 62–75. [CrossRef]
30. Sun, D.; Huo, J.; Chen, H.; Dong, Z.; Ren, R. Experimental study of fretting fatigue in dovetail assembly considering temperature effect based on damage mechanics method. *Eng. Fail. Anal.* **2022**, *131*, 105812. [CrossRef]
31. Liu, Y.; Yuan, H. A hierarchical mechanism-informed neural network approach for assessing fretting fatigue of dovetail joints. *Int. J. Fatigue* **2023**, *168*, 107453. [CrossRef]
32. Mirsayar, M.M. A generalized criterion for fatigue crack growth in additively manufactured materials–Build orientation and geometry effects. *Int. J. Fatigue* **2021**, *145*, 106099. [CrossRef]
33. Huo, J.; Yang, B.; Ren, R.; Dong, J. Research on fretting fatigue life estimation model considering plastic effect. *J. Braz. Soc. Mech. Sci. Eng.* **2022**, *44*, 112. [CrossRef]

**Disclaimer/Publisher's Note:** The statements, opinions and data contained in all publications are solely those of the individual author(s) and contributor(s) and not of MDPI and/or the editor(s). MDPI and/or the editor(s) disclaim responsibility for any injury to people or property resulting from any ideas, methods, instructions or products referred to in the content.

# Article

# Mathematical Model for Estimating the Sound Absorption Coefficient in Grid Network Structures

Takamasa Satoh [1], Shuichi Sakamoto [2,*], Takunari Isobe [3], Kenta Iizuka [3] and Kastsuhiko Tasaki [3]

[1] FUKOKU Co., Ltd., 6 Showa Chiyoda-machi, Oura-gun, Gunma 370-0873, Japan
[2] Department of Engineering, Niigata University, Ikarashi 2-nocho 8050, Nishi-ku, Niigata 950-2181, Japan
[3] Graduate School of Science and Technology, Niigata University, Ikarashi 2-nocho 8050, Nishi-ku, Niigata 950-2181, Japan
* Correspondence: sakamoto@eng.niigata-u.ac.jp; Tel.: +81-25-262-7003

**Abstract:** Although grid network structures are often not necessarily intended to absorb sound, the gaps between the rods that make up the grid network are expected to have a sound absorption effect. In this study, the one-dimensional transfer matrix method was used to develop a simple mathematical model for accurately estimating the sound absorption coefficient of a grid network structure. The gaps in the grid network structure were approximated as the clearance between two parallel planes, and analysis units were derived to consider the exact geometry of the layers. The characteristic impedance and propagation constant were determined for the approximated gaps and treated as a one-dimensional transfer matrix. The transfer matrix obtained for each layer was used to calculate the sound absorption coefficient. The samples were fabricated from light-curing resin by using a Form2 3D printer from Formlabs. The measurement results showed that a sound absorption coefficient of 0.81 was obtained at the peak when seven layers were stacked. A sensitivity analysis was carried out to investigate the influence of the rod diameter and pitch. The simulated values tended to be close to the experimental values. The above results indicate that the mathematical model used to calculate the sound absorption coefficient is sufficiently accurate to predict the sound absorption coefficient for practical application.

**Keywords:** sound absorption coefficient; porous material; grid network structure; transfer matrix method

## 1. Introduction

Grid network structures formed by a large number of round or square rods are often used as screens in machine openings and in natural or forced ventilation inlets and outlets. Various experimental studies have investigated the fluid flow characteristics in grid network structures, such as the oscillatory pressure drop when airflow is applied to a mesh screen [1], the use of a wire mesh as a catalyst for soot mitigation [2] and the pressure loss of wire mesh filters [3].

Yamamoto et al. [4] numerically analyzed the acoustic properties of a frame structure, which is similar to a grid network structure. Satoh et al. [5] compared the predicted and experimental results for sound waves incident perpendicular to a group of cylinder axes, including the sound absorption coefficient. In the grid network structure, such sound waves traverse a direction where the size of the gap between the cylinders varies continuously, which is considered to facilitate a high sound absorption coefficient.

In general, thin grid network materials are often treated as acoustically transparent objects. Therefore, they are used in practical applications or experiments to hold acoustic materials. However, if the grid network has a low aperture ratio, an air layer behind it or several stacked layers, its sound absorption may be significant. In a previous study, Iizuka et al. [6] demonstrated the sound absorption effect of laminated wire mesh structures. Dias and Monaragala [7] performed a similar study on the sound absorption of knitted structures and on metamaterials consisting of laminated thin sheets with apertures [8–10]. Dias and Monaragala [7] carried out a similar study on the sound absorption of knitted structures.

Although grid network structures are often not necessarily intended to absorb sound, the gaps between the rods that make up the grid network are expected to have a sound absorption effect. Therefore, predicting acoustic properties such as the sound absorption coefficient of a grid network structure is useful in noise engineering because it is a common engineering shape. Being able to predict the sound absorption effect based on the geometric dimensions of the grid network structure and the physical properties of the gas may help in the development of a compact and simple sound absorption mechanism for various applications. It is expected to provide basic knowledge for many applications, such as filters, sifters, and screens. It can also provide insights into the noise control properties of heaters, heat exchangers, catalysts, and other devices. Furthermore, the grid network structure can function as a reflection barrier for electromagnetic waves and as a vibration damper; thus, it can be used as a device that combines these functions with sound reduction.

Although the finite element method [11,12] is useful in the analysis of acoustic metamaterials, the simpler transfer matrix method was used in this work. The one-dimensional transfer matrix method was used to develop a simple mathematical model for accurately estimating the sound absorption coefficient of a grid network structure. The propagation constants and characteristic impedances were derived by using the Navier–Stokes equations and other equations to consider the viscosity of the air. In addition, a simulated analysis was performed to account for the continuous change in the cross-sectional area when the sound waves are incident perpendicular to the rods in a group of cylinder axes [5]. Moreover, a simulated analysis corresponding to the continuous change in the cross-sectional area is attempted by geometrically estimating the surface area of the wall surface and the volume of the void portion in each segmented element correctly and applying it to a two-plane approximation. Finally, the mathematical model was validated by 3D-printing samples and comparing the measured sound absorption coefficients with the predicted values. The measurement results showed that a sound absorption coefficient of 0.81 was obtained at the peak when seven layers were stacked. A sensitivity analysis was performed to evaluate the effects of parameters such as the diameter and pitch on the predicted values.

## 2. Experimental Validation

### 2.1. Measurement Samples

Figure 1 shows a schematic of the test samples used for the experimental measurements. Figure 2 shows a photograph of the samples. Table 1 presents the sample specifications. The samples were fabricated from acrylic-based light-curing resin by using a Form2 3D printer from Formlabs Inc., (Somerville, MA, USA). The rods comprising the grid network had a length of 25.7 mm, and the diameter and pitch were varied to evaluate their effects on the sound absorption coefficient. Four samples were produced with different dimensions. The 3D printer used a layer pitch of 0.025 mm, which allows a fabrication error of less than 0.05 mm in rod diameter. However, during the initial fabrication runs, the fabrication error of the rod diameters was biased to be several tens of micrometers smaller. To compensate for this, the diameter of the rod on the computer-aided design (CAD) drawing was increased by several tens of micrometers. This resulted in a finished product with more accurate dimensions, as shown in Table 1. The room temperature at the time of fabrication was 20 °C, and the time required to fabricate one sample was approximately 7 h.

The samples were inserted into a sample holder made of aluminum alloy with internal dimensions of 25.7 mm per side for measurement of the sound absorption coefficient. The end faces of the rods and the gap between the bottoms of the sample holder and sample were filled with Vaseline to eliminate factors other than the sample that could contribute to sound absorption.

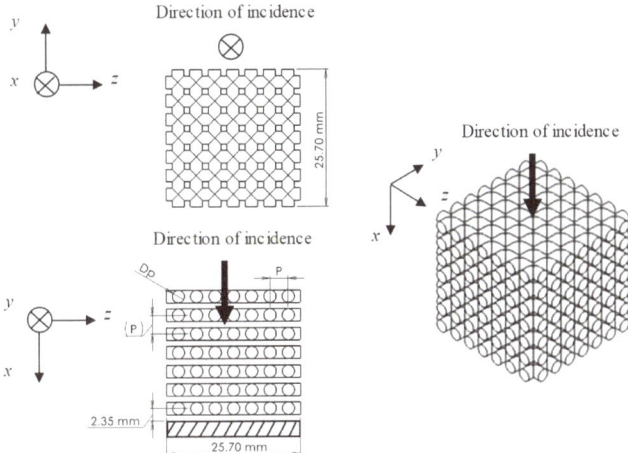

**Figure 1.** Schematic of a test sample. $D_p$: rod diameter; $P$: rod pitch; and $N_m$: number of grids.

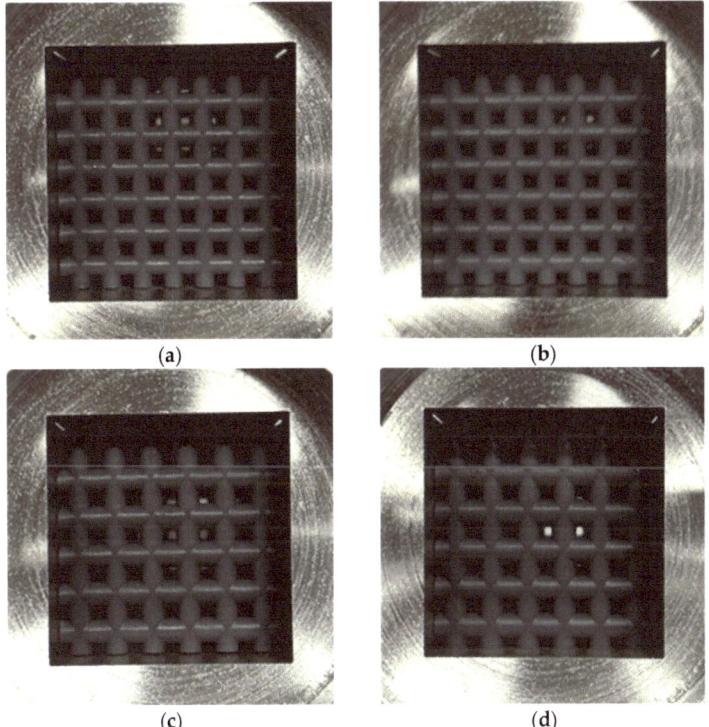

**Figure 2.** Test samples in sample holder: (**a**) $D_p$ = 2.3 mm, $P$ = 3.5 mm, $N_m$ = 64; (**b**) $D_p$ = 2.5 mm, $P$ = 3.5 mm, $N_m$ = 64; (**c**) $D_p$ = 2.8 mm, $P$ = 4.2 mm, $N_m$ = 49; (**d**) $D_p$ = 3.0 mm, $P$ = 4.2 mm, and $N_m$ = 49.

Table 1. Specifications of test samples.

| Diameter of Rods [mm] (Measured) | Pitch of Rods [mm] | Height of Test Sample [mm] | Number of Grids $N_m$ | Number of Layers $N_l$ | Correspondence to Figure |
|---|---|---|---|---|---|
| 2.3 (2.31) | 3.5 | 25.7 | 64 | 7 | Figure 2a |
| 2.5 (2.51) | 3.5 | 25.7 | 64 | 7 | Figure 2b |
| 2.8 (2.80) | 4.2 | 25.7 | 49 | 6 | Figure 2c |
| 3.0 (3.02) | 4.2 | 25.7 | 49 | 6 | Figure 2d |

### 2.2. Measurement Equipment

A Brüel & Kjær Type 4206 (Brüel & Kjær Sound & Vibration Measurement, Nærum, Denmark)two-microphone acoustic impedance tube system was used to measure the sound absorption coefficient. Figure 3a shows the measurement system. Sound waves with a sinusoidal signal were generated by a signal generator built into the fast Fourier transform (FFT) analyzer and were radiated into the impedance tube by a loudspeaker. The transfer function between the sound pressure signals of two microphones attached to the impedance tube was measured by a FFT analyzer. An Onosokki DS-3000 FFT (Ono Sokki Co., Ltd., Yokohama, Japan) analyzer was used for the measurement. The measured transfer function was then used to calculate the normal incident sound absorption coefficient in accordance with ISO 10534-2 [13]. The derivation of the normal incident sound absorption coefficient is described in detail in ISO 10534-2.

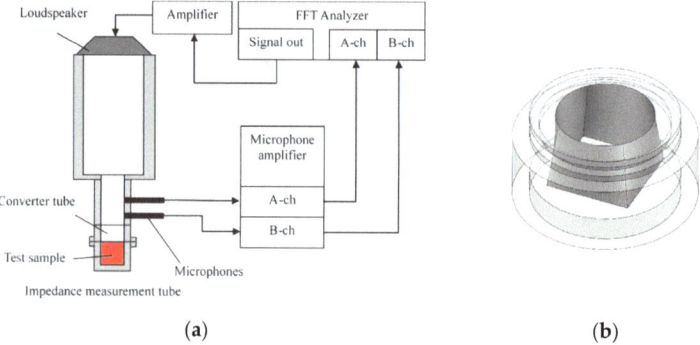

**Figure 3.** Schema of absorption coefficient measurement: (**a**) two-microphone impedance tube and (**b**) converter tube to convert the cross-sectional shape from circular to square.

The critical frequency at which a plane wave forms depends on the inner diameter of the impedance tube. In this study, impedance tubes with an inner diameter of 29 mm were used because the sound absorption coefficient is not high at low frequencies. However, a rectangular tube with side lengths of 25.7 mm was used as the sample holder to ensure uniform rod lengths. Thus, a conversion tube was used to smoothly change the cross-sectional shape of the impedance tube from circular to square. Figure 3b shows a perspective view of the conversion tube. The diagonal length of the square cross-section was approximately 36.3 mm, so the limiting frequency of the plane wave was determined as approximately 5400 Hz.

### 3. Simulated Analysis

#### 3.1. Transfer Matrix of an Acoustic Element

The transfer matrix method was used to calculate the simulated value of the sound absorption coefficient of a grid network structure [14,15]. Figure 4 shows the Cartesian coordinate system for a gap of thickness $b$ between two parallel planes 1 and 2. The transfer matrix method can be used to calculate the sound pressure and volume velocity for such a gap based on the one-dimensional wave equation. The relationship between the sound pressure $p_1$ and volume velocity $Su_1$ at plane 1 and the sound pressure $p_2$ and volume velocity $Su_2$ at plane 2 can be expressed by using the four-terminal constant $t$:

$$\begin{bmatrix} p_1 \\ Su_1 \end{bmatrix} = \begin{bmatrix} \cosh \gamma l & \frac{Z_c}{S} \sinh \gamma l \\ \frac{S}{Z_c} \sinh \gamma l & \cosh \gamma l \end{bmatrix} \begin{bmatrix} p_2 \\ Su_2 \end{bmatrix} = \begin{bmatrix} t_{11} & t_{12} \\ t_{21} & t_{22} \end{bmatrix} \begin{bmatrix} p_2 \\ Su_2 \end{bmatrix} \quad (1)$$

where $Z_c$ is the characteristic impedance, $\gamma$ is the propagation constant, $S$ is the area of the gap and $l$ is the length between planes 1 and 2.

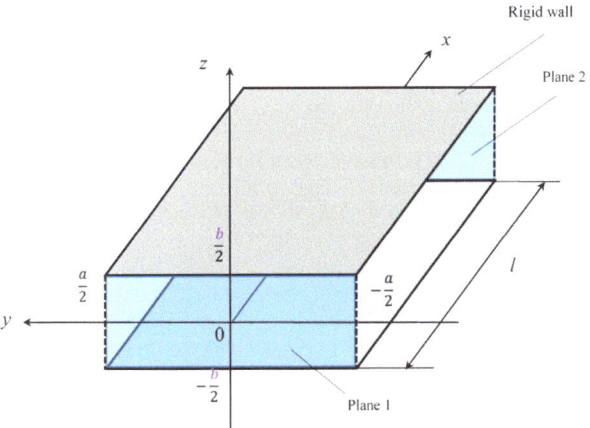

**Figure 4.** Cartesian coordinate system for the parallel clearance between a pair of planes.

*3.2. Propagation Constant and Characteristic Impedance Considering Attenuation*

The propagation constant and characteristic acoustic impedance of small tubes considering attenuation due to air viscosity have been studied for tubes with circular [16–18] and equilateral-triangle cross-sections [19].

In this study, Stinson's method [19] was applied to consider the attenuation of sound waves by the loss of friction due to the viscosity of the boundary layer near the wall. The propagation constant $\gamma$ and characteristic acoustic impedance $Z_c$ of a gap between two surfaces, as shown in Figure 4, can be derived by solving the Navier–Stokes equations, continuity equation, gas state equation, energy equation and dissipative function representing heat transfer. Here, air was assumed to be a compressible fluid, and the air viscosity was assumed to be constant.

The propagation constant $\gamma$ can be expressed as follows:

$$\gamma = \sqrt{\frac{\kappa - (\kappa - 1) B(s\sigma)}{B(s)}}, \quad B(x) = 1 - \frac{\tanh x}{x}, \quad s = \frac{b}{2} \sqrt{\frac{\rho_0 \omega}{\mu}} \quad (2)$$

where $\kappa$ is the specific heat ratio, $\sigma$ is the square root of the Prandtl number (0.8677), $\mu$ is the viscosity (1.869 × 10$^{-5}$ Pa·s), $\omega$ is the angular frequency and $\rho_0$ is the density of the gas between two planes. Then, the characteristic impedance $Z_c$ can be expressed as follows:

$$Z_c = \frac{p^+}{u^+} \quad (3)$$

$u^+$ and $p^+$ are the particle velocity and sound pressure, respectively, of the travelling wave:

$$\overline{u}^+ = c_0 B(s) \frac{j}{\kappa} \left( -\frac{\gamma}{k} \right) \beta e^{-\gamma x} \quad (4)$$

$$p^+ = P_s \beta e^{-\gamma x} \quad (5)$$

where $k$ is the wavenumber and $P_s$ is the atmospheric pressure ($1.013 \times 10^5$ Pa).

*3.3. Mathematical Model of the Grid Network Structure*

The transfer matrix method was used to calculate the sound absorption coefficient of the grid network structure when the sound wave is incident perpendicular to the axial direction of the rods of the grid network structure. Three analysis units were considered to represent different positions in the grid network structure in relation to the tube wall.

3.3.1. Analysis Unit Surrounded by Four Rods

Figure 5 shows a model of the grid network structure. The first analysis unit (i.e., Unit I) represents an area surrounded by four rods (red rectangle). To obtain the transfer matrix, Unit I is divided into $n$ equal layers in the direction of sound incidence (i.e., $x$-axis direction), as shown in Figure 6. The rod diameter is $D_p$, so the thickness of one layer is $D_p/n$. For the actual analysis, 100 divisions were used because this is when the simulated values converged sufficiently. Then, the clearance between two parallel planes can be approximated by using Allard's method [20] so that the volume of the void space and the surface area of the rod portion are each equal. Figure 7 approximates a single layer as two parallel planes. The surface area of a wall constituted by one rod is $S_1$. As shown in Figure 8, the central angle $\theta_n$ of the arc constituting $S_1$ can be expressed as

$$\theta_n = \cos^{-1}\left\{ 1 - \left( \frac{x}{r} + \frac{D_p}{nr} \right) \right\} - \cos^{-1}\left( 1 - \frac{x}{r} \right) \quad (6)$$

where $r$ is the radius of the rod.

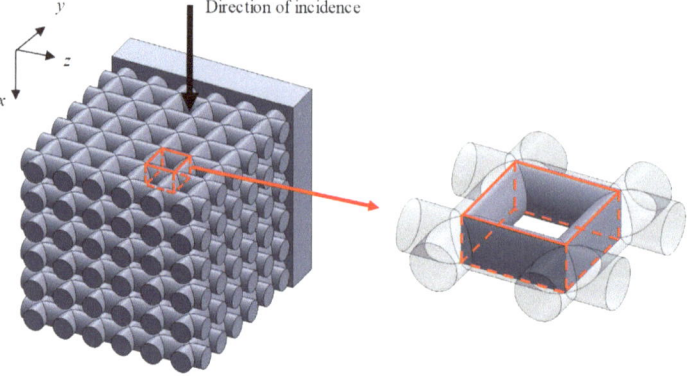

**Figure 5.** Analysis unit (Unit I) of the grid network structure.

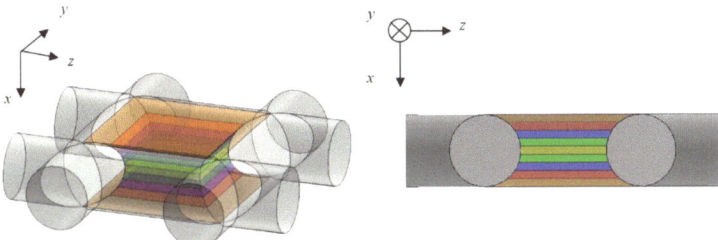

**Figure 6.** Analysis unit (Unit I) divided into layers of equal thickness in the *x*-axis direction.

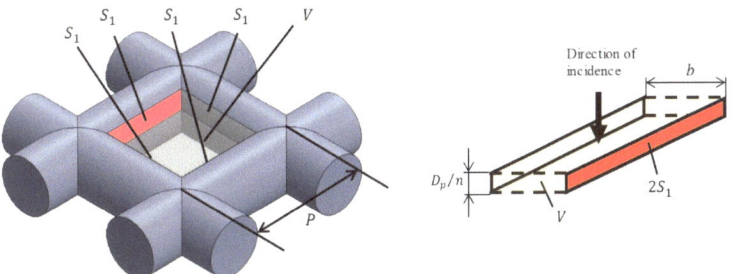

**Figure 7.** Approximation of analysis unit (Unit I) as the clearance between two planes.

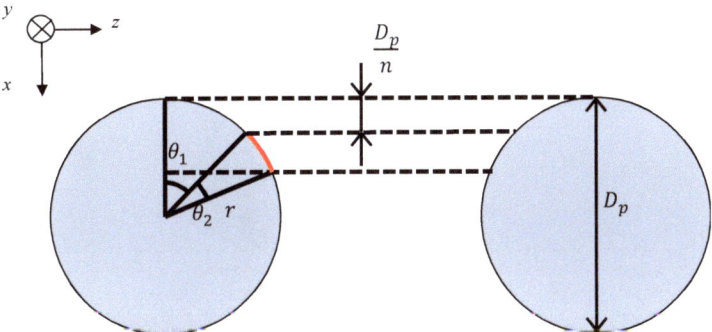

**Figure 8.** Derivation of surface area $S_1$.

Then, the surface area $S_1$ is given by

$$S_1 = \int (P - 2r\sin\theta) r d\theta \tag{7}$$

where $P$ is the pitch of the rod. The volume $V$ of the void space is given by

$$V = \int \left(P - 2\sqrt{r^2 - x^2}\right)^2 dx \tag{8}$$

The thickness in the direction of the sound wave incidence is equal to $D_p/n$, which is the thickness of one layer. The distance $b$ between the two planes can be expressed as

$$b = \frac{V}{2S_1} \tag{9}$$

where the volume between two parallel planes is $V$. An open-ended correction was added to the grid network structure because the voids can be regarded as orifices in the $x$-axis direction. For the cross-section perpendicular to the direction of sound wave incidence, the orifice was assumed to be in the $x$-axis direction. Then, an open-ended correction length of 0.4 times the radius of a circle equivalent to the area of the smallest gap is applied [21,22]:

$$\Delta l = \frac{P - D_p}{\sqrt{\pi}} \times 0.4 \tag{10}$$

The open-ended correction length is added to the length $l$ of the relevant layer.

### 3.3.2. Analysis Unit Surrounded by Three Rods and Tube Wall

Figure 9 shows an analysis unit surrounded by three rods and a tube wall (Unit II). Figure 10 shows the approximation of Unit II as two parallel planes. The lateral surface area $S_1$ is derived in the same manner as in Unit I. The lateral surface area $S_2$ is derived in the same manner as in Unit I. The side area $S_2$ of the rod in contact with the tube wall is derived as follows:

$$S_2 = \int \left\{ \frac{L_t - (\sqrt{N_p} - 1)P}{2} - r\sin\theta \right\} r d\theta \tag{11}$$

where $L_t$ is the length of one side of the sample holder (i.e., $L_t$ = 25.7 mm).

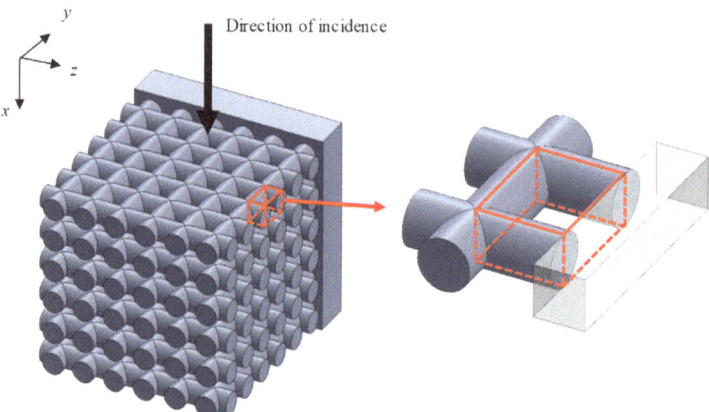

**Figure 9.** Analysis unit (Unit II) of the grid network structure.

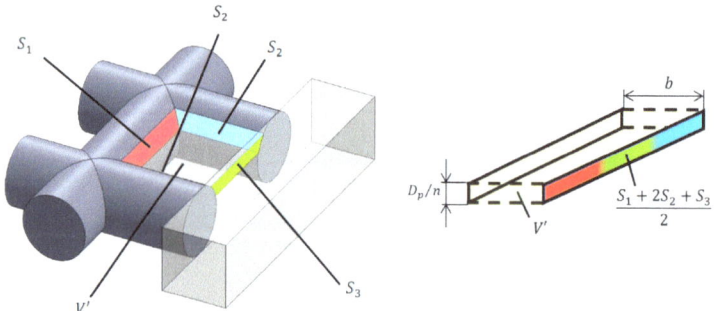

**Figure 10.** Approximation of Unit II as the clearance between two planes.

Figure 11 shows the area $S_3$ of the inner wall of the sample holder, which is the rectangular area comprising the pitch between the rods and layer thickness minus the area of overlap between the rectangle and end faces of the rods. In other words, $A_{ar}$ (area enclosed by the red border) can be calculated as follows:

$$A_{ar} = \int_{x1}^{x2} \sqrt{r^2 - x^2} dx \tag{12}$$

Then, the surface area $S_3$ can be expressed as follows:

$$S_3 = \frac{D_p}{n} \times P - 2A_{ar} \tag{13}$$

The volume $V'$ in Figure 12 can be expressed as follows:

$$V' = \int \left\{ \frac{L_t - (\sqrt{N_p} - 1)P}{2} - \sqrt{r^2 - x^2} \right\} \left( P - 2\sqrt{r^2 - x^2} \right) dx \tag{14}$$

Then, the distance $b$ between two planes is derived as follows:

$$b = \frac{V'}{\frac{S_1 + 2S_2 + S_3}{2}} \tag{15}$$

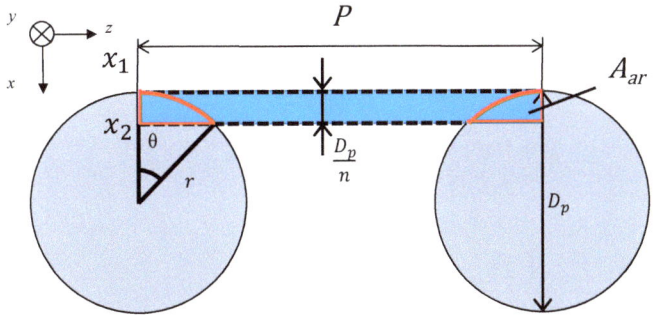

**Figure 11.** Derivation of surface area $S_3$.

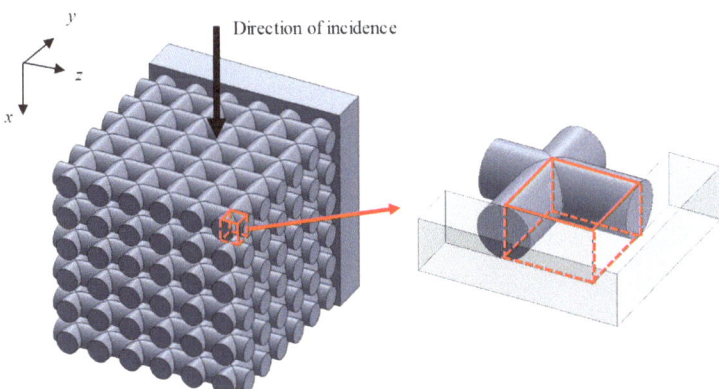

**Figure 12.** Analysis unit (Unit III) of the grid network structure.

### 3.3.3. Analysis Unit Surrounded by Two Rods and Tube Wall

Figure 12 shows an analysis unit surrounded by two rods and a tube wall (Unit III). Figure 13 shows the approximation of Unit III as two parallel planes. The lateral area $S_2$ of the rod tangential to the tube wall is derived in the same way as in Unit II. The area of the tube wall $S_3{}'$ is derived as follows:

$$S_3{}' = \frac{D_p}{n} \times \frac{L_t - (\sqrt{N_p}-1)P}{2} - A_{ar} \qquad (16)$$

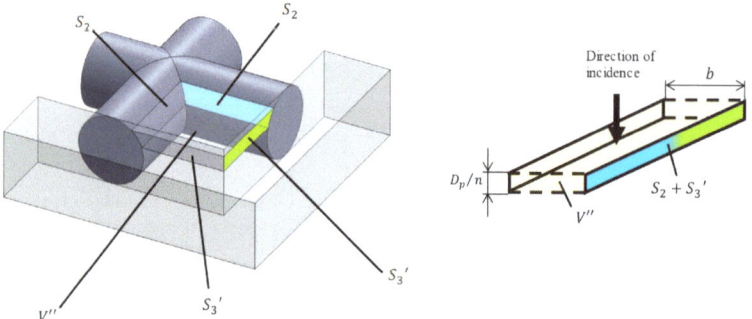

**Figure 13.** Approximation of Unit III as the clearance between two planes.

The volume $V''$ in Figure 13 can be expressed as follows:

$$V'' = \int \left\{ \frac{L_t - (\sqrt{N_p}-1)P}{2} - \sqrt{r^2 - x^2} \right\}^2 dx \qquad (17)$$

Then, the distance $b$ between the two planes can be derived as follows:

$$b = \frac{V''}{S_2 + S_3{}'} \qquad (18)$$

### 3.4. Transfer Matrix of the Grid Network Structure
#### 3.4.1. Transfer Matrix of Analysis Units

The two parallel planes of each layer in an analysis unit correspond to the planes in Figure 4. Thus, Equations (2) and (3) can be used to determine the propagation constant and characteristic impedance for each layer. The propagation constant and characteristic impedance can then be substituted into Equation (1) to obtain the transfer matrices $T_1, T_2, T_3, \ldots, T_n$ for the layers, where the subscript 1 represents the layer at the plane incident to the sound wave, and the subscript $n$ represents the number of layers. These transfer matrices can then be cascaded, as shown in Figure 14a, to obtain the transfer matrix for an analysis unit.

Here, $T_u$ is the transfer matrix of Unit I, $T_w$ is the transfer matrix of Unit II and $T_f$ is the transfer matrix of Unit III.

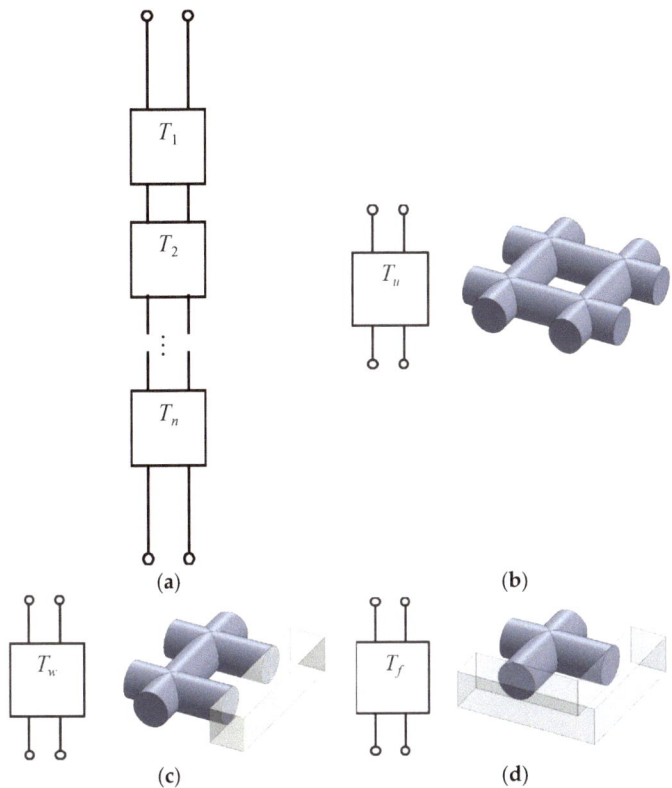

**Figure 14.** Equivalent circuit analysis: (**a**) cascade connecting $T_n$, (**b**) summation of all $T_n$ for Unit I, (**c**) summation of all $T_n$ for Unit II, and (**d**) summation of all $T_n$ for Unit III.

3.4.2. Transmission Matrix of the Whole Sample

Figure 15 shows that the transfer matrices of each analysis unit are connected in parallel along the y-axis and then along the z-axis. This allows the transfer matrix $T_l$ per grid network layer to be derived.

As shown in Figures 15 and 16, the transfer matrices $T_M$ and $T_N$ are calculated for each analysis unit in parallel in the y-axis direction and for one row in the y-axis direction, respectively. The number of parallel connections of $T_w$ in $T_M$ and $T_u$ in $T_N$ is one less than the number of rods in the y-axis direction. As shown in Figure 17, the calculated $T_M$ and $T_N$ are connected in parallel in the z-axis direction. Then, the transfer matrix $T_l$ of the entire sample can be calculated. In this case, the number of $T_N$ in parallel is one less than the number of rods in the z-axis direction.

Next, the transfer matrix $T_{all}$ for the entire sample is derived. As shown in Figure 18, the transfer matrix $T_l$ for one grid network layer and the transfer matrix $T_a$ corresponding to the air layer are connected in an alternating cascade. This allows the entire transfer matrix to be derived.

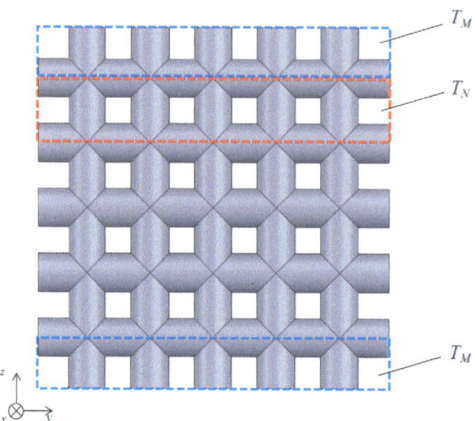

**Figure 15.** Parallel connection of transfer matrices in a grid network structure.

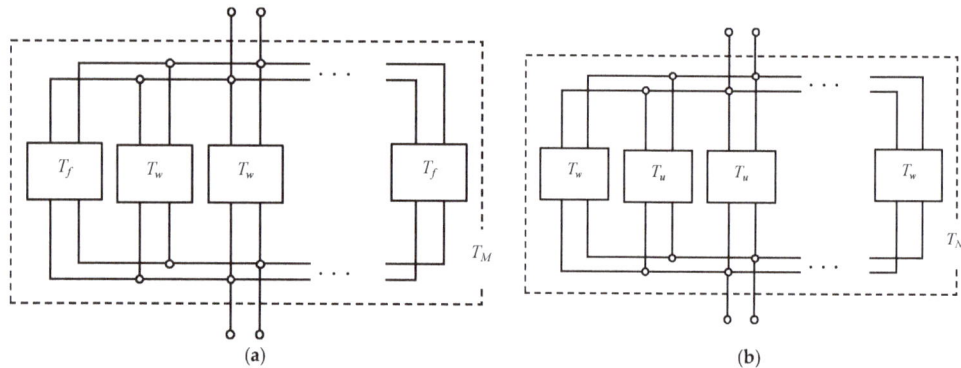

**Figure 16.** Equivalent circuit of parallel connections in the $y$-axis direction: (**a**) $T_w$ and $T_f$ and (**b**) $T_u$ and $T_w$.

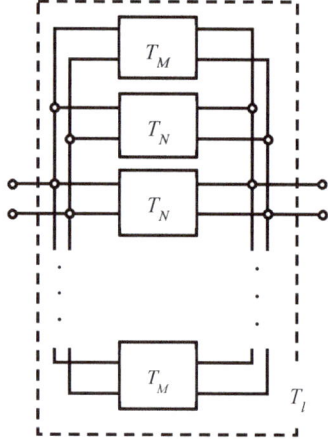

**Figure 17.** Equivalent circuit of the whole sample as parallel connections of $T_M$ and $T_N$ in the z-axis direction.

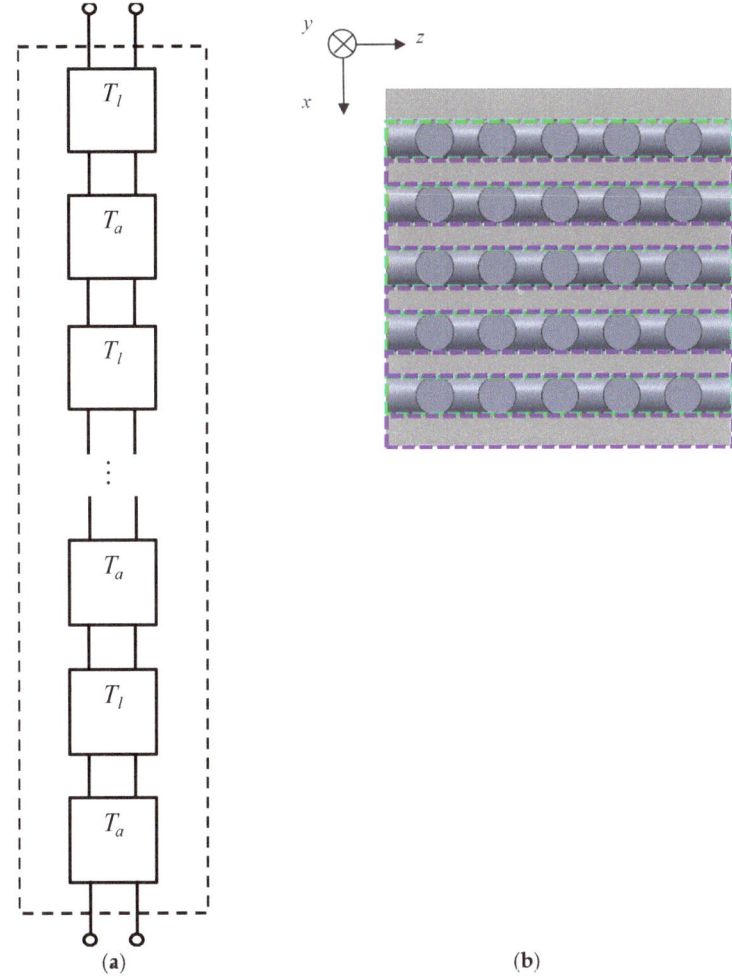

**Figure 18.** Equivalent circuit of analysis unit: (**a**) cascade connecting $T_l$, $T_a$ and $T_{all}$; and (**b**) cross-section of the grid network structure.

### 3.5. Derivation of the Sound Absorption Coefficient

The transfer matrix can be used to derive the sound absorption coefficient. The end of the gap was assumed to be a rigid wall, so the particle velocity $u_2 = 0$. Then, Equation (1) can be transformed as follows:

$$\begin{bmatrix} p_1 \\ Su_1 \end{bmatrix} = \begin{bmatrix} t_{11} & t_{12} \\ t_{21} & t_{22} \end{bmatrix} \begin{bmatrix} p_2 \\ 0 \end{bmatrix} = \begin{bmatrix} t_{11}p_2 \\ t_{21}p_2 \end{bmatrix} \tag{19}$$

As shown in Figure 4, $p_0$ and $u_0$ are the sound pressure and particle velocity, respectively, just outside plane 1. If $p_0 = p_1$ and $S_0 u_0 = S u_1$, then Equation (19) can be used to obtain the specific acoustic impedance $Z_0$ looking inside the sample from the plane of incidence:

$$Z_0 = \frac{p_0}{u_0} = \frac{p_0}{u_0 S_0} S_0 = \frac{p_1}{u_1 S} S_0 = \frac{t_{11}}{t_{21}} S_0 \tag{20}$$

The relationship between the specific acoustic impedance $Z_0$ and reflectance $R$ is expressed as follows:

$$R = \frac{Z_0 - \rho_0 c_0}{Z_0 + \rho_0 c_0} \tag{21}$$

Equation (21) can then be used to obtain the sound absorption coefficient $\alpha$:

$$\alpha = 1 - |R|^2 \tag{22}$$

## 4. Results

### 4.1. Experimental and Simulated Values of Sound Absorption Coefficient

The experimental and simulated sound absorption coefficients were compared for each sample. Figure 19 shows the experimental and simulated values for the rod diameters $D_p$ = 2.3 and 2.5 mm and pitch $P$ = 3.5 mm. Figure 20 shows the results for the rod diameters $D_p$ = 2.8 and 3.0 mm and pitch $P$ = 4.2 mm.

For all samples, the experimental and simulated values generally showed similar trends. The simulated values showed high prediction accuracy, but tended to be lower than the experimental values at peak frequencies. A similar tendency was previously observed by Sakamoto et al. [23] when the one-dimensional transfer matrix method was used to estimate the sound absorption coefficient for other cross-sectional shapes. The difference between the simulated and experimental values may be attributed to the simulated calculation underestimating the sound wave attenuation. The reason for this can be that the particle velocity distribution of the sound wave between the two surfaces was averaged in this simulated analysis.

When sound waves are incident to a gap, the distribution of the $x$-axial component of the particle velocity in the $z$-axis direction within the gap generally has a distribution with respect to the distance from the wall. In contrast, the present simulated analysis assumed the $x$-axial component of the particle velocity in the gap between two planes to have a uniform distribution in the $z$-axis direction. This assumption may have led to the difference between the experimental and simulated values.

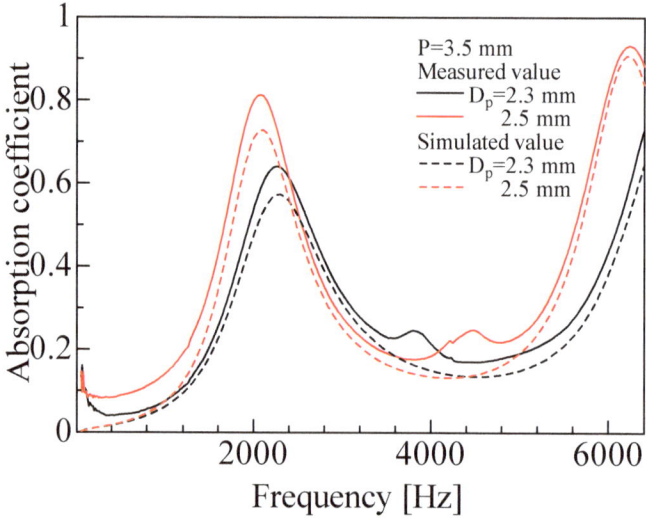

**Figure 19.** Comparison between experimental and simulated values ($D_p$ = 2.3, 2.5 mm, $P$ = 3.5 mm). The rod diameters used in the calculations are the measured values given in Table 1.

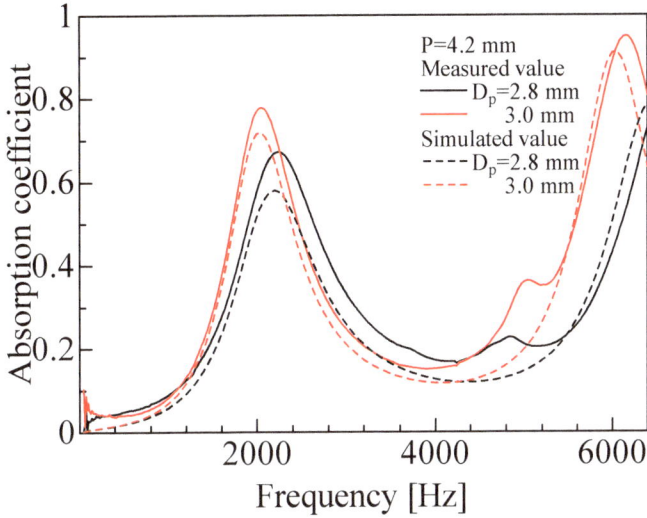

**Figure 20.** Comparison between experimental and simulated values ($D_p$ = 2.8, 3.0 mm, $P$ = 4.2 mm). The rod diameters used in the calculations are the measured values given in Table 1.

*4.2. Sensitivity Analysis*

In this section, simulation results are presented for different wire diameters and pitches.

For the sensitivity analysis, the sample shown in Figure 2b (i.e., $D_p$ = 2.5 mm, pitch $P$ = 3.5 mm, number of grid network layers $N_1$ = 7) was used as the baseline. Figure 21 shows the calculated results when the rod diameter $D_p$ was varied. The corresponding experimental result is shown for reference. When $D_p$ was incremented 0.1 mm larger than the actual value (i.e., $D_p$ = 2.6 mm), the simulated and experimental values were almost identical. Therefore, the difference between the simulated and experimental values can be attributed to a difference in the rod diameter of approximately 0.1 mm. One way to reduce the difference between the experimental and simulated values to an acceptable level is to adjust the rod diameter used in the simulated calculation.

Figure 21 also shows that the peak value of the sound absorption coefficient increased with an increasing rod diameter, and the peak frequency tended to shift lower. This may be attributed to the increasing influence of the boundary layer as the gap becomes smaller. In general, the peak becomes sharper as the aperture ratio decreases. Here, the sound absorption peak also became sharper as the rod diameter increased (i.e., as the aperture ratio decreased).

Figure 22 shows the calculated results when the pitch $P$ was varied. The size of the sample tube was changed to match the change in pitch. A smaller pitch tended to increase the peak value of the sound absorption coefficient. This may be because a smaller pitch decreases the gap, which again increases the influence of the boundary layer.

Figure 23 shows the calculated results when the number of grid network layers $N_l$ was varied. The peak value increased with the number of layers, and the peak frequency tended to shift lower. This may be because an increase in the number of grid network layers corresponds to an increase in the thickness of the sound-absorbing material. For a single grid network layer, the sound absorption coefficient was difficult to confirm.

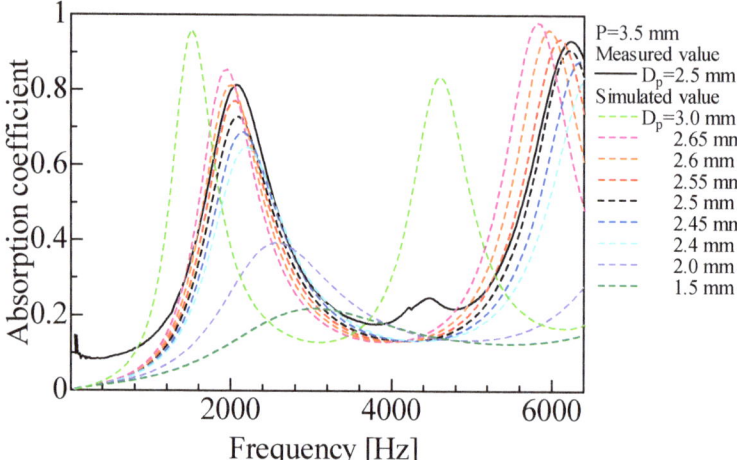

**Figure 21.** Comparison between experimental and simulated values with changing rod diameter ($D_p \approx$ 1.5–3.0 mm, $P$ = 3.5 mm).

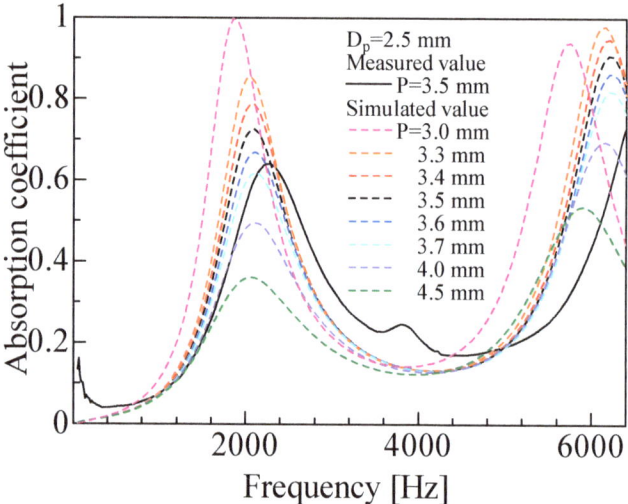

**Figure 22.** Comparison between experimental and simulated values with changing pitch ($P \approx$ 3.3–3.7 mm, $D_p$ = 2.5 mm).

Figure 24 shows the calculation results for one grid network layer when the thickness of the back air layer was varied. Increasing the thickness of the back air layer decreased the peak value of the sound absorption coefficient, and the peak frequency tended to shift lower. This confirmed that a sound absorption effect could be achieved by the back air layer even when there was only one grid network layer.

These results show that the parameters used in the simulated analysis can be varied to explore their effects on the sound absorption coefficient. The comparison between the experimental and simulated results show that the observed trends for the changes in the sound absorption coefficient are reasonable.

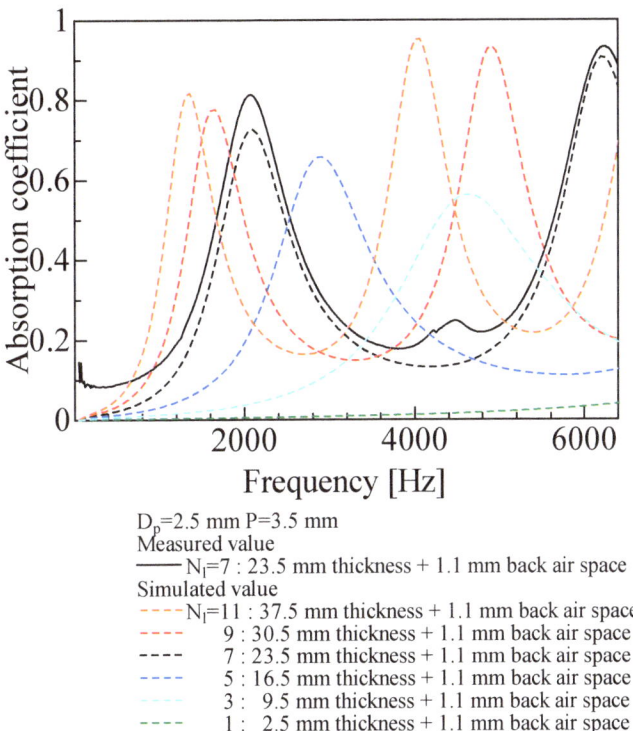

**Figure 23.** Comparison between experimental and simulated values with changing number of layers ($N_l \approx$ 1–9).

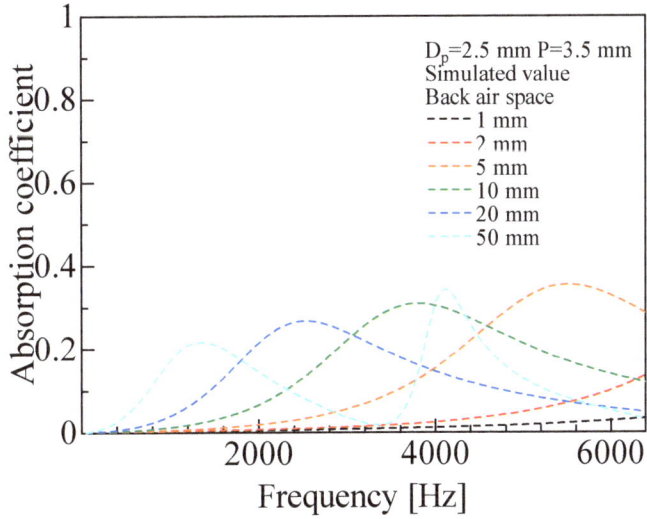

**Figure 24.** Simulated values with a changing length of back air space (1–9 mm) ($N_l = 1$, $D_p = 2.5$ mm, $P = 3.5$ mm).

## 5. Conclusions

A mathematical model of a grid network structure consisting of rods was developed to estimate the sound absorption coefficient, and the simulated results were compared with experimental measurements. The gaps in the grid network structure were approximated as the clearance between two parallel planes, and analysis units were derived to consider the exact geometry of the layers. The characteristic impedance and propagation constant were determined for the approximated gaps and treated as a one-dimensional transfer matrix. The transfer matrix obtained for each layer was used to calculate the sound absorption coefficient.

The simulated values tended to be close to the experimental values, but generally underestimated them. This may be because the simulated analysis underestimated the sound wave attenuation. For the peak frequency, the simulated analysis showed high prediction accuracy for the peak frequency. The measurement results show that the rod diameter of 2.5 mm, pitch of 3.5 mm, and the presence of seven layers resulted in a sound absorption coefficient of 0.81 at the first peak.

A sensitivity analysis was carried out to investigate the influence of the rod diameter and pitch. The peak sound absorption coefficient tended to increase with an increasing rod diameter and decreasing pitch. This was attributed to the increasing influence of the boundary layer as the gaps become smaller. The above results indicate that the mathematical model used to calculate the sound absorption coefficient is sufficiently accurate to predict the sound absorption coefficient for practical applications. Thus, it was found that the difference between the simulated and experimental values was approximately equivalent to 0.1 mm in diameter. Therefore, it is possible to reduce the difference between the experimental and calculated values to a practically acceptable level by adjusting the diameter of the rod.

**Author Contributions:** Conceptualization, S.S.; data curation, T.I., T.S. and K.T.; formal analysis, T.S., T.I., K.I. and K.T.; project administration, S.S.; software, K.I.; supervision, S.S. All authors have read and agreed to the published version of the manuscript.

**Funding:** This research was funded by FUKOKU Co., Ltd.

**Institutional Review Board Statement:** Not applicable.

**Informed Consent Statement:** Not applicable.

**Data Availability Statement:** Data is contained within the article.

**Acknowledgments:** This study was conducted in collaboration with FUKOKU Co., Ltd.

**Conflicts of Interest:** The authors declare no conflict of interest.

## References

1. Zhao, T.S.; Cheng, P. Oscillatory pressure drops through a woven-screen packed column subjected to a cyclic flow. *Cryogenics* **1996**, *36*, 333–341. [CrossRef]
2. Banús, E.D.; Sanz, O.; Milt, V.G.; Miró, E.E.; Montes, M. Development of a stacked wire-mesh structure for diesel soot combustion. *Chem. Eng. J.* **2014**, *246*, 353–365. [CrossRef]
3. Sun, H.; Bu, S.; Luan, Y. A high-precision method for calculating the pressure drop across wire mesh filters. *Chem. Eng. Sci.* **2015**, *127*, 143–150. [CrossRef]
4. Yamamoto, T.; Maruyama, S.; Terada, K.; Izui, K.; Nishiwaki, S. A generalized macroscopic model for sound-absorbing poroelastic media using the homogenization method. *Comput. Methods Appl. Mech. Eng.* **2011**, *200*, 251–264. [CrossRef]
5. Satoh, T.; Sakamoto, S.; Unai, S.; Isobe, T.; Iizuka, K.; Tasaki, K.; Nitta, I.; Shintani, T. Sound-absorption coefficient of a pin-holder structure for sound waves incident in the direction perpendicular to the cylinder's axis. *Noise Control Eng. J.* **2022**, *70*, 136–149. [CrossRef]
6. Iizuka, K.; Sakamoto, S.; Sato, T.; Tasaki, K.; Nitta, I.; Mizuno, C. Theoretical Estimation and Experiment of Sound Absorption Coefficient for Mesh Structure. In Proceedings of the JSME, No.227-1, Paper No. C032, Kanazawa, Japan, 5 March 2022.
7. Dias, T.; Monaragala, R. Sound absorption in knitted structures for interior noise reduction in automobiles. *Meas. Sci. Technol.* **2006**, *17*, 2499. [CrossRef]

8. Gao, N.; Tang, L.; Deng, J.; Lu, K.; Hou, H.; Chen, K. Design, fabrication and sound absorption test of composite porous metamaterial with embedding I-plates into porous polyurethane sponge. *Appl. Acoust.* **2021**, *175*, 107845. [CrossRef]
9. Gao, N.; Wu, J.; Lu, K.; Zhong, H. Hybrid composite meta-porous structure for improving and broadening sound absorption. *Mech. Syst. Signal Process.* **2021**, *154*, 107504. [CrossRef]
10. Gao, N.; Wang, B.; Lu, K.; Hou, H. Teaching-learning-based optimization of an ultra-broadband parallel sound absorber. *Appl. Acoust.* **2021**, *178*, 107969. [CrossRef]
11. Yang, F.; Wang, E.; Shen, X.; Zhang, X.; Yin, Q.; Wang, X.; Yang, X.; Shen, C.; Peng, W. Optimal Design of Acoustic Metamaterial of Multiple Parallel Hexagonal Helmholtz Resonators by Combination of Finite Element Simulation and Cuckoo Search Algorithm. *Materials* **2022**, *15*, 6450. [CrossRef] [PubMed]
12. Yang, X.; Yang, F.; Shen, X.; Wang, E.; Zhang, X.; Shen, C.; Peng, W. Development of Adjustable Parallel Helmholtz Acoustic Metamaterial for Broad Low-Frequency Sound Absorption Band. *Materials* **2022**, *15*, 5938. [CrossRef] [PubMed]
13. ISO 10534-2; Acoustics—Determination of Sound Absorption Coefficient and Impedance in Impedance Tubes—Part 2: Transfer-Function Method. International Organization for Standardization: Geneva, Switzerland, 1998.
14. Sasao, H. A guide to acoustic analysis by Excel-Analysis of an acoustic structural characteristic-(4) Analysis of the duct system silencer by Excel. *J. Soc. Heat. Air-Cond. Sanit. Eng. Jpn.* **2007**, *81*, 51–58. (In Japanese)
15. Suyama, E.; Hirata, M. Attenuation Constant of Plane Wave in a Tube: Acoustic Characteristic Analysis of Silencing Systems Based on Assuming of Plane Wave Propagation with Frictional Dissipation Part 1. *J. Acoust. Soc. Jpn.* **1979**, *35*, 165–170. (In Japanese)
16. Sakamoto, S.; Hoshino, A.; Sutou, K.; Sato, T. Estimating Sound-Absorption Coefficient and Transmission Loss by the Di-mentions of Bundle of Narrow Holes (Comparison between Theoretical Analysis and Experiments). *Trans. Jpn. Soc. Mech. Eng. Ser. C* **2013**, *79*, 4164–4176. (In Japanese) [CrossRef]
17. Tijdeman, H. On the propagation of sound waves in cylindrical tubes. *J. Sound Vib.* **1975**, *39*, 1–33. [CrossRef]
18. Stinson, M.R. The propagation of plane sound waves in narrow and wide circular tubes, and generalization to uniform tubes of arbitrary cross—Sectional shape. *J. Acoust. Soc. Am.* **1991**, *89*, 550–558. [CrossRef]
19. Stinson, M.R.; Champou, Y. Propagation of sound and the assignment of shape factors in model porous materials having simple pore geometries. *J. Acoust. Soc. Am.* **1992**, *91*, 685–695. [CrossRef]
20. Allard, J.F.; Atalla, N. *Propagation of Sound in Porous Media: Modeling Sound Absorbing Materials*, 2nd ed.; Wiley: Hoboken, NJ, USA, 2009; pp. 45–54.
21. Bolt, H.R.; Labate, S.; Ingård, U. The acoustic reactance of small circular orifices. *J. Acoust. Soc. Am.* **1949**, *21*, 94–97. [CrossRef]
22. Benade, A.H. Measured End Corrections for Woodwind Tone—Holes. *J. Acoust. Soc. Am.* **1967**, *41*, 1609. [CrossRef]
23. Sakamoto, S.; Higuchi, K.; Saito, K.; Koseki, S. Theoretical analysis for sound-absorbing materials using layered narrow clearances between two planes. *J. Adv. Mech. Des. Syst. Manuf.* **2014**, *8*, JAMDSM0036. [CrossRef]

**Disclaimer/Publisher's Note:** The statements, opinions and data contained in all publications are solely those of the individual author(s) and contributor(s) and not of MDPI and/or the editor(s). MDPI and/or the editor(s) disclaim responsibility for any injury to people or property resulting from any ideas, methods, instructions or products referred to in the content.

# Article

# Estimation of the Acoustic Properties of the Random Packing Structures of Granular Materials: Estimation of the Sound Absorption Coefficient Based on Micro-CT Scan Data

Shuichi Sakamoto [1,*], Kyosuke Suzuki [2], Kentaro Toda [2] and Shotaro Seino [2]

[1] Department of Engineering, Niigata University, Ikarashi 2-no-cho 8050, Nishi-ku, Niigata 950-2181, Japan
[2] Graduate School of Science and Technology, Niigata University, Ikarashi 2-no-cho 8050, Nishi-ku, Niigata 950-2181, Japan
* Correspondence: sakamoto@eng.niigata-u.ac.jp; Tel.: +81-25-262-7003

**Abstract:** In this study, the sound absorption properties of randomly packed granular materials were estimated. Generally, it is difficult to construct a general mathematical model for the arrangement of randomly packed granular materials. Therefore, in this study, an attempt was made to estimate the sound absorption coefficient using a theoretical analysis by introducing data from computed tomography (CT) scans, as the tomographic images of CT scans correspond to the slicing and elemental division of packing structures. In the theoretical analysis, the propagation constants and characteristic impedances in the voids were obtained by approximating each tomographic image as a void between two parallel planes. The derived propagation constants and characteristic impedances were then treated as a one-dimensional transfer matrix in the propagation of sound waves, and the transfer matrix method was used to calculate the normal incident sound absorption coefficient. The theoretical value of the sound absorption coefficient was derived using the effective density to which the measured tortuosity was applied. As a result, for the theoretical values considering the tortuosity, in many cases, the theoretical values were close to the measured values. For the theoretical values, when both the surface area and tortuosity were considered, the peak sound absorption frequency moved to a lower frequency and was in general agreement with the measured values.

**Keywords:** granular material; random packing; micro-CT scan; tortuosity

## 1. Introduction

Granular packing structures have been used for noise reduction purposes, such as low-noise pavements [1], ballasted tracks [2] and consolidated expanded clay granulates [3], due to their acoustic properties. These structures have continuous voids and exhibit acoustic properties based on the same principle of porous materials, and their acoustic properties vary depending on their layer thickness, grain size, and filling structure. Therefore, from an engineering point of view, it is useful to predict the acoustic properties of granular packing structures through calculations based on their grain size, packing structure, and gas physical properties.

Various studies have been performed on the sound absorption properties of the regular packing structures of granular materials. For example, while there are experimental studies on the acoustic properties of loosely packed lattices of granular materials [4], predicting the acoustic properties of face-centered cubic lattices [5], and fundamental studies on the acoustic properties of granular materials [6], such as sound absorption by simple cubic and hexagonal lattices [7] and by hexagonal most dense and face-centered cubic lattices [8], there are few examples of theoretical calculations on random packing, although experimental studies [9] have been previously performed. Additionally, numerical analyses have been performed on several regular packing structures using commercial software [10] (e.g., a numerical analysis of pseudo-random packing combined with multiple regular packings) [11] and on the sound absorption characteristics of random close packing using the discrete element method (DEM) [12]. However, the software used for the theoretical analyses in these studies is expensive and computationally time-consuming. Generally, when granular materials are packed, the structure is random packing. Therefore, it is scientifically significant to analyze random packing rather than regular packing.

In this study, the sound absorption properties of randomly packed granular materials were estimated. Generally, it is difficult to construct a general mathematical model for the arrangement of randomly packed granular materials. Therefore, in this study, an attempt was made to estimate the sound absorption coefficient using a theoretical analysis by introducing data from computed tomography (CT) scans, as the tomographic images of CT scans correspond to the slicing and elemental division of packing structures. To estimate the propagation constants and characteristic impedances of voids, we considered each tomographic image as a void between two parallel planes using theoretical calculations. The transfer matrix method was used to calculate the normal incident sound absorption coefficient, which was estimated from the propagation constants and characteristic impedances generated from sound waves travelling through the medium. This method allows the sound absorption coefficient to be calculated based on CT scan images of granular materials. Moreover, the used program is simple, the calculation time is short, and the calculation can be performed using a common PC. By applying this method, it may be possible to calculate the sound absorption coefficients of granular sound-absorbing materials with irregular shapes, not limited to the glass beads used in this measurement.

Tortuosity, one of the Biot parameters, is a parameter that expresses the complexity of the sound waves propagating in a structure, and it was measured in this study. As a result, by combining the measured tortuosity with an effective density, it is possible to calculate the theoretical value of the sound absorption coefficient.

Regarding the measurement of the sound absorption coefficient, a two-microphone impedance tube was used to measure the normal incident sound absorption coefficient of each sample. The results of a comparison between the experimental measurements and theoretical values were also reported.

## 2. Samples and Measuring Equipment Used for Measuring the Sound Absorption Coefficient

### 2.1. Sample for Measuring the Sound Absorption Coefficient

In this study, random packing was used as a packing structure for granular materials. Figure 1a–c shows the samples used to measure the sound absorption coefficient. Glass beads with diameters of 1 mm, 2 mm, and 4 mm were used in the experiments.

(a)

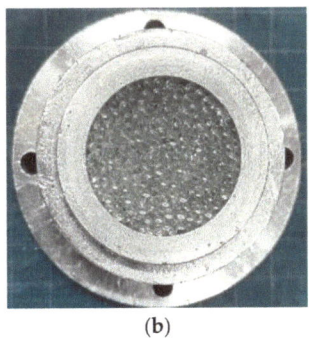
(b)

(c)

**Figure 1.** Test samples of random packing granules: (**a**) $d$ = 1 mm; (**b**) $d$ = 2 mm; (**c**) $d$ = 4 mm.

A sample holder tube by aluminum alloy with an inner diameter of 29 mm was filled with glass beads.

## 2.2. Equipment for Measuring the Sound Absorption Coefficient

A Brüel & Kjær Type 4206 two-microphone impedance tube (Brüel & Kjær Sound & Vibration Measurement A/S, Skodsborgvej 307 DK-2850, Nærum, Denmark) was used to measure the sound absorption coefficient, and Figure 2 shows the configuration of the apparatus. As shown in the figure, a sample was enclosed in the impedance tube, a sinusoidal signal was output by the signal generator DS-3000 with a built-in fast Fourier transform (FFT) analyzer fabricated by Ono Sokki (Yokohama, Japan), and the transfer function between the sound pressure signals of the two microphones attached to the impedance tube was measured by the FFT analyzer. The measured transfer function was used to calculate the normal incident sound absorption coefficient in accordance with ISO 10534-2. The critical frequency for the formation of plane waves differs based on the inner diameter of the acoustic tube. Small tubes with an inner diameter of 29 mm were used because the sound absorption coefficient in the low-frequency range was not high enough for the samples used in this study. Therefore, the measurement range was 500–6400 Hz.

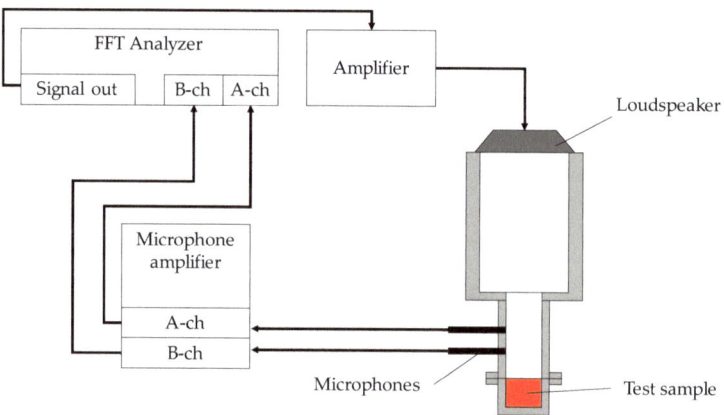

**Figure 2.** Scheme of two microphone impedance tubes for the sound absorption coefficient measurement.

## 3. Methods and Results of Measuring Tortuosity

### 3.1. Tortuosity Measurement Sample

As indicated in the previous report and in Figure 3, tortuosity is one of the Biot parameters [8]. In this study, the tortuosity of randomly packed structures was considered and was experimentally measured using ultrasonic sensors. In general, the tortuosity $\alpha_\infty$ can be expressed using the speed of sound $c_0$ in air and the apparent speed of sound $c$ in the filling structure, as shown in Equation (1) [13]:

$$\alpha_\infty = \left(\frac{c_0}{c}\right)^2 \qquad (1)$$

Tortuosity was measured using the same measuring equipment as previously reported [8]. Glass beads with a grain size of $d$ = 2 mm were used for the measurements, as the tortuosity is independent of grain size.

Ultrasonic sensors with center frequencies of 32.7 kHz, 58 kHz, 110 kHz, 150 kHz, 200 kHz, and 300 kHz were used. In the initial phase, we measured the tortuosity $\alpha_\infty$ at each frequency to determine the performance of the random packing structure. The inverse of the square root of the frequency used for the measurement was then taken as the value on the horizontal axis, and the tortuosity $\alpha_\infty$ at each frequency (determined by the meas-urement) was plotted as the value on the vertical axis. A linear approximation of these points was performed using the least-squares method, yielding a straight line that rose steadily to the right. The extreme value of the tortuosity at infinite frequency in the approximated straight line was obtained, where the *y*-intercept of the graph was the tortuosity $\alpha_\infty$ of the packing structure.

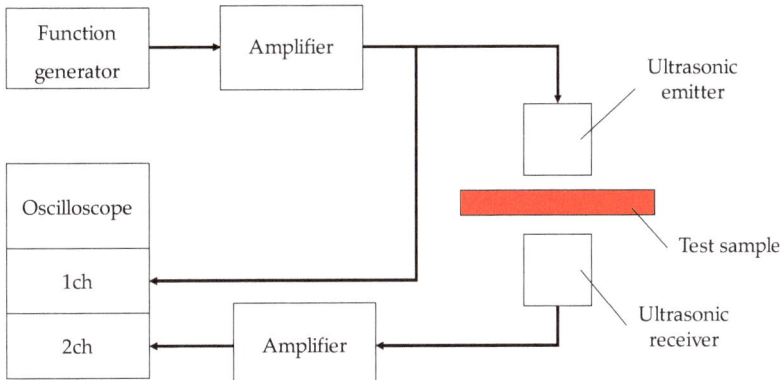

**Figure 3.** Configuration diagram of measurement of tortuosity.

*3.2. Tortuosity Measurement*

The signal-to-noise ratio (S/N ratio) of the ultrasonic sensor was low at high frequencies due to the low conversion efficiency and high sound wave attenuation. A total of 150 measurements were synchronously added to increase the S/N ratio and improve the measurement accuracy. The signals were measured with a resolution of 16 bits.

Table 1 shows the results of the tortuosity measurements, and Figure 4 shows a graphical representation. This was confirmed by the fact that the y-intercept of the approximation line indicated the complexity of each packing configuration. The measurement results show that the tortuosity was $\alpha_\infty$ = 1.45.

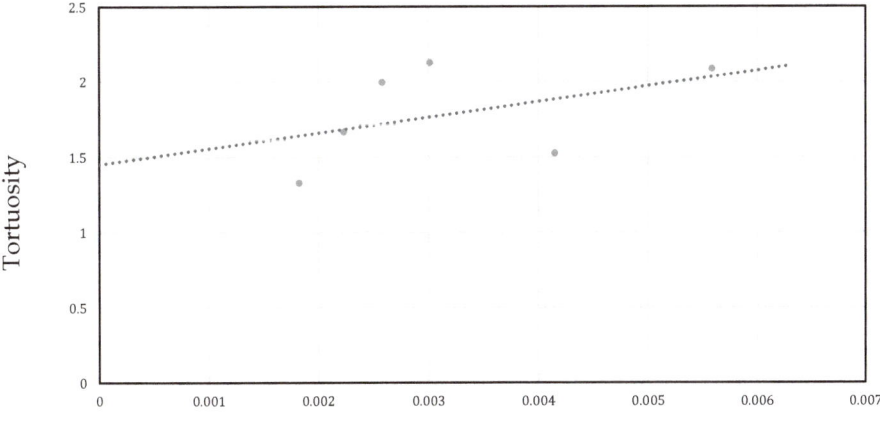

**Figure 4.** Results of the tortuosity measurements.

Table 1. Result of the tortuosity measurement.

| Frequency [kHz] | Inverse of the Square Root of Frequency [1/√Hz] | Distance between Sensors [mm] | Transmission Time [ms] | Tortuosity |
|---|---|---|---|---|
| 32.7 | 0.00559 | 395 | 1.229 | 2.09 |
| 58 | 0.004152 | 395 | 1.199 | 1.53 |
| 110 | 0.003015 | 345 | 1.044 | 2.13 |
| 150 | 0.002582 | 204 | 0.626 | 2.00 |
| 200 | 0.002236 | 204 | 0.618 | 1.67 |
| 300 | 0.001826 | 204 | 0.615 | 1.33 |
| ∞ | 0 | - | - | 1.45 |

## 4. Theoretical Analysis

### 4.1. Analysis Unit and Element Division of the Packing Structure

In random packing, there is no periodicity in the arrangement of granular materials, making it difficult to develop mathematical models. Therefore, this study considered the use of geometric information on the measured particles, and the method adopted for this purpose was micro-CT scanning.

The cross-sectional image was binarized, as shown in Figure 5a. The sphere boundaries were then clarified, as shown in Figure 5b. The cross-sectional images of the micro-CT scan were taken over a rectangular area of 20 mm in the $x$-direction and 13 mm square in the $y$-$z$ plane. The instrument used was an MCT225 Metrology CT, manufactured by NIKON Corp (Tokyo, Japan). The tomograms were taken to be sliced in the $y$-$z$ plane, which was perpendicular to the direction of the sound wave incidence ($x$-direction). The CT scan data for $d$ = 2 mm were also used for the analysis of other grain sizes, as the arrangement of grains was independent of the grain size.

As shown in Figure 5c, the cross-section of the sphere in the tomogram was approximated to a cylindrical shape of length $l$. Subsequently, as shown in Figure 5d, it was approximated as a clearance between two planes, where the surface area $S_n$ of the grain and the volume $V_n$ of the clearance were equal. The number of images used for the theoretical analysis $n$ was 2000, 1000, and 500 for the grain size $d$ = 1 mm, $d$ = 2 mm, and $d$ = 4 mm, respectively, for 20 mm in the $x$-direction. Thus, the thickness $l$ of the element in Figure 5c corresponded to the pitch in the $x$-direction of the image, which was 10 μm, 20 μm, and 40 μm for the grain sizes of $d$ = 1 mm, 2 mm, and 4 mm, respectively. Thus, for each grain size $d$ = 1 mm, 2 mm, and 4 mm, the number of analysis units was $n$ = 100 in the packing structure partitioning method [7,8], as shown in Figure 5c, which was a value at which the theoretical values of the normal incident sound absorption coefficient sufficiently converged.

For each image, the cross-sectional area of the void and the sum of the circumferences of the cross-sections of all of the spheres in the image were calculated. Multiplying the cross-sectional area of the void by the image pitch $l$ provided the volume of the void $V_n$, as shown in Figure 5c. Similarly, multiplying the total circumference of the cross-section by $l$ yielded $S_n$, as shown in Figure 5c. Thus, as shown in Figure 5d, using Equation (2), the thickness of the clearance between the two planes, $b_n$, could be obtained for a single image with a thickness $l$. Here, $F_n$ is the correction factor for obtaining the true surface area, explained in the next section.

$$b_n = \frac{2V_n}{F_n S_n} \quad (2)$$

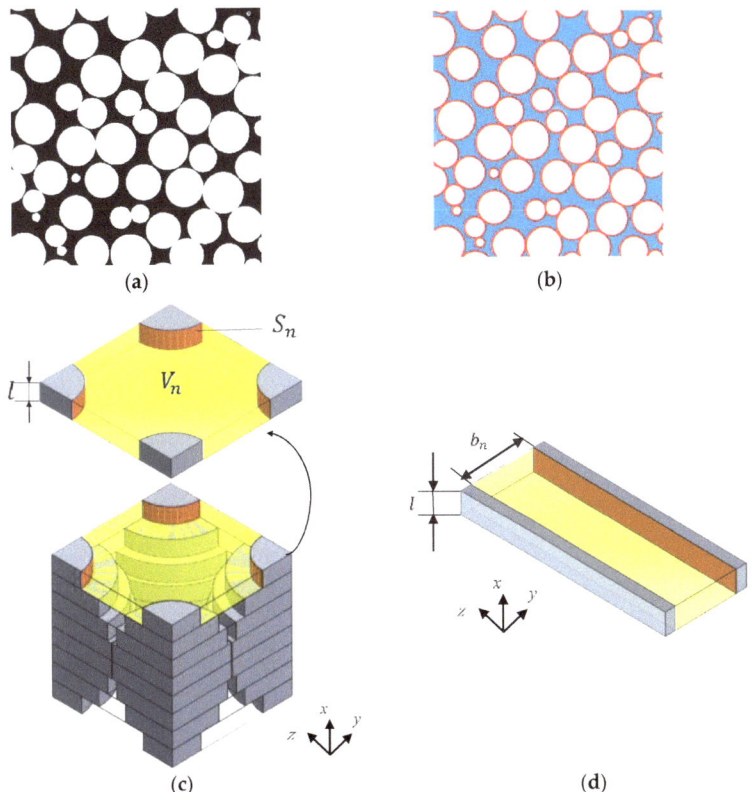

**Figure 5.** Divided element approximated to the clearance between the two planes: (**a**) binarization of the cross-sectional image; (**b**) calculation of the circumference of the sphere and the cross-sectional area of the clearance; (**c**) divided element; (**d**) approximated clearance between the two planes.

### 4.2. Numerical Analysis of the Tomograms in Random Packing

Using the method shown in Figure 6a, the surface area of the granular material used in the analysis was estimated to be smaller than the surface area of a true sphere. Thus, a correction for the surface area was made, as shown in Figure 6b and below.

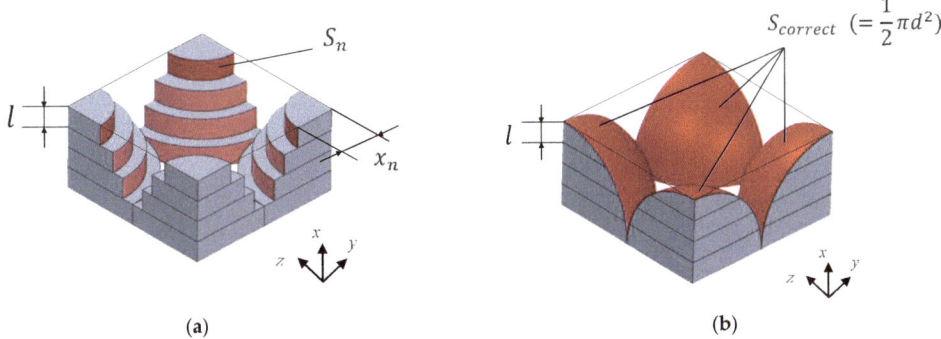

**Figure 6.** Divided element for the half sphere: (**a**) cylindrically approximated surface area; (**b**) true surface area.

In Figure 6a, the radius $x_n$ of the cylindrical element in the $n$th division is expressed as in Equation (3), where $k$ is the number of divisions of the hemisphere in the $x$-direction.

$$x_n = \sqrt{\frac{d^2}{4} - \left\{(n-1)\frac{d}{2k}\right\}^2} \quad (3)$$

Using the above equation, the area $S_n$ of the cylindrical element could be expressed as in Equation (4):

$$S_n = \pi x_n \frac{d}{k} \quad (4)$$

When the number of divisions $k = 1$, the ratio between the sum of the areas $S_n$ of the divided cylindrical elements obtained from the CT images and the true surface area $S_{correct}$ ($= 1/2 \cdot \pi d^2$) of the hemisphere is shown in the following equation, where $F_1$ is the correction factor.

$$F_1 = \frac{S_{correct}}{\lim_{k \to 1} \sum_{n=1}^{k} S_n} = 1 \quad (5)$$

As shown in the above equation, it is geometrically clear that when the number of divisions $k = 1$, the correction factor $F_1$ is unity (i.e., they are equal). However, when the number of divisions $k$ is infinite, the correction factor $F_\infty$ asymptotically approaches ~1.273.

$$F_\infty = \frac{S_{correct}}{\lim_{k \to \infty} \sum_{n=1}^{k} S_n} \cong 1.273 \quad (6)$$

Furthermore, the correction factor $F_{50}$ for $k = 50$ is ~1.259.

$$F_{50} = \frac{S_{correct}}{\lim_{k \to 50} \sum_{n=1}^{k} S_n} \cong 1.259 \quad (7)$$

Therefore, for the resolution of the CT scan used in this study, an area equal to the surface area of the true sphere could be used for the analysis by multiplying the correction factor $F_{50} = 1.259$ for $k = 50$ by $S_n$ in Equation (2).

### 4.3. Propagation Constants and Characteristic Impedance Considering the Tortuosity

Our investigation was based on Stinson [14] and Allard [15] descriptions of investigative procedures. The Cartesian coordinate system was used, as shown in Figure 7, and through a three-dimensional analysis using the Navier–Stokes equations, the gas equation of state, continuity equation, energy equation, and dissipative function representing the heat transfer, effective density $\rho_s$, and compressibility $C_s$ were derived, as shown in Equations (8) and (9), respectively [14]. For several atmospheric properties, including the density ($\rho_0$) of air, $\lambda_s$ is the parameter of mediation, $b_n$ is the clearance thickness between the two planes, $\omega$ is the angular frequency, $\eta$ is the viscosity of air, $\kappa$ is the specific heat ratio of air, $P_0$ is the atmospheric pressure, and $N_{pr}$ is the Prandtl number.

$$\rho_s = \rho_0 \left[1 - \frac{\tanh(\sqrt{j\lambda_s})}{\sqrt{j\lambda_s}}\right]^{-1}, \lambda_s = \frac{b_n}{2}\sqrt{\frac{\omega \rho_0}{\eta}} \quad (8)$$

$$C_s = \left(\frac{1}{\kappa P_0}\right)\left\{1 + (\kappa - 1)\left[\frac{\tanh(\sqrt{jN_{pr}\lambda_s})}{\sqrt{jN_{pr}\lambda_s}}\right]\right\} \quad (9)$$

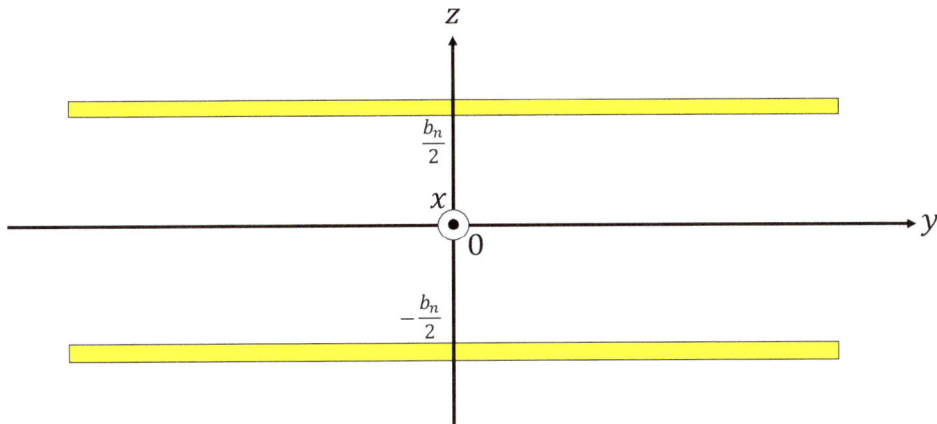

**Figure 7.** Coordinates for the clearance between the two planes.

Using the effective density $\rho_s$ multiplied by the tortuosity $\alpha_\infty$, the propagation constant and characteristic impedance considering the tortuosity could be obtained [15]. Therefore, the propagation constant $\gamma$ and characteristic impedance $Z_c$ when considering the tortuosity $\alpha_\infty$ could be expressed using the effective density $\rho_s$ and compression ratio $C_s$, as follows [15]:

$$\gamma = j\omega\sqrt{\alpha_\infty \rho_s C_s} \tag{10}$$

$$Z_c = \sqrt{\frac{\alpha_\infty \rho_s}{C_s}} \tag{11}$$

*4.4. Transfer Matrix*

Using a one-dimensional wave equation, the transfer matrix method was used to derive the clearance between the two planes for the sound pressure and volume velocity. Figure 8 shows a schematic representation of one component of the *x*-direction clearance between the two planes indicated in Figure 7. Using the characteristic impedance, propagation constants, and cross-sectional area *S* of the clearance, as well as the length *l* of the divided element derived in the previous section, the transfer matrix $T_n$ and four-terminal constants *A~D* of the acoustic tube element could be calculated, as shown in Equation (12):

$$T_n = \begin{bmatrix} \cosh(\gamma l) & \frac{Z_c}{S}\sinh(\gamma l) \\ \frac{S}{Z_c}\sinh(\gamma l) & \cosh(\gamma l) \end{bmatrix} = \begin{bmatrix} A & B \\ C & D \end{bmatrix} \tag{12}$$

Plane 1 is the plane of incidence of the sound wave, and Plane 2 is the plane of transmission of the sound wave. If the sound pressure and particle velocity at Plane 1 are $p_1$ and $u_1$, respectively, and the sound pressure and particle velocity at Plane 2 are $p_2$ and $u_2$, respectively, the transfer matrix can be expressed as in Equation (13):

$$\begin{bmatrix} p_1 \\ Su_1 \end{bmatrix} = \begin{bmatrix} A & B \\ C & D \end{bmatrix}\begin{bmatrix} p_2 \\ Su_2 \end{bmatrix} \tag{13}$$

As shown in Figure 9, an equivalent circuit was used to construct the transfer matrices $T_{\text{unit}}$ and $T_{\text{top}}$ for the analysis unit at the top end of the sample, respectively, by cascading the transfer matrices of each divided element.

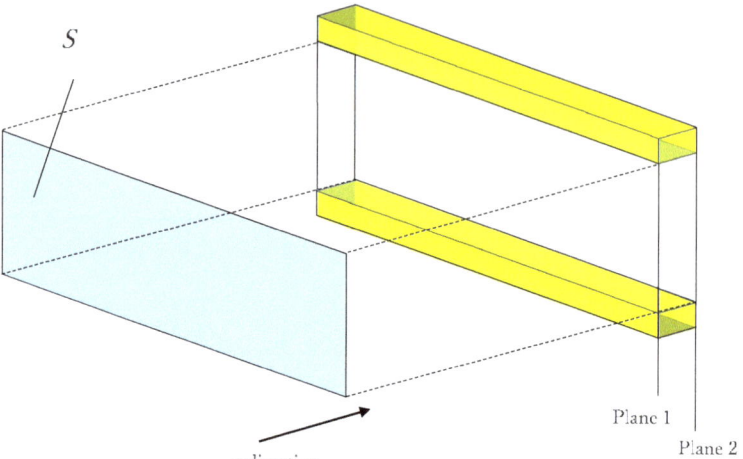

**Figure 8.** Cross-sectional shape at any position of the analysis unit (sound incident area and aperture area).

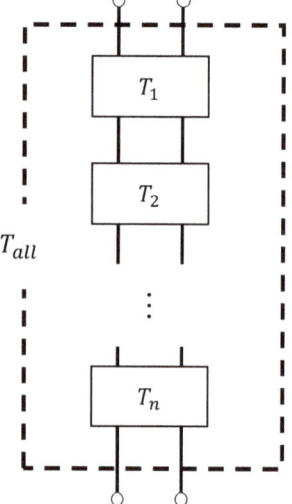

**Figure 9.** Equivalent circuit in the analysis (cascade connection of the transfer matrix of each element).

### 4.5. Vertical Incident Sound Absorption Coefficient

The transfer matrix, described in Section 4.4, was used to calculate the sound absorption coefficient. Since the end of the sample used in this study was a rigid wall, the particle velocity $u_2 = 0$, and Equation (13) could be transformed into Equation (14):

$$\begin{bmatrix} p_1 \\ Su_1 \end{bmatrix} = \begin{bmatrix} A & B \\ C & D \end{bmatrix} \begin{bmatrix} p_2 \\ 0 \end{bmatrix} = \begin{bmatrix} Ap_2 \\ Cp_2 \end{bmatrix} \quad (14)$$

If the sound pressure and particle velocity immediately outside Plane 1 are denoted by $p_0$ and $u_0$, respectively, the specific acoustic impedance $Z_0$ looking inward from the plane of incidence of the sample can be expressed as follows:

$$Z_0 = \frac{p_0}{u_0} \quad (15)$$

Therefore, through $p_0 = p_1$ and Equation (15), the specific acoustic impedance $Z_0$ of the sample can be expressed as follows:

$$Z_0 = \frac{p_0}{u_0} = \frac{A}{C}S \tag{16}$$

The relationship between the specific acoustic impedance $Z_0$ and reflectance $R$ can be expressed as follows:

$$R = \frac{Z_0 - \rho_0 c_0}{Z_0 + \rho_0 c_0} \tag{17}$$

The following relationship between the sound absorption coefficient, the reflectance, and Equation (17) provides the theoretical value of the sample's normal incident sound absorption coefficient $\alpha$.

$$\alpha = 1 - |R|^2 \tag{18}$$

## 5. Comparison of the Measured and Theoretical Values

The measured and theoretical values of the normal incident sound absorption coefficient were compared for each sample with a varying particle diameter in a random packing structure. First, the results were demonstrated, as shown in Figure 10a–c. Figure 10a–c show comparisons for the particle diameters $d$ = 1 mm, $d$ = 2 mm, and $d$ = 4 mm, respectively.

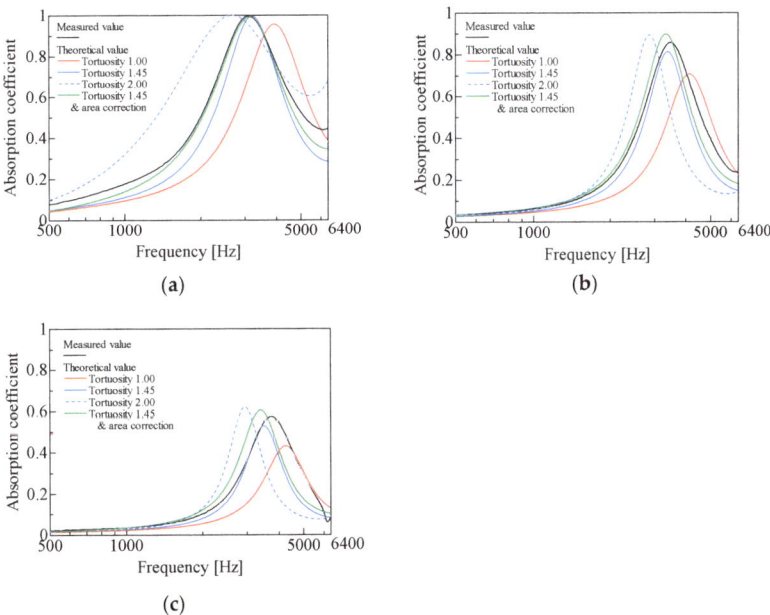

**Figure 10.** The experiment in comparison with calculation: (**a**) $d$ = 1 mm; (**b**) $d$ = 2 mm; (**c**) $d$ = 4 mm.

First, a comparison was performed between the measured values in each graph and the theoretical values without considering the tortuosity. In all of the cases shown in Figure 10a–c, the theoretical values without considering the tortuosity appeared at a frequency higher than the peak frequency of the measured values. In all cases, the theoretical values without considering the tortuosity had lower peak sound absorption values than the measured values. The differences between the theoretical and measured values could be attributed to the fact that tortuosity was not considered.

Attention was then drawn to the theoretical values considering the tortuosity. Considering the tortuosity, the peak frequency moved lower in all cases, as shown in Figure 10a–c, and the peak sound absorption value increased in all cases.

As a result, the difference between the theoretical and measured values decreased when the tortuosity was considered. A tortuosity greater than unity means that the path of the sound waves propagating in the sample is longer, which has the same effect as an increase in the sample thickness. Thus, it can be considered that the peak frequency, considering the theoretical value, moved to the lower frequency side. For the same reasons mentioned above, the peak sound absorption values in the theoretical values could be considered to have increased due to the tortuosity consideration. The theoretical values for a tortuosity of 2.00 were also demonstrated for reference.

Theoretical values were considered with the surface area correction described in Section 4.2 after considering tortuosity. For the particle diameters $d = 1$ mm and $d = 2$ mm shown in Figure 10a,b, the theoretical and measured values for both the peak frequency and peak sound absorption agreed well when the tortuosity and surface area corrections were considered. The reason for this result was that the surface area correction reduced the void thickness $b_n$ between the two planes in Equation (2), resulting in a larger proportion of the boundary layer in the voids and a larger calculated attenuation of the sound waves by viscosity. However, for the grain size $d = 4$ mm, as shown in Figure 10c, the peak frequency of the theoretical value was even lower than the measured value when considering the tortuosity and surface area corrections, and the reasons for this result are discussed below.

There are three possible reasons for the different degrees of agreement between the measured and theoretical values depending on the particle diameter. As a note, it is generally accepted that for equal packing structures, the paths of sound waves are analogous irrespective of the particle diameter; therefore, the same tortuosity can be applied. Thus, tortuosity is not considered the cause of the abovementioned results.

The first possible reason is that the method used to estimate the attenuation of sound waves in clearance [14] is more inaccurate for larger gaps (grain size) [16].

The second possible reason is the difference between the CT scan image and the sample used to measure the sound absorption coefficient. In the CT scan image, only a part of the particle was captured at the edge of the image, as shown in Figure 5a. In the experiment, the particles were spherical even at the edge, and there was a wall on the inner circumference of the sample tube. Thus, the difference between the CT scan image of the particles at the edge and the sample used for the measurement of the sound absorption coefficient could be a reason for the difference between the theoretical and measured values.

The third possible reason is an error in the layer thickness when the granular material was filled. To perform the experiment with a layer thickness of 20 mm, the granular material was added to the measuring tube and then slid off at a position of 20 mm from the rigid wall at the bottom of the measuring tube. Therefore, there were actually scattered cavities with a layer thickness of less than 20 mm. The effect of this phenomenon is more pronounced for larger particle diameters. In general, the smaller the sample thickness, the higher the peak frequency moves and the lower the peak sound absorption value. Therefore, in the case of a particle diameter of $d = 4$ mm in Figure 10c, it can be considered that the peak frequency of the measured value moved to a higher frequency due to levelling.

In summary, the obtained difference between the experimental and theoretical values (red line) was because the tortuosity and actual surface area were not considered. In other words, the obtained theoretical value based on both the tortuosity and actual surface area agreed well with the experimental value. The theoretical value that considered the tortuosity and surface area correction was less sharp (i.e., slightly higher than the experimental value at the base of the sound absorption curve). This result suggests that the attenuation of sound waves was slightly overestimated when considering the two factors. However, the policy of considering the tortuosity and surface area correction is suggested to be correct.

## 6. Conclusions

A theoretical calculation of the sound absorption coefficient at a normal incidence for the random packing of granular materials was performed using tomography images taken by CT scan. The measured tortuosity was considered in the derivation of the propagation constant and characteristic impedance. In the theoretical analysis, a surface area correction was performed to make the cross-sectional images closer to reality. Theoretical values with and without considering tortuosity, theoretical values with considering both tortuosity and surface area correction, and experimental values were compared, and the following results were obtained.

1. It is difficult to construct a mathematical model for random packing, as it has no structure periodicity. Therefore, the sound absorption coefficient was estimated using a theoretical analysis based on cross-sectional CT scan images.
2. For the theoretical values considering the tortuosity, the peak sound absorption values were higher, and the peak frequency moved to lower frequencies compared with the case without considering the tortuosity. As a result, in all cases, the theoretical values were closer to the measured values. Therefore, the measured tortuosity values are reasonable.
3. Regarding the theoretical values, when both the surface area and tortuosity were considered, the peak sound absorption frequency moved to a lower frequency compared with the theoretical value when only considering tortuosity, and was in general agreement with the measured values for the particle diameters of $d = 1$ mm and $d = 2$ mm. Therefore, the method of estimating the vertical incident sound absorption coefficient using computed tomographic images is useful. Moreover, this model can be applied even if the material changes, provided the granular material can be considered rigid. Additionally, the model can be applied without problems for general granular densities and grain sizes in the order of mm.

**Author Contributions:** Conceptualization, S.S. (Shuichi Sakamoto); Data curation, K.T. and S.S. (Shotaro Seino); Formal analysis, K.S.; Project administration, S.S. (Shuichi Sakamoto); Software, K.S.; Supervision, S.S. (Shuichi Sakamoto). All authors have read and agreed to the published version of the manuscript.

**Funding:** This work was supported by Japan Society for the Promotion of Science (JSPS) KAKENHI Grant No. 20K04359.

**Institutional Review Board Statement:** Not applicable.

**Conflicts of Interest:** The authors declare no conflict of interest.

## References

1. Sandberg, U. Low noise road surfaces—A state-of-the-art review. *J. Acoust. Soc. Jpn.* **1999**, *20*, 1–17. [CrossRef]
2. Zhang, X.; Thompson, D.; Jeong, H.; Squicciarini, G. The effects of ballast on the sound radiation from railway track. *J. Sound Vib.* **2017**, *339*, 137–150. [CrossRef]
3. Vašina, M.; Hughes, D.C.; Horoshenkov, K.V.; Lapčík, L., Jr. The acoustical properties of consolidated expanded clay granulates. *Appl. Acoust.* **2006**, *67*, 787–796. [CrossRef]
4. Voronina, N.N.; Horoshenkov, K.V. A new empirical model for the acoustic properties of loose granular media. *Appl. Acoust.* **2003**, *64*, 415–432. [CrossRef]
5. Dung, V.V.; Panneton, R.; Gagné, R. Prediction of effective properties and sound absorption of random close packings of monodisperse spherical particles: Multiscale approach. *J. Acoust. Soc. Am.* **2019**, *145*, 3606–3624. [CrossRef] [PubMed]
6. Sakamoto, S.; Tsutsumi, Y.; Yanagimoto, K.; Watanabe, S. Study for Acoustic Characteristics Variation of Granular Material by Water Content. *Trans. Jpn. Soc. Mech. Eng.* **2009**, *75*, 2515–2520. [CrossRef]
7. Sakamoto, S.; Ii, K.; Katayama, I.; Suzuki, K. Measurement and Theoretical Analysis of Sound Absorption of Simple Cubic and Hexagonal Lattice Granules. *Noise Control. Eng. J.* **2021**, *69*, 401–410. [CrossRef]
8. Sakamoto, S.; Suzuki, K.; Toda, K.; Seino, S. Mathematical Models and Experiments on the Acoustic Properties of Granular Packing Structures (Measurement of tortuosity in hexagonal close-packed and face-centered cubic lattices). *Materials* **2022**, *15*, 7393. [CrossRef] [PubMed]
9. Sakamoto, S.; Sakuma, Y.; Yanagimoto, K.; Watanabe, S. Basic study for the acoustic characteristics of granular material (Normal incidence absorption coefficient for multilayer with different grain diameters). *Trans. Jpn. Soc. Mech. Eng.* **2008**, *74*, 2240–2245. [CrossRef]

10. Gasser, S.; Paun, F.; Bréchet, Y. Absorptive properties of rigid porous media: Application to face centered cubic sphere packing. *J. Acoust. Soc. Am.* **2005**, *117*, 2090–2099. [CrossRef] [PubMed]
11. Zieliński, T.G. Microstructure-based calculations and experimental results for sound absorbing porous layers of randomly packed rigid spherical beads. *J. Appl. Phys.* **2014**, *116*, 034905. [CrossRef]
12. Lee, C.-Y.; Leamy, M.J.; Nadler, J.H. Acoustic absorption calculation in irreducible porous media: A unified computational approach. *J. Acoust. Soc. Am.* **2009**, *126*, 1862–1870. [CrossRef] [PubMed]
13. Allard, J.F.; Castagnede, B.; Henry, M. Evaluation of tortuosity in acoustic porous materials saturated by air. *Rev. Sci. Instrum.* **1994**, *65*, 754. [CrossRef]
14. Stinson, M.R.; Champou, Y. Propagation of sound and the assignment of shape factors in model porous materials having simple pore geometries. *J. Acoust. Soc. Am.* **1992**, *91*, 685–695. [CrossRef]
15. Allard, J.F. Propagation of sound in Porous Media Modeling Sound Absorbing Materials. *J. Acoust. Soc. Am.* **1994**, *95*, 2785. [CrossRef]
16. Sakamoto, S.; Higuchi, K.; Saito, K.; Koseki, S. Theoretical analysis for sound-absorbing materials using layered narrow clearances between two planes. *J. Adv. Mech. Des. Syst. Manuf.* **2014**, *8*, 16. [CrossRef]

**Disclaimer/Publisher's Note:** The statements, opinions and data contained in all publications are solely those of the individual author(s) and contributor(s) and not of MDPI and/or the editor(s). MDPI and/or the editor(s) disclaim responsibility for any injury to people or property resulting from any ideas, methods, instructions or products referred to in the content.

Article

# Mathematical Models and Experiments on the Acoustic Properties of Granular Packing Structures (Measurement of Tortuosity in Hexagonal Close-Packed and Face-Centered Cubic Lattices)

**Shuichi Sakamoto [1,\*], Kyosuke Suzuki [2], Kentaro Toda [2] and Shotaro Seino [2]**

1 Department of Engineering, Niigata University, Ikarashi 2-no-cho 8050, Nishi-ku, Niigata 950-2181, Japan
2 Graduate School of Science and Technology, Niigata University, Ikarashi 2-no-cho 8050, Nishi-ku, Niigata 950-2181, Japan
\* Correspondence: sakamoto@eng.niigata-u.ac.jp; Tel.: +81-25-262-7003

**Abstract:** In this study, the sound absorption characteristics of hexagonal close-packed and face-centered cubic lattices were estimated by theoretical analysis. Propagation constants and characteristic impedances were obtained by dividing each structure into elements perpendicular to the incident direction of sound waves and by approximating each element to a clearance between two parallel planes. Consequently, the propagation constant and the characteristic impedance were treated as a one-dimensional transfer matrix in the propagation of sound waves, and the normal incident sound absorption coefficient was calculated by the transfer matrix method. The theoretical value of the sound absorption coefficient was derived by using the effective density applied to the measured tortuosity. As a result, the theoretical value was becoming closer to the measured value. Therefore, the measured tortuosity is reasonable.

**Keywords:** sound absorption coefficient; hexagonal close-packed lattice; face-centered cubic lattice; tortuosity

**Citation:** Sakamoto, S.; Suzuki, K.; Toda, K.; Seino, S. Mathematical Models and Experiments on the Acoustic Properties of Granular Packing Structures (Measurement of Tortuosity in Hexagonal Close-Packed and Face-Centered Cubic Lattices). *Materials* **2022**, *15*, 7393. https://doi.org/10.3390/ma15207393

Academic Editor: Martin Vašina

Received: 6 September 2022
Accepted: 19 October 2022
Published: 21 October 2022

**Publisher's Note:** MDPI stays neutral with regard to jurisdictional claims in published maps and institutional affiliations.

**Copyright:** © 2022 by the authors. Licensee MDPI, Basel, Switzerland. This article is an open access article distributed under the terms and conditions of the Creative Commons Attribution (CC BY) license (https://creativecommons.org/licenses/by/4.0/).

## 1. Introduction

The structures filled with granular materials are used to reduce noise, such as in low-noise pavements [1] and ballast tracks [2], owing to their acoustic characteristics. These continuous clearance structures exhibit acoustic properties similar to those of porous materials, and their acoustic properties change depending on the layer thickness, particle size, and packing structure. Therefore, the prediction of the acoustic properties of the structures filled with granular materials based on the particle size, packing structure, and physical properties of gases is useful for their engineering applications.

Various studies have been conducted on the sound absorption characteristics of granular packing structures, including experimental studies on the acoustic properties of loosely packed granular materials [3], the prediction of the acoustic properties of face-centered cubic structures [4], the numerical analysis of multiple regularly packed structures using commercial software [5], and the study of the sound absorption characteristics of random close-packed materials [6]. In addition, the acoustic properties of granular materials [7], the sound absorption coefficient in a powder bed [8,9], and sound absorption due to gaps in the packing structure of simple cubic and hexagonal lattices [10] have been studied.

In this study, the sound absorption characteristics of hexagonal close-packed and face-centered cubic lattices were estimated by theoretical analysis. Propagation constants and characteristic impedances were obtained by dividing each structure into elements perpendicular to the incident direction of sound waves and approximating each element to a clearance between two parallel planes. Consequently, the propagation constant and the characteristic impedance were treated as a one-dimensional transfer matrix in the propagation of sound waves, and the normal incident sound absorption coefficient was calculated by the transfer matrix method. This method measures the sound absorption coefficient using a one-dimensional (1D) wave equation from simplified geometrical information of the granular material. The program and the number of calculations are so simple that they cannot be compared to the finite element method (FEM), for example, and the calculation can be completed in a fraction of a second on a standard PC.

We also measured the tortuosity, which is used to represent the complexity of the path through which sound waves propagate in the structure. The theoretical value of the sound absorption coefficient was derived by using the effective density applied to the measured tortuosity which is one of the Biot parameters. The drawback of this method may be that the theoretical value includes a measurement of tortuosity, but instead, a simplified method for measuring tortuosity is proposed.

In the measurement of the sound absorption coefficient, the normal incident sound absorption coefficient of each sample was measured using a two-microphone impedance tube. The comparison between experimental and theoretical values was reported.

## 2. Samples and Measuring Device Used to Measure the Sound Absorption Coefficient

### 2.1. Transmission Loss Measurement

In this study, two types of packing structures, hexagonal close-packed and face-centered cubic lattices, were investigated. The samples used for the measurement of sound absorption are shown in Figure 1a–d, and the sample specifications are shown in Table 1. Stainless steel spheres with diameters of $d$ = 4 mm and 8 mm were used as particles.

In the preparation of the measurement sample, the granular material was packed regularly with the sample holder. The sample holder was regularly arranged with a hemispherical convex part on the wall, and a regular packing structure was formed by placing a granular body in a predetermined position. The sample holder was fabricated in photocurable resin using a photocurable 3D printer Form 2 manufactured by Formlabs Inc. (Somerville, MA, USA).

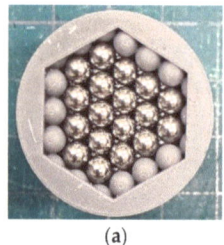   

(a) (b) (c) (d)

**Figure 1.** Test samples: (**a**) Hexagonal close-packed lattice ($d$ = 4 mm); (**b**) Hexagonal close-packed lattice ($d$ = 8 mm); (**c**) Face-centered cubic lattice ($d$ = 4 mm); (**d**) Face-centered cubic lattice ($d$ = 8 mm).

**Table 1.** Properties of test samples.

| Packing Structure | Diameter [mm] | Length [mm] | Aperture Ratio of Sample Holder | Filling Rate | Measured Tortuosity | Correspondence to Figure |
|---|---|---|---|---|---|---|
| Hexagonal close-packed | 4 | 27 | 0.57 | 0.74 | 1.44 | 1a |
|  | 8 | 27 | 0.67 | 0.74 | 1.44 | 1b |
| Face-centered cubic | 4 | 22 | 0.58 | 0.74 | 1.43 | 1c |
|  | 8 | 21 | 0.85 | 0.74 | 1.43 | 1d |

## 2.2. Measurement Equipment for Sound Absorption Coefficient

A 4206-type two-microphone impedance tube made by Brüel and Kjær was used to measure the sound absorption coefficient. The configuration of the device is shown in Figure 2. The sample was enclosed in an impedance tube, and a sine wave signal was output by an internal signal generator in the DS-3000 fast Fourier transform (FFT) analyzer manufactured by Ono Sokki, and the transfer function between the sound pressure signals of two microphones attached to the impedance tube was measured by an FFT analyzer. Using the measured transfer function, the normal incident sound absorption coefficient was calculated in accordance with ISO 10534-2. The critical frequency of the plane wave depended on the inner diameter of the acoustic tube. In this study, a small tube with an inner diameter of 29 mm was used because the sound absorption coefficient in the low-frequency range was not high. Therefore, the measurement was performed in the range of 500–6400 Hz.

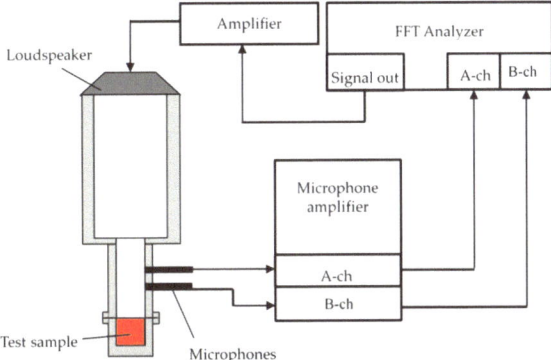

**Figure 2.** Configuration diagram of a two-microphone impedance tube for the absorption coefficient measurement.

## 3. Measurement Method and Results of Tortuosity

### 3.1. Overview of Tortuosity

Tortuosity, which is a Biot parameter, is the ratio of the average length of voids in the poroelastic material to the thickness of the sound-absorbing material. When a sound wave passes through a sound-absorbing material with a complicated internal clearance structure, the tortuosity represents the complexity of the path of the sound wave. In this study, the tortuosity for each packed structure was measured using an ultrasonic sensor. In general, the tortuosity $\alpha_\infty$ is expressed as Equation (1) using the sound velocity $c_0$ in the air and the apparent sound velocity $c$ in the packed structure [11]:

$$\alpha_\infty = \left(\frac{c_0}{c}\right)^2 \tag{1}$$

### 3.2. Tortuosity Measurement

The tortuosity was calculated by Equation (1) from the square of the ratio of two sound velocities: the sound velocity in the air without a sample and the sound velocity transmitted through the sample. Ultrasonic waves were output from an ultrasonic sensor on the transmitter side using a sinusoidal signal generated by a signal generator. Ultrasonic waves propagated in the sample were measured using an ultrasonic sensor on the receiving side, and the waveform was observed using an oscilloscope. The propagation time of ultrasonic waves from the transmitter to the receiver was calculated from the comparison between the original waveform obtained by the signal generator and that of ultrasonic waves propagated in the sample, and the apparent sound velocity $c$ in the sample in Equation (1) was calculated.

Figure 3 shows the configuration of the tortuosity measurement device. Ultrasonic sensors with center frequencies of 32.7 kHz, 40 kHz, 58 kHz, 110 kHz, 150 kHz, 200 kHz, and 300 kHz were used. First, the tortuosity $\alpha_\infty$ at each frequency was measured for each packed structure. Then, the reciprocal of the square root of the frequency used for the measurement was determined as the value of the horizontal axis, and the tortuosity $\alpha_\infty$ at each frequency obtained by the measurement was plotted as the value of the vertical axis. The linear approximation of these point clouds leads to a soaring straight line using the least-squares method. When the frequency of the approximate line is set to infinity, the limiting value of the tortuosity, i.e., the $y$-intercept of the graph, becomes the tortuosity $\alpha_\infty$ of the packing structure.

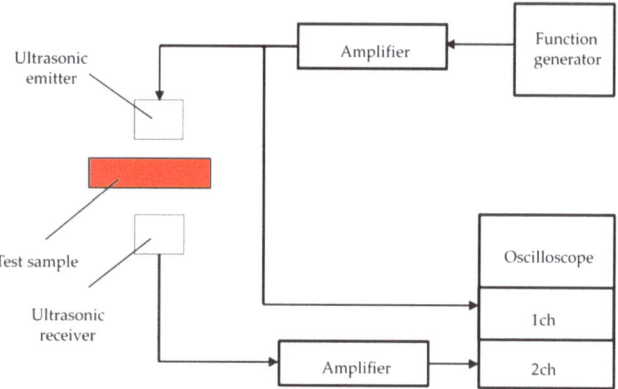

**Figure 3.** Configuration diagram of the tortuosity measurement.

At higher frequencies, ultrasonic sensors have a lower signal-to-noise ratio (S/N ratio or SNR) due to a lower conversion efficiency and a higher attenuation of sound waves in the air. To improve the S/N ratio and measurement accuracy, 150 measurements were added synchronously. The signal was measured in the amplitude direction with a resolution of 16 bits.

Table 2 and Figure 4 show the results of tortuosity measurements. As mentioned above, the $y$-intercept of the approximate straight line is the tortuosity $\alpha_\infty$ of each packing structure. At 40 kHz, the deviation was excluded for the hexagonal structure. The tortuosity of each packed structure was 1.44 for the hexagonal close-packed structure and 1.43 for the face-centered cubic lattice.

Generally, measuring tortuosity requires expensive equipment, and outsourcing the measurement is not cost-effective. In addition, experimental methods with very high ultrasonic frequencies make it difficult to measure tortuosity on rigid granular objects such as those used in this study (where the rigid surfaces have a significant size relative to the wavelength) because ultrasonic waves are reflected.

To avoid the abovementioned problems, this apparatus uses several ultrasonic transducers with low frequencies. Other devices are very common. This simple method of measuring tortuosity may be useful for rigid granular materials such as those used in this study.

Table 2. Measurement results of the tortuosity.

| Frequency [kHz] | Inverse of the Square Root of Frequency [1/√Hz] | Distance between Sensors [mm] | Transmission Time [ms] | Tortuosity Hexagonal Close-Packed | Tortuosity Face-Centered Cubic |
|---|---|---|---|---|---|
| 32.7 | 0.00559 | 395 | 1.229 | 4.78 | 4.48 |
| 40 | 0.005 | 395 | - | - | 3.35 |
| 58 | 0.004152 | 395 | 1.199 | 4.31 | 3.98 |
| 110 | 0.003015 | 345 | 1.044 | 3.59 | 3.36 |
| 150 | 0.002582 | 204 | 0.626 | 3.11 | 2.84 |
| 200 | 0.002236 | 204 | 0.618 | 2.56 | 2.67 |
| 300 | 0.001826 | 204 | 0.615 | 2.56 | 1.93 |
| ∞ | 0 | - | - | 1.44 | 1.43 |

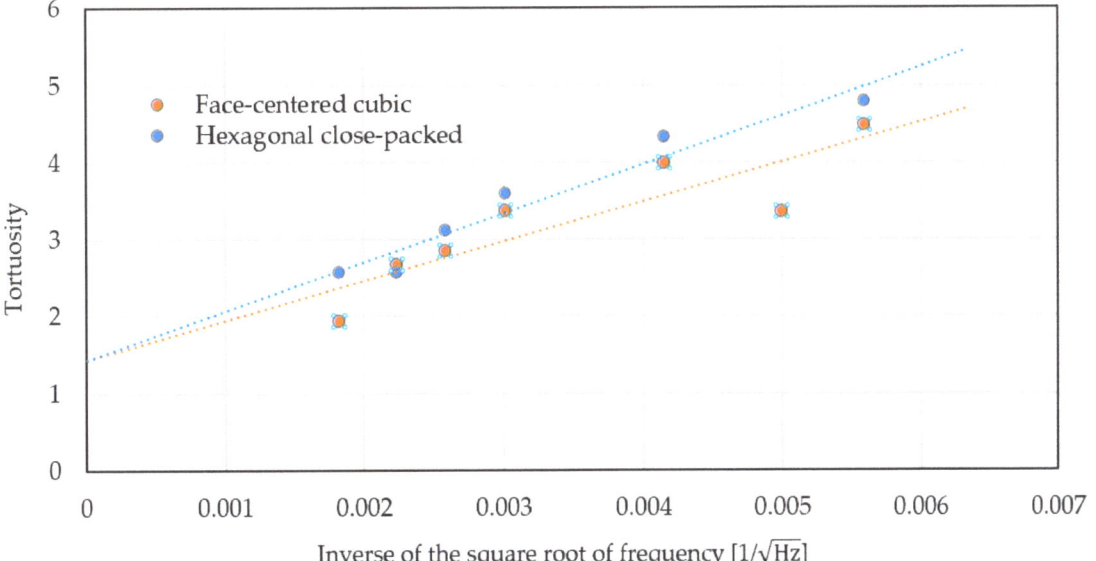

Figure 4. Measurement results of the tortuosity.

## 4. Theoretical Analysis

### 4.1. Analysis Units and Element Division

The gap in the packed structure of granular materials was analyzed by the transfer matrix method based on the one-dimensional wave equation. In this section, the outline of the theoretical analysis is explained. Figure 5 shows the analysis units for each packing structure. In these structures, the cross-sectional shape changes periodically in the $x$-, $y$-, and $z$-axis directions. The range, whereby the change in the cross-sectional shape was completed in one cycle, was used as the analysis unit. Namely, the analysis unit was defined as the area surrounded by the broken line in Figure 5. In addition, the analysis unit located at the upper edge of the sample in Figure 5a,b (i.e., the incident surface of the sound wave) in the $x$ direction is shaped, as shown in Figure 6a,b, respectively.

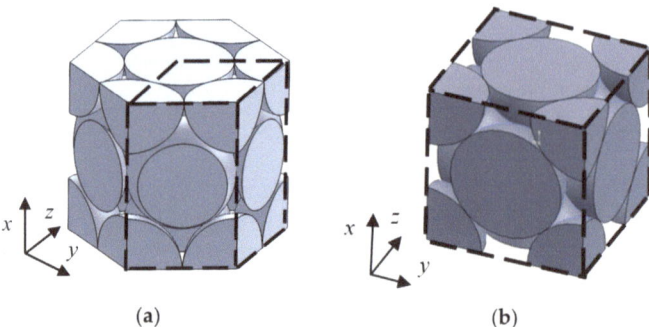

**Figure 5.** Analysis unit in each structure: (**a**) Hexagonal close-packed lattice; (**b**) Face-centered cubic lattice.

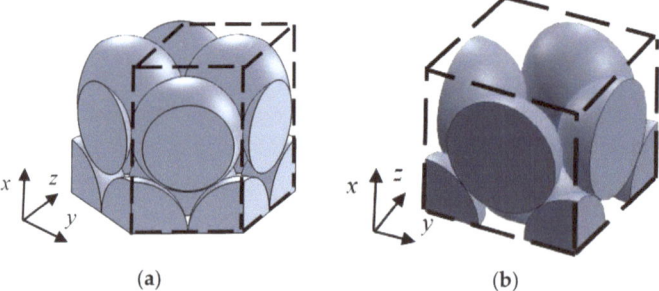

**Figure 6.** Analysis unit at the top of the sample: (**a**) Hexagonal close-packed lattice; (**b**) Face-centered cubic lattice.

In the packing structure of the granular materials, the shape of the cross-section changes continuously in the plane perpendicular to the direction of sound wave propagation. Therefore, the analysis unit was divided into 100 parts in the direction perpendicular to the $x$-axis. Figure 7a shows the analysis unit division method for the face-centered cubic lattice shown in Figure 5b. The analysis units shown in Figures 5a and 6a,b were also divided, as in the case of Figure 7a. The number of divisions of the analysis unit, $n = 100$, is the value at which the theoretical value of the normal incident sound absorption coefficient converges.

For the clearance between particles in the divided element shown in Figure 7a, the real surface area $S_n$ and real volume $V_n$ of the clearance were geometrically calculated and approximated to the thickness of the clearance between two planes $b_n$, as shown in Figure 7b and Equation (2).

For the divided elements in contact with the sample holder, as shown in Figure 8a,b, the area $S_h$ of the wall of the sample holder was also taken into consideration, and it was approximated to the clearance between two planes. As shown in Figure 8a,b, when the clearance thickness is set to $b_n'$, $S_n$, $S_h$, and $V_n$ are expressed as Equation (3):

$$b_n = \frac{2V_n}{S_n} \quad (2)$$

$$b_n' = \frac{2V_n}{S_n + S_h} \quad (3)$$

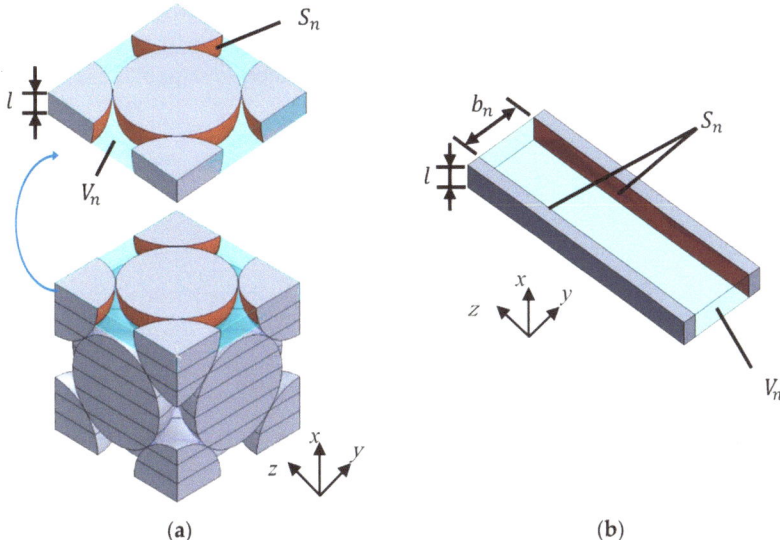

(a) (b)

**Figure 7.** Divided element approximated to the clearance between two planes: (**a**) Divided element (face-centered cubic); (**b**) Approximated clearance between two planes.

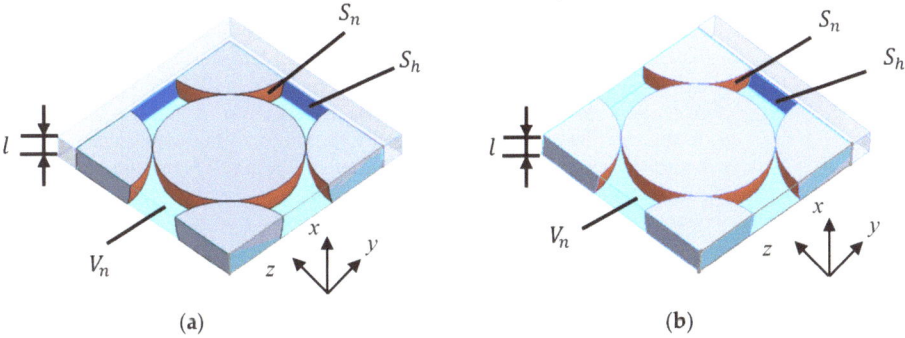

(a) (b)

**Figure 8.** Approximated clearance between two planes at the element in contact with the sample holder: (**a**) Divided element in contact with the sample holder on two sides; (**b**) Divided element in contact with the sample holder on one side.

*4.2. Derivation of the Surface Area of a Sphere in a Divided Element*

In this section, we describe the derivation method of the surface area $S_n$ of spheres in the divided element for the hexagonal close-packed and face-centered cubic lattices used in Equations (2) and (3).

Using the radius $r$ of the sphere and the length $Z_{unit}$ of the dividing element in the $x$-axis direction, the surface area $S_n$ of the sphere at the dividing element can be obtained by Equation (4):

$$S_n = 2\pi r \times Z_{unit} \quad (4)$$

As shown in Figure 8, the area $S_h$ of the blue portion of the sample holder wall was considered for the divided element in contact with the sample holder.

*4.3. Derivation of the Volume of the Clearance in a Divided Element*

In this section, we describe the derivation method of the volume $V_n$ of clearances in the divided element of the hexagonal close-packed and face-centered cubic lattice structures.

$V_n$, for the hexagonal close-packed structure, shows the clearance in the range surrounded by the broken line in Figure 5a and is derived geometrically using the dimensions of the radius $r$ of the sphere and the length $Z_{unit}$ of the divided element in the $x$-axis direction, as follows:

$$V_n = 2\sqrt{3}r^2 \times Z_{unit} - 2\int \pi\left(r^2 - x^2\right)dx \tag{5}$$

$V_n$, for the face-centered cubic structure, shows the clearance in the range surrounded by the broken line in Figure 5b, which is geometrically derived as follows:

$$V_n = 8r^2 \times Z_{unit} - 4\int \pi\left(r^2 - x^2\right)dx \tag{6}$$

*4.4. Propagation Constant and Characteristic Impedance Considering Tortuosity*

Propagation constants and characteristic impedances in the clearance between two planes approximated in Section 4.1 were obtained by considering the attenuation of sound waves. Propagation constants and characteristic impedances considering the viscosity of the air in tubes were studied by Tijdeman [12] and Stinson [13] for circular tubes, Stinson [14] for equilateral triangles, and Beltman [15] for rectangular tubes. In addition, Allard [16] considered the effect of tortuosity. In this study, the methods reported by Stinson [14] and Allard [16] were applied.

The Cartesian coordinate system was used, as shown in Figure 9, and the effective density $\rho_s$ and compressibility $C_s$ were obtained from a three-dimensional analysis using Equations (7) and (8) [14], respectively, using the Navier–Stokes equations, the equation of state of gas, the continuity equation, the energy equation, and the dissipation function representing heat transfer, where $\rho_0$ is the density of the air, $\lambda_s$ is an intermediary variable, $b_n$ is the clearance thickness between two planes, $\omega$ is the angular frequency, $\eta$ is the viscosity of the air, $\kappa$ is the specific heat ratio of the air, $P_0$ is the atmospheric pressure, and $N_{pr}$ is the Prandtl number.

$$\rho_s = \rho_0\left[1 - \frac{\tanh\left(\sqrt{j}\lambda_s\right)}{\sqrt{j}\lambda_s}\right]^{-1}, \lambda_s = \frac{b_n}{2}\sqrt{\frac{\omega\rho_0}{\eta}} \tag{7}$$

$$C_s = \left(\frac{1}{\kappa P_0}\right)\left\{1 + (\kappa - 1)\left[\frac{\tanh\left(\sqrt{jN_{pr}}\lambda_s\right)}{\sqrt{jN_{pr}}\lambda_s}\right]\right\} \tag{8}$$

By using the effective density $\rho_s$ multiplied by the tortuosity $\alpha_\infty$, the propagation constant and characteristic impedance considering the tortuosity can be obtained [16]. Therefore, the propagation constant $\gamma$ and the characteristic impedance $Z_c$ can be expressed by the following equations [16] in terms of the effective density $\rho_s$ and compressibility $C_s$ when the tortuosity $\alpha_\infty$ is considered:

$$\gamma = j\omega\sqrt{\alpha_\infty \rho_s C_s} \tag{9}$$

$$Z_c = \sqrt{\frac{\alpha_\infty \rho_s}{C_s}} \tag{10}$$

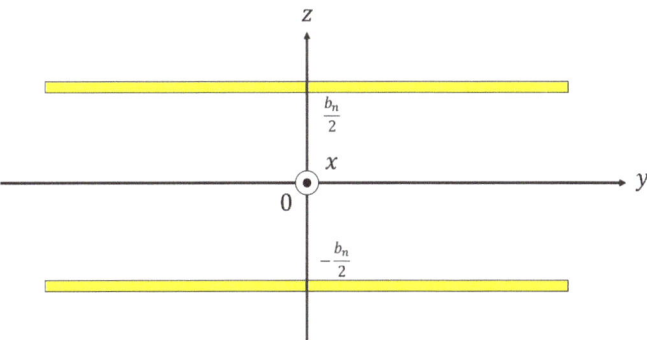

**Figure 9.** Cartesian coordinate system for the parallel clearance between the two planes.

*4.5. Transfer Matrix*

The clearance between two planes was analyzed by the transfer matrix method with respect to sound pressure and volumetric velocity based on the one-dimensional wave equation. Figure 10 shows a schematic diagram showing one element in the $x$-direction for the clearance between two planes shown in Figure 9. Using the characteristic impedance, the propagation constant; the cross-sectional area of the clearance, $S$; and the length of the divided element, $l$, obtained in Sections 4.1–4.4 the transfer matrix, $T_n$; and the four-terminal constants of the acoustic tube element, $A$ to $D$, can be calculated using Equation (11):

$$T_n = \begin{bmatrix} \cosh(\gamma l) & \frac{Z_c}{S}\sinh(\gamma l) \\ \frac{S}{Z_c}\sinh(\gamma l) & \cosh(\gamma l) \end{bmatrix} = \begin{bmatrix} A & B \\ C & D \end{bmatrix} \quad (11)$$

Plane 1 is the incident surface of sound waves, and Plane 2 is the transmission surface of sound waves. The sound pressure and the particle velocity can be expressed as $p_1$ and $u_1$ at Plane 1 and $p_2$ and $u_2$ at Plane 2, respectively, and the transfer matrix can be expressed as Equation (12):

$$\begin{bmatrix} p_1 \\ Su_1 \end{bmatrix} = \begin{bmatrix} A & B \\ C & D \end{bmatrix} \begin{bmatrix} p_2 \\ Su_2 \end{bmatrix} \quad (12)$$

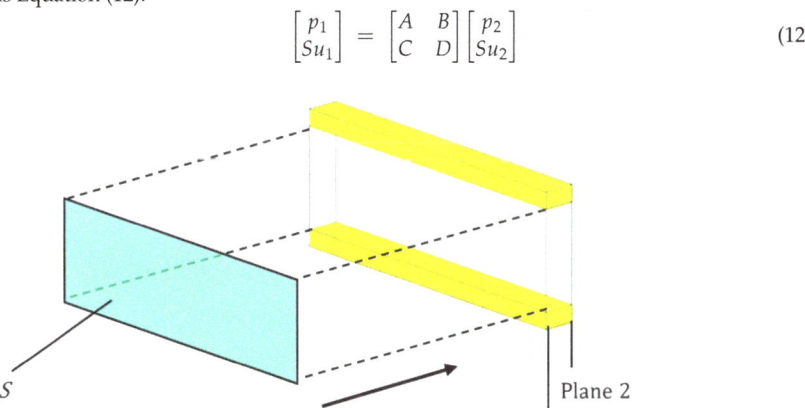

**Figure 10.** Sound incident area, incident plane, and transmission plane of approximated clearance between two planes in Figures 7b and 9.

When Equation (12) was applied to the clearance between the two planes obtained in Sections 4.2 and 4.3, the transfer matrix was obtained at each divided element. Since each divided element is continuous in the $x$-axis direction, the transfer matrix $T_{unit}$ and $T_{top}$ of the analysis unit and the analysis unit at the upper edge of the sample were obtained by cascading the transfer matrix of each divided element based on the equivalent circuit shown in Figure 11.

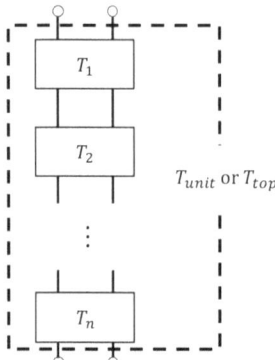

**Figure 11.** Equivalent circuit in the analysis unit (cascade connecting the transfer matrix of each element).

Next, as shown in Figure 12, the transfer matrix corresponding to the analysis unit was connected based on the equivalent circuit of the whole sample, and the transfer matrix of the whole sample was derived.

First, the analysis unit that was arranged in the $x$-axis direction, i.e., the incident direction of sound waves, was cascaded. Next, the transfer matrix of the whole sample was obtained by connecting the cascade-connected transfer matrix that aligned on the $y$–$z$ plane, the plane perpendicular to the incident direction of the sound wave, in parallel.

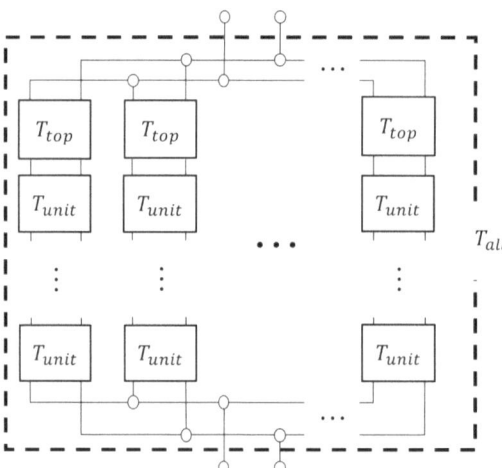

**Figure 12.** Equivalent circuit of the whole sample (parallel connection of the cascaded $T_{unit}$).

### 4.6. Normal Incident Sound Absorption Coefficient

The sound absorption coefficient was calculated from the transfer matrix $T_{all}$ obtained in Section 4.5. Since the end of the specimen used in this study was a rigid wall, the particle velocity $u_2 = 0$, Equation (12), can be deformed as shown in Equation (13), yielding Equation (14):

$$\begin{bmatrix} p_1 \\ Su_1 \end{bmatrix} = \begin{bmatrix} A & B \\ C & D \end{bmatrix} \begin{bmatrix} p_2 \\ 0 \end{bmatrix} \tag{13}$$

$$\begin{bmatrix} p_1 \\ Su_1 \end{bmatrix} = \begin{bmatrix} Ap_2 \\ Cp_2 \end{bmatrix} \qquad (14)$$

When the sound pressure and the particle velocity just outside of Plane 1 are $p_0$ and $u_0$, the specific acoustic impedance $Z_0$ from the incident surface of the sample is expressed as follows:

$$Z_0 = \frac{p_0}{u_0} \qquad (15)$$

Therefore, according to $p_0 = p_1$, $S_0 u_0 = S u_1$, and Equation (15), the specific acoustic impedance $Z_0$ of the sample is expressed as

$$Z_0 = \frac{p_0}{u_0} = \frac{p_0}{u_0 S_0} S_0 = \frac{p_1}{u_1 S} S_0 = \frac{A}{C} S_0 \qquad (16)$$

where $S/S_0$ is the aperture ratio of the sample shown in Table 1.

The relation between the specific acoustic impedance $Z_0$ and the reflectance $R$ is expressed by the following equation:

$$R = \frac{Z_0 - \rho_0 c_0}{Z_0 + \rho_0 c_0} \qquad (17)$$

According to Equation (17), the theoretical value of the normal incident sound absorption coefficient $\alpha$ of the sample is expressed as follows:

$$\alpha = 1 - |R|^2 \qquad (18)$$

## 5. Comparison between Measured and Theoretical Values

In this section, the theoretical value of the normal incident sound absorption coefficient is compared with the measured value for each sample. Three theoretical values are presented for Figure 13a–d: first, for the case where the tortuosity is not considered (unity); for the tortuosity measured in Section 3.2 (hexagonal close-packed lattice: 1.44, face-centered cubic lattice: 1.43); and for the sound absorption coefficient which derived from the tortuosity obtained by the numerical analysis in [5] (HCP: 1.65, FCC: 1.66). Here, the tortuosity measured in Section 3.2 shows smaller values than those obtained from the numerical analysis in [5].

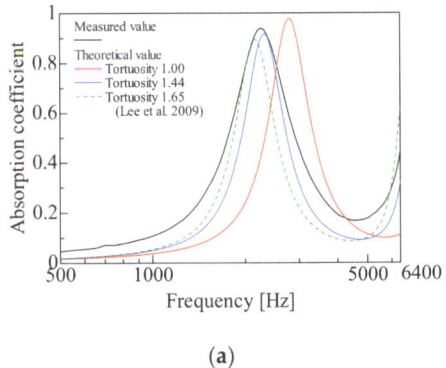
(a)

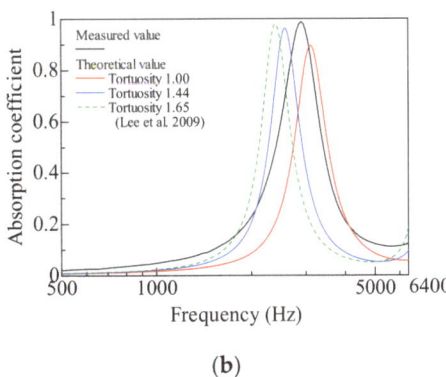
(b)

Figure 13. Cont.

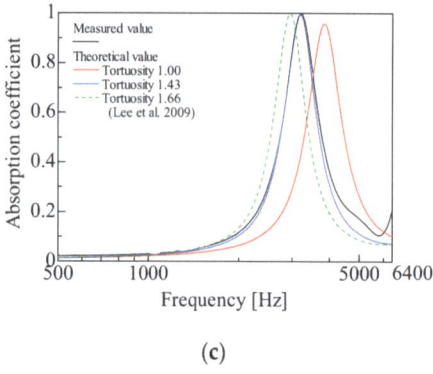 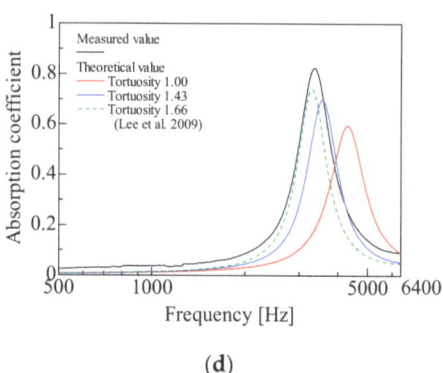

(c)                     (d)

**Figure 13.** Comparison between the experimental and calculated values (considering tortuosity in Section 3.2 and tortuosity in Reference (Lee et al. 2009) [5]) of the peak frequency and the sound absorption coefficient: (**a**) Hexagonal close-packed lattice ($d$ = 4 mm); (**b**) Hexagonal close-packed lattice ($d$ = 8 mm); (**c**) Face-centered cubic lattice ($d$ = 4 mm); and (**d**) Face-centered cubic lattice ($d$ = 8 mm).

Figure 13a,b show the hexagonal close-packed structure, and Figure 13c,d show the comparison of the face-centered cubic lattice with grain sizes of $d$ = 4 mm and $d$ = 8 mm, respectively.

First, we compared the measured values with theoretical values without considering the tortuosity. As shown in Figure 13a–d, the peak frequency of the theoretical value, which did not consider the tortuosity, compared with the measured value that appeared on the high-frequency side. Except for Figure 13a, the theoretical sound absorption peak values, without considering the tortuosity, were lower than the measured values.

Next, we focused on the theoretical values that considered the tortuosity. The peak frequency shifted to a lower frequency in Figure 13a–d, and the peak sound absorption value increased in all cases (except for Figure 13a).

As a result, the difference between the theoretical and measured values decreased when the tortuosity was considered. A larger tortuosity than unity means a longer path of sound waves that propagate in the sample, similar to the increase in sample thickness. Therefore, it is considered that the peak frequency of the theoretical value moved to the low-frequency side.

In general, the sound absorption coefficient of porous materials is greatly influenced by thickness [17,18]. In other words, the sound absorption peak occurs at a frequency where the thickness of the porous sound-absorbing material corresponds to one-quarter wavelength of the sound wave. Conversely, if the apparent sound velocity in the material decreases because of boundary layer friction or tortuosity, the peak sound absorption frequency will decrease, as this is equivalent to an increase in the apparent thickness of the material. That is, an increase in tortuosity decreases the peak sound absorption frequency and increases the sound absorption curve in most cases. For the same reason, in Figure 13b–d, it is considered that the theoretical sound absorption peak value increased, owing to the consideration of the tortuosity.

Next, we discuss the tortuosity in each sample. As shown in Section 3, the measured tortuosities in the hexagonal close-packed and face-centered cubic lattices were 1.44 and 1.43, respectively.

By considering one of the two types of tortuosities, the theoretical values of the peak frequency and the sound absorption peak were in good agreement with the measured values for the hexagonal close-packed lattice with a grain size $d$ = 4 mm, as shown in Figure 13a,b. Here, the theoretical values, considering the measured tortuosity in Section 3.2, are closer to the experimental values. However, when the grain size of the hexagonal close-packed structure was $d = 8$ mm, the peak frequency shifted to a lower frequency with the consideration of either of the tortuosities, but it was lower than the peak frequency side of the measured value because the division and assembly of the sample holder led to errors in the measured values. For the measured hexagonal close-packed structure with a grain size of $d = 8$ mm, the number of spheres used was small. To avoid interference between the particles during the sample assembly, the holder was divided into three parts perpendicular to the incident direction of the sound wave. Dimensional errors between the divided holders may have caused gaps in the spheres, thereby affecting the measured values. On the other hand, for the grain size $d = 4$ mm, the accuracy of the theoretical value was improved by considering the measured tortuosity in Section 3.2.

In the case of the face-centered cubic lattice with a grain size of $d = 4$ mm, as shown in Figure 13c, the theoretical values were fairly close to the measured values when either tortuosity was considered. When the two types of tortuosities were compared, the theoretical values from the tortuosity measured in Section 3.2 agreed very well with the measured values. When the grain size of the face-centered cubic lattice was $d = 8$ mm, as shown in Figure 13d, by considering either of the tortuosities, the theoretical values of both the peak frequency and the sound absorption peak approached the measured values, but the degree of agreement was smaller than that in Figure 13c. In terms of the agreement between theoretical (applying the tortuosity in Section 3.2) and experimental values, the case with a grain size of 4 mm (Figure 13a,c) is better than the case with a grain size of 8 mm (Figure 13b,d). This is because the method used to estimate the attenuation of sound waves in gaps [14] is more accurate for smaller gaps [19].

## 6. Conclusions

The normal incidence sound absorption coefficients of hexagonal close-packed and face-centered cubic lattices, which are typical packing structures of granular materials, were theoretically calculated. In the derivation of the propagation constant and the characteristic impedance, the measured tortuosity was taken into account. The results were compared with the experimental data. The following conclusions were obtained:

1. In both packing structures, the real area of the granular surface and the real volume of the clearance were obtained geometrically and analyzed theoretically.
2. In both packing structures, the peak frequency tended to appear at a higher frequency than the measured value when the tortuosity was not considered.
3. In the theoretical sound absorption, the peak value was higher when the tortuosity was considered compared to that without the consideration of the tortuosity (the peak frequency moved to a lower frequency) As a result, the theoretical value was becoming closer to the measured value.

**Author Contributions:** Conceptualization, S.S. (Shuichi Sakamoto); Data curation, K.T. and S.S. (Shotaro Seino); Formal analysis, K.S.; Project administration, S.S. (Shuichi Sakamoto); Software, K.S.; Supervision, S.S. (Shuichi Sakamoto). All authors have read and agreed to the published version of the manuscript.

**Funding:** This work was supported by Japan Society for the Promotion of Science (JSPS) KAKENHI Grant No. 20K04359.

**Institutional Review Board Statement:** Not applicable.

**Conflicts of Interest:** The authors declare no conflict of interest.

## References

1. Sandberg, U. Low noise road surfaces—A state-of-the-art review. *J. Acoust. Soc. Jpn.* **1999**, *20*, 1–17. [CrossRef]
2. Zhang, X.; Thompson, D.; Jeong, H.; Squicciarini, G. The effects of ballast on the sound radiation from railway track. *J. Sound Vib.* **2017**, *339*, 137–150. [CrossRef]
3. Voronina, N.N.; Horoshenkov, K.V. A new empirical model for the acoustic properties of loose granular media. *Appl. Acoust.* **2003**, *64*, 415–432. [CrossRef]

4. Gasser, S.; Paun, F.; Bréchet, Y. Absorptive properties of rigid porous media: Application to face centered cubic sphere packing. *J. Acoust. Soc. Am.* **2005**, *117*, 2090–2099. [CrossRef] [PubMed]
5. Lee, C.-Y.; Leamy, M.J.; Nadler, J.H. Acoustic absorption calculation in irreducible porous media: A unified computational approach. *J. Acoust. Soc. Am.* **2009**, *126*, 1862–1870. [CrossRef] [PubMed]
6. Dung, V.V.; Panneton, R.; Gagné, R. Prediction of effective properties and sound absorption of random close packings of monodisperse spherical particles: Multiscale approach. *J. Acoust. Soc. Am.* **2019**, *145*, 3606–3624. [CrossRef]
7. Sakamoto, S.; Sakuma, Y.; Yanagimoto, K.; Watanabe, S. Basic Study for the Acoustic Characteristics of Granular Material (Normal Incidence Absorption Coefficient for Multilayer with Different Grain Diameters). *J. Environ. Eng.* **2012**, *7*, 12–22. [CrossRef]
8. Sakamoto, S.; Yamaguchi, K.; Ii, K.; Takakura, R.; Nakamura, Y.; Suzuki, R. Theoretical and experiment analysis on the sound absorption characteristics of a layer of fine lightweight powder. *J. Acoust. Soc. Am.* **2019**, *146*, 2253–2262. [CrossRef]
9. Sakamoto, S.; Takakura, R.; Suzuki, R.; Katayama, I.; Saito, R.; Suzuki, K. Theoretical and Experimental Analyses of Acoustic Characteristics of Fine-grain Powder Considering Longitudinal Vibration and Boundary Layer Viscosity. *J. Acoust. Soc. Am.* **2021**, *149*, 1030–1040. [CrossRef]
10. Sakamoto, S.; Ii, K.; Katayama, I.; Suzuki, K. Measurement and Theoretical Analysis of Sound Absorption of Simple Cubic and Hexagonal Lattice Granules. *Noise Control Eng. J.* **2021**, *69*, 401–410. [CrossRef]
11. Allard, J.F.; Castagnede, B.; Henry, M. Evaluation of tortuosity in acoustic porous materials saturated by air. *Rev. Sci. Instrum.* **1994**, *65*, 754. [CrossRef]
12. Tijdeman, H. On the propagation of sound waves in cylindrical tubes. *J. Sound Vib.* **1975**, *39*, 1–33. [CrossRef]
13. Stinson, M.R. The propagation of plane sound waves in narrow and wide circular tubes, and generalization to uniform tubes of arbitrary cross-sectional shape. *J. Acoust. Soc. Am.* **1991**, *89*, 550–558. [CrossRef]
14. Stinson, M.R.; Champou, Y. Propagation of sound and the assignment of shape factors in model porous materials having simple pore geometries. *J. Acoust. Soc. Am.* **1992**, *91*, 685–695. [CrossRef]
15. Beltman, W.M.; van der Hoogt, P.J.M.; Spiering, R.M.E.J.; Tijdeman, H. Implementation and experimental validation of a new viscothermal acoustic finite element for acousto-elastic problems. *J. Sound Vib.* **1998**, *216*, 159–185. [CrossRef]
16. Allard, J.F. Propagation of sound in Porous Media Modeling Sound Absorbing Materials. *J. Acoust. Soc. Am.* **1994**, *95*, 2785. [CrossRef]
17. Tabana, E.; Khavanina, A.; Jafarib, A.J.; Faridanc, M.; Tabrizid, A.K. Experimental and mathematical survey of sound absorption performance of date palm fibers. *Heliyon* **2019**, *5*, e01977. [CrossRef] [PubMed]
18. Duan, C.; Cui, G.; Xu, X.; Liu, P. Sound absorption characteristics of a high-temperature sintering porous ceramic material. *Appl. Acoust.* **2012**, *73*, 865–871. [CrossRef]
19. Sakamoto, S.; Higuchi, K.; Saito, K.; Koseki, S. Theoretical analysis for sound-absorbing materials using layered narrow clearances between two planes. *J. Adv. Mech. Des. Syst. Manuf.* **2014**, *8*, 16. [CrossRef]

Article

# Optimal Design of Acoustic Metamaterial of Multiple Parallel Hexagonal Helmholtz Resonators by Combination of Finite Element Simulation and Cuckoo Search Algorithm

Fei Yang [1], Enshuai Wang [1], Xinmin Shen [1,*], Xiaonan Zhang [1,*], Qin Yin [1], Xinqing Wang [1], Xiaocui Yang [2,3], Cheng Shen [3] and Wenqiang Peng [4]

1 College of Field Engineering, Army Engineering University of PLA, Nanjing 210007, China
2 Engineering Training Center, Nanjing Vocational University of Industry Technology, Nanjing 210023, China
3 MIIT Key Laboratory of Multifunctional Lightweight Materials and Structures (MLMS), Nanjing University of Aeronautics and Astronautics, Nanjing 210016, China
4 College of Aerospace Science and Engineering, National University of Defense Technology, Changsha 410073, China
* Correspondence: xmshen2014@aeu.edu.cn (X.S.); zxn8206@163.com (X.Z.); Tel.: +86-025-8082-1451 (X.S.)

**Abstract:** To achieve the broadband sound absorption at low frequencies within a limited space, an optimal design of joint simulation method incorporating the finite element simulation and cuckoo search algorithm was proposed. An acoustic metamaterial of multiple parallel hexagonal Helmholtz resonators with sub-wavelength dimensions was designed and optimized in this research. First, the initial geometric parameters of the investigated acoustic metamaterials were confirmed according to the actual noise reduction requirements to reduce the optimization burden and improve the optimization efficiency. Then, the acoustic metamaterial with the various depths of the necks was optimized by the joint simulation method, which combined the finite element simulation and the cuckoo search algorithm. The experimental sample was prepared using the 3D printer according to the obtained optimal parameters. The simulation results and experimental results exhibited excellent consistency. Compared with the derived sound absorption coefficients by theoretical modeling, those achieved in the finite element simulation were closer to the experimental results, which also verified the accuracy of this optimal design method. The results proved that the optimal design method was applicable to the achievement of broadband sound absorption with different low frequency ranges, which provided a novel method for the development and application of acoustic metamaterials.

**Keywords:** broadband sound absorption; optimal design; acoustic metamaterial; hexagonal Helmholtz resonators; joint simulation method; finite element simulation; cuckoo search algorithm

## 1. Introduction

The prevalence of low-frequency noise pollution in the modern industrial production and living has seriously affected the daily life and work of people [1]. The traditional acoustic materials, such as porous media [2] and microperforated panel [3], are commonly utilized to reduce the impact of noise, but the low frequency noise is very difficult to eliminate due to the long wavelength [4]. To block low frequency noises, the sizes of acoustic materials have to be larger than a quarter of their wavelength, but the traditional acoustic materials are hard to satisfy the special requirements of noise reduction due to the limitation of occupied space [5,6]. Thus, the novel composites with special physical characteristics have become the research focus in the fields of noise reduction and sound absorption [7]. For example, the acoustic metamaterial with subwavelength physical properties that traditional acoustic materials do not possess has attracted great attention among various scholars, which can provide novel methods to reduce the low frequency noise effectively.

In the latest years, the Helmholtz resonators with extended aperture depth have been a favorable choice for the low frequency noise absorption [8–12]. Jimenez et al. [8] proposed realistic panels which were made of Helmholtz resonator bricks, and the perfect absorption of sound at the frequency of 338.5 Hz was achieved. The acoustic metamaterial absorber composed of thick-necked embedded Helmholtz resonators was developed by Zhang and Xin [9] to achieve the perfect sound absorption at the 150 Hz. Sharafkhani [10] presented a multi-band sound absorber based on the Helmholtz resonators, and it could achieve the perfect absorptions at 100 Hz, 200 Hz, and 300 Hz simultaneously. However, the relatively narrow absorption bandwidth of a single Helmholtz resonator cannot satisfy practical engineering applications. Moreover, the material thickness is usually limited to small installation spaces. To overcome these problems, the prominent strategy involved in the embedded multiple individual single Helmholtz resonators with various sizes were proposed in this research. The overall bandwidth of the entire acoustic material is expanded by overlapping the absorption bandwidth of each single Helmholtz resonator.

For the practical applications, the design of acoustic materials with the desired sound absorption effect is the most important goal. Thus, substantial research has been carried out on optimizing the geometric parameters of the acoustical materials [13–17]. Gao et al. [13] proposed a hybrid design of unit cells consisting of multiple labyrinthine channels, and a genetic algorithm was used to optimize the absorption performance so that it could have best absorption energy in the frequency range of interest. Yan et al. [14] adopted the particle swarm algorithm to optimize the depth of the honeycomb core cavity of the honeycomb micro-perforated plate (HMPP). The optimization results showed that the HMPP structure had the multiple peak absorption coefficients in the range of 0–3500 Hz, and the absorption coefficient was above 0.85 in a wide frequency band. Gao et al. [15] had optimized the ultra-broadband parallel sound absorber (UBPSA) using a teaching–learning-based optimization algorithm, in which geometric parameters were utilized as the optimization variables. The optimized UBPSA showed the high absorption effect in the frequency range of 200–1715 Hz. Currently, most studies on the optimization of acoustic materials have adopted the acoustic–electric analogy or transfer matrix method for theoretical modelling, followed by the combination of the intelligent algorithms for optimization, which have greatly reduced the optimization time. However, these methods have not been able to accurately optimize the acoustic metamaterial. Due to the immature research on the acoustic absorption mechanism of acoustic metamaterial [13–17], a number of simplifications are usually taken into the process of the theoretical modeling, resulting in large errors between the theoretical results and experimental results. Hence, the accuracy of the optimization results may not be precise. To overcome these shortcomings, the joint simulation method incorporating the finite element simulation and cuckoo search algorithm was adopted in this study, which had attempted to improve the accuracy of the optimization result.

In this study, the design method of acoustic metamaterial of multiple parallel hexagonal Helmholtz resonators with sub-wavelength dimensions was discussed. The optimized acoustic metamaterials could achieve the ideal broadband sound absorption effect at a low frequency range in a small space. First, in order to ease the optimization burden and speed up the optimization process, the overall thickness, aperture size, and side length of the cavity for the acoustic metamaterial were determined depending on the application scene and processing difficulty. The depth of neck was initially selected by the two-dimensional finite element analysis. Then, a joint simulation method incorporating finite element analysis and cuckoo search algorithm was conducted to obtain the optimal geometric parameters, which satisfied the sound absorption requirement. The optimization took the depth of neck as the optimization parameter and the total absorbed sound energy at the target frequency range as the optimization target. Finally, samples were prepared to obtain experimental data according to the optimized parameters. The optimized parameters were also substituted into a theoretical model to obtain the theoretical data. By comparing the theoretical data, simulation data, and experimental data, the accuracy and reliability of the proposed optimization method were verified. In addition, the applicability of this optimal design method for the different conditions was also further discussed, which provided a novel method for the development of similar acoustic metamaterial.

## 2. Initial Structural Design

### 2.1. Design Objective

A noise test was conducted on a workshop under the operation mode. The test results indicated that major frequencies of the noise were concentrated in the range of 487–675 Hz. The acoustic materials would exhibit a better sound absorption performance in the simulation environments than in the reality of applications due to the fabrication errors and actual installation conditions. Thus, the simulation target was set as slightly larger in the design stage. To realize all the actual sound absorption coefficients larger than 0.8 in the interested frequency range, sound absorption coefficients of the designed acoustic metamaterial were required to be above 0.85 in the frequency range of 420–700 Hz during the simulation process.

### 2.2. Geometric Parameters Design

In order to reduce the noise and improve the indoor acoustic environment in the given workshop, the acoustic metamaterial of multiple parallel hexagonal Helmholtz resonators was proposed in this research, as shown in Figure 1.

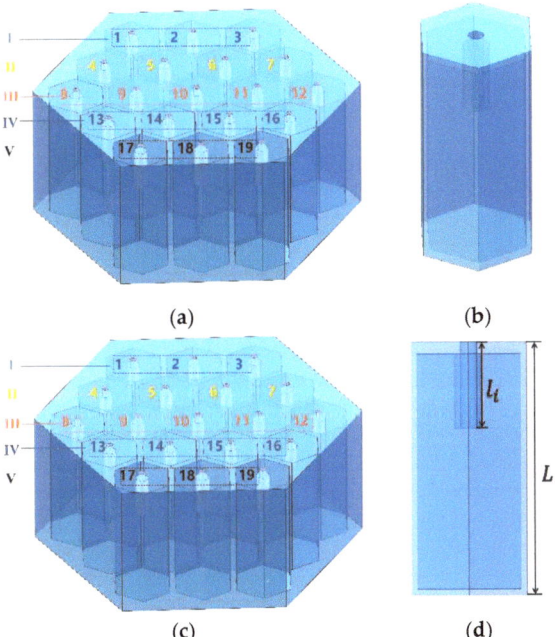

**Figure 1.** The designed acoustic metamaterial of multiple parallel hexagonal Helmholtz resonators. (**a**) Schematic diagram of the whole structure; (**b**) Structure of a single hexagonal Helmholtz resonator; (**c**) Top view of the single hexagonal Helmholtz resonator; (**d**) Main view of the single hexagonal Helmholtz resonator.

The schematic diagram of the whole structure for the designed acoustic metamaterial is shown in Figure 1a. The acoustic metamaterial consisted of an array of nineteen hexagonal Helmholtz resonators, each of which was a hexagonal cavity structure that contained a microporous plate embedded in an extended neck and rear cavity, as shown in Figure 1b.

The nineteen single hexagonal Helmholtz resonators were divided into five groups to reduce the computational burden, with structures 1–3 as the first group, structures 4–7 as the second group, structures 8–12 as the third group, structures 13–16 as the fourth group, and structures 17–19 as the fifth group, as shown in Figure 1a. Figure 1c,d showed details of a single hexagonal Helmholtz resonator, where $a$ is the side length of a single hexagonal resonator; $t$ is the thickness of wall; $d_i$ is the diameter of the micropore; $l_i$ is the depth of the neck; $L$ is the total thickness of the acoustic metamaterial.

The various restrictions would be placed on the data of design variables to meet the certain usage requirements in the subsequent optimization process. Excessive variables would lead to an exponential increase in the calculation effort for the simulation and the optimization, so a certain size design should need to be predetermined.

To ensure the low frequency sound absorption, the perforation rate should not be too high. Meanwhile, taking the widest possible sound absorption band into consideration, the perforation rate should not be too low [18–20]. Therefore, the perforation rate of the five groups was selected to be in the range of 0.5–0.9% [20], and the size of the perforation was ensured to be kept different among the various groups. Therefore, for the five groups of hexagonal Helmholtz resonator in Figure 1a, the derived diameters of micropore were 4.47 mm, 3.54 mm, 3.75 mm, 4.74 mm, and 5.16 mm, respectively, which corresponded to the perforation rate of 0.6%, 0.5%, 0.7%, 0.9%, and 0.8% successively. Because the designed acoustic metamaterial would be installed between the building decorative panels and the roof, the thickness of acoustic metamaterial should not be too large, and here the total thickness $L$ was limited to be 40 mm. To reduce the overall structure mass, the thickness of wall $t$ was taken to be 2 mm. The side length $a$ of the single Helmholtz resonator was set to be 10 mm by taking into account the requirement of standing wave tube test for actual sound absorption coefficients of the investigated acoustic metamaterial.

With these three already determined structural parameters, the ideal sound absorption effect was achieved by adjusting the depth of neck for each single structure. Without taking the mutual coupling effect among the various Helmholtz resonators into account, each single structure corresponded to generate one resonant frequency. The depth of neck was adjusted to allow nineteen resonant frequencies at around 420–700 Hz. The resonant frequencies were 425 Hz, 440 Hz, 455 Hz, 470 Hz, 485 Hz, 500 Hz, 515 Hz, 530 Hz, 545 Hz, 560 Hz, 575 Hz, 590 Hz, 605 Hz, 620 Hz, 635 Hz, 650 Hz, 665 Hz, 680 Hz, and 695 Hz as an interval of 15 Hz was taken for the division, which were aimed to be obtained by hexagonal Helmholtz resonator of 4, 5, 6, 7, 8, 9, 10, 11, 12, 1, 2, 3, 13, 14, 15, 16, 17, 18, and 19, respectively, in Figure 1a.

To visually illustrate the design process of initial value of the depth of neck, structure 4, corresponding to the resonant frequency 425 Hz, was taken as an example. First, a cylindrical representative hexagonal structure with the approximate boundary conditions was chosen to achieve the desired sound absorption properties of acoustic materials [21,22]. The radius of cylinder is equal to $\sqrt{S/\pi}$, where $S$ is the cross-sectional area of the hexagon in Figure 1c and is calculated as $\frac{3\sqrt{3}}{2}a^2$. In order to decrease the computational time and improve the research efficiency, the rotational 3D model could be further simplified to a 2D model for the finite element analysis, as shown in Figure 2. Afterwards, through setting the parametric scan of the depth of neck in the range of 10–15 mm with the interval of 0.5 mm to obtain the absorption curve in the frequency range of 420–430 Hz, an approximate range of value was obtained as [14.5 mm, 15 mm]. To further confirm the value of the depth of neck, the parameter scan of the depth of neck was performed in the approximate range of value [14.5 mm, 15 mm] with the interval of 0.1 mm. Finally, the initial depth of neck of structure 4 was determined to be 14.7 mm, which could obtain a resonant frequency 425 Hz in theory. By analogy, the initial geometric parameters of overall structure were obtained and summarized in Table 1.

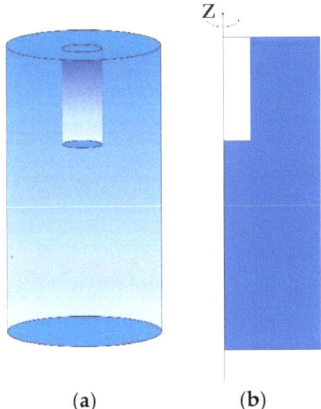

**Figure 2.** The single Helmholtz resonator. (**a**) Schematic diagram of the cylindrical representative hexagonal structure; (**b**) Schematic diagram of the 2D rotationally symmetric model for the finite element analysis.

**Table 1.** The initial geometric parameters of the proposed acoustic metamaterial.

| Group | Serial Number | Thickness/mm | Diameter of Micropore/mm | Depth of Neck/mm |
|---|---|---|---|---|
| I | 1 | | 4.47 | 13.0 |
|   | 2 | | | 12.1 |
|   | 3 | | | 11.2 |
| II | 4 | | 3.54 | 14.7 |
|    | 5 | | | 13.4 |
|    | 6 | | | 12.3 |
|    | 7 | | | 11.3 |
| III | 8 | 40 | 3.75 | 12.1 |
|     | 9 | | | 11.2 |
|     | 10 | | | 10.3 |
|     | 11 | | | 9.5 |
|     | 12 | | | 8.8 |
| IV | 13 | | 4.74 | 12.3 |
|    | 14 | | | 11.4 |
|    | 15 | | | 10.7 |
|    | 16 | | | 10.0 |
| V | 17 | | 5.16 | 11.8 |
|   | 18 | | | 11.0 |
|   | 19 | | | 10.3 |

According to the initial geometric parameters in Table 1, the finite element simulation model of the designed acoustic metamaterial of multiple parallel hexagonal Helmholtz resonators was obtained, and distributions of its sound absorption coefficients are shown in Figure 3. Judging from Figure 3, the sound absorption coefficients of the initial structure were larger than 0.85 in the frequency range of 496–686 Hz, which was far from the desired target frequency band 420–700 Hz. Therefore, the further optimization of geometric parameters for the designed acoustic metamaterial of multiple parallel hexagonal Helmholtz resonators was essentially required to satisfy the noise reduction for the given conditions in this research.

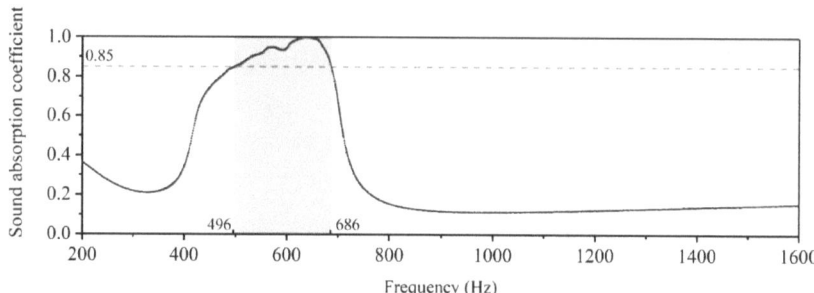

**Figure 3.** Sound absorption coefficient of the initial structure gained by finite element simulation. The gray area represents the frequency range that meets the design requirements.

## 3. Optimization of Acoustic Metamaterial

### 3.1. Establishment of Finite Element Simulation Model

Finite element simulation was a prerequisite for the optimization of the acoustic metamaterial by using optimization algorithms. In this study, the numerical model of acoustic metamaterial was constructed by COMSOL [23–25]. The constructed finite element model based on the thermo-viscous acoustic module is shown in Figure 4.

**Figure 4.** Finite element simulation model. (**a**) Finite element model of the acoustic metamaterial of multiple parallel hexagonal Helmholtz resonators. (**b**) Finite element mesh division of the proposed acoustic metamaterial.

The components of the numerical model were identified in the figure. The background pressure field was simulated as a plane wave sound field with the sound pressure value of 1 Pa and the sound velocity of 343 m/s. The input sound waves were incident vertically on the surface of the acoustic metamaterial with the negative direction of the z-axis. The perfect match layer was added at the end of the background pressure field for full absorption of the reflected acoustic waves, which prevented the acoustic reflection from affecting the calculation results. The thickness of the perfect match layer was 1.5 times of that of background pressure field, and their sizes were determined by total thickness of the acoustic absorber and the required frequency range simultaneously. Generally speaking, their sizes would be larger when the total thickness of the acoustic absorber was larger or the analyzed frequency range was in a lower region. In this study, thickness of the background was 8 mm, and that of the perfect match layer was 12 mm. The model material was set as air, and the material properties used the default values in the COMSOL.

Moreover, in the numerical model, the mesh of the perfect match layer was divided into 6 layers through sweeping, and the rest were tetrahedral meshes, as shown in Figure 4b. Each wavelength must be divided into at least six meshes to ensure computational accuracy. Therefore, the utilized parameters of finite element simulation model in this research were as follows: size of the biggest unit, 3.5 mm; size of the smallest unit, 0.15 mm; the maximum growth rate of neighboring unit, 1.35; curvature factor, 0.3; resolution ratio of the narrow area, 0.85; mesh type, free tetrahedron mesh; layer number of the boundary area, 5; stretching factor of the boundary layer, 1.2; regulatory factor of the thickness of the boundary layer, 1. Furthermore, the frequency domain solver was chosen for the calculation, and the investigated frequency range was selected as 200–1600 Hz.

### 3.2. Optimization Algorithm

The cuckoo search algorithm [26–32] was chosen as the optimization algorithm in this research, which performed a global search by simulating the parasitic brood behavior of the cuckoo nests using Lévy flight. The algorithm had an excellent global optimization seeking capability. Currently, various studies and applications have demonstrated that the cuckoo search algorithm is very effective in solving scheduling problems and combinatorial optimization problems. Therefore, it was quite suitable for solving the optimization problems of acoustic metamaterial under the limited conditions.

The acoustic metamaterial was optimized by the cuckoo search algorithm in order to achieve an absorption coefficient in the target frequency range 420–700 Hz greater than 0.85. Therefore, the optimization process was divided into the following five steps.

(1) Set the initial parameters for the algorithm. In the current optimization process, the number of host nest populations was set as $N = 20$, the maximum number of iterations was $N\_iterTotal = 100$, and the maximum probability of discovery was $p_a = 0.25$.

(2) Calculate the fitness function of the population individual. In this research, sound absorption performance within the target frequency range of the acoustic metamaterial of multiple parallel hexagonal Helmholtz resonators was the research object and the depth of neck of each resonator was the optimization parameter. To ensure that the resulted absorption curve could maintain a high absorption coefficient in the target frequency range, the fitness value function $\max(\alpha)$ was chosen as the total amount of sound energy absorbed at the target frequency range, as shown in Equation (1).

$$\max(\alpha) = \int_{f_0}^{f_1} \alpha(f) df \tag{1}$$

The corresponding discrete form of the fitness function was as follows.

$$\max(\alpha) = \sum_{i=1}^{n} \alpha_i \Delta f \tag{2}$$

In Equations (1) and (2), $\alpha(f)$ is the sound absorption coefficient for the corresponding frequency $f$; $\Delta f$ is the given adjacent frequency interval; $f_0$ and $f_1$ are the upper and lower limit of the target frequency band, respectively; $n$ is the number of frequency intervals.

(3) Update all nests according to Equation (3).

$$x_i^{t+1} = x_i^t + \partial \oplus L(\lambda) \tag{3}$$

Here, $x_i^{t+1}$ is the $i$th bird nest at the $t$th iteration; $\partial$ denotes step control amount, and $L(\lambda) = t^{-\lambda}(1 < \lambda \leq 3)$ is the Lévy random search path. Then, better individuals were selected by assessing fitness.

(4) For all nests $x_i(i = 1, \cdots, N)$, a random number $r_i \in [0,1]$ is generated and compared to the probability $p_a$. If $r_i > p_a$, then $x_i^{t+1}$ is randomly changed, otherwise there will be no change. The adaptation of the nests before and after the random change was evaluated, retaining the nest with better adaptation as the final $x_i^{t+1}$. Afterwards, the return step (2) was iterated.

(5) When all the absorption coefficients at the target frequency range were above 0.85, or it had reached the maximum number of iterations, the iteration ended, and the current optimal individual was output.

The optimization process of the cuckoo search algorithm for the acoustic metamaterial of multiple parallel hexagonal Helmholtz resonators is shown in Figure 5.

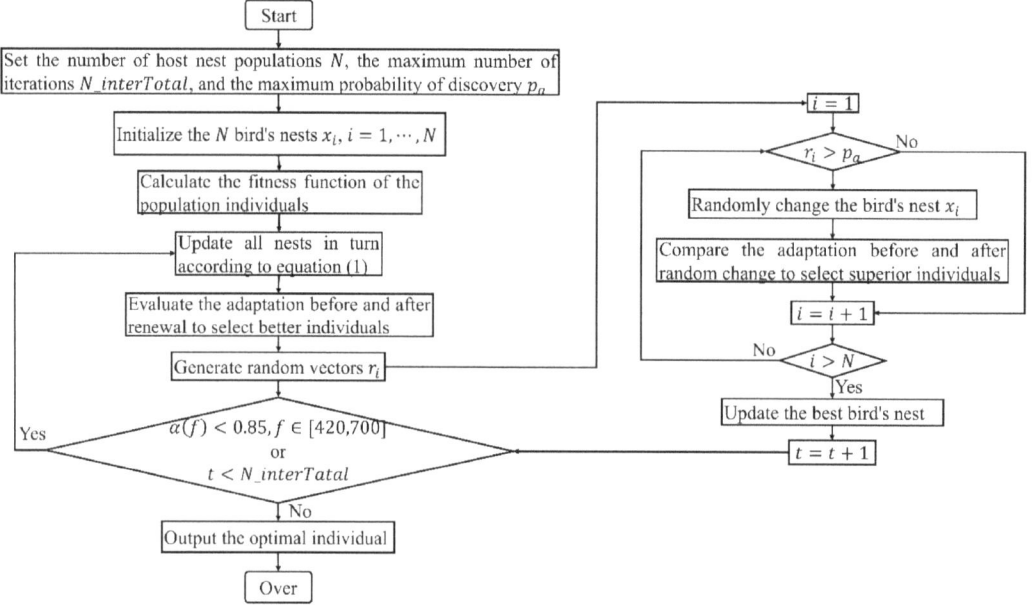

**Figure 5.** The optimization process of cuckoo search algorithm for the acoustic metamaterial of multiple parallel hexagonal Helmholtz resonators.

*3.3. Interactive Operations with MATLAB and COMSOL*

COMSOL had a powerful multi-physics field simulation capability and weak ability to optimize physical field parameters by using the intelligent algorithms. However, COMSOL's powerful interface compatibility could construct a joint simulation platform with MATLAB to realize the joint simulation of algorithms and physical fields.

First, the acoustic finite element simulation model was constructed in COMSOL and the other physical parameters, except these optimization parameters were set. The project was exported to an m-file by using the 'Save As' option. Next, MATLAB was opened by "COMSOL with MATLAB" and it operated the COMSOL automatically in the background. The cuckoo search algorithm was combined with the m-file of the acoustic finite element simulation model in MATLAB to realize the control of the algorithm on the physical field simulation. The process of using the joint simulation platform for algorithm and physical field simulation was summarized and is shown in Figure 6.

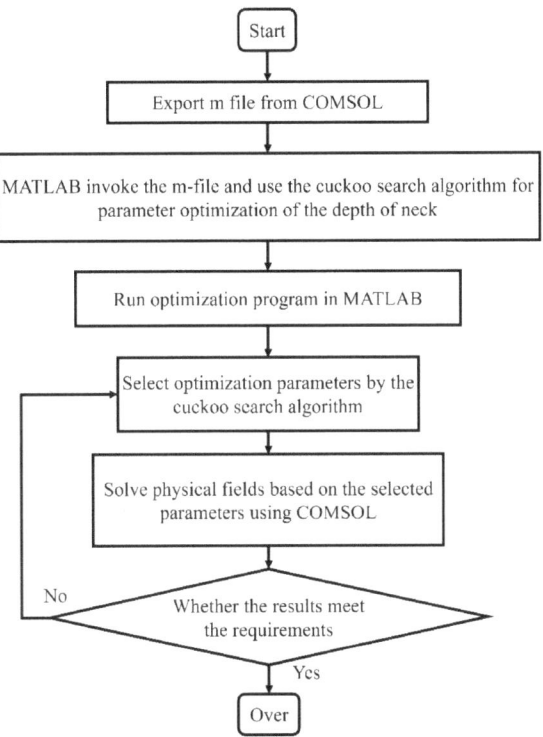

**Figure 6.** The process of the joint simulation platform for algorithm and physical field simulation.

## 4. Results and Discussion

*4.1. Optimization Results*

The geometric parameters for the optimized acoustic metamaterial of multiple parallel hexagonal Helmholtz resonators are summarized and shown in Table 2. The optimal sound absorption coefficients with their corresponding nineteen single Helmholtz resonator structures are shown in Figure 7a. It could be judged from Figure 7a that the sound absorption curve met the design requirement of a sound absorption coefficient above 0.85 in the frequency range of 418–709 Hz. The absorption peaks were 0.92917, 0.89134, 0.95469, 0.95203, 0.89158, 0.88892, 0.89092, 0.93914, and 0.91841 corresponding to resonant frequency approximately 432 Hz, 463 Hz, 538 Hz, 545 Hz, 571 Hz, 600 Hz, 633 Hz, 660 Hz, and 694 Hz, respectively. The overall thickness of the acoustic material was only 40 mm, which was near 1/20 of the wavelength of the first peak absorption ($\lambda = c/f = 343/432 = 0.794$ m), meaning that the acoustic metamaterial had the ability to absorb the sound in the sub-wavelength range. The acoustic metamaterial had better sound absorption performance after the optimization. The absorption peaks of the combined structure were larger than the absorption peaks of single Helmholtz resonators, which indicated that the absorption curve of the combined structure was not a simple superposition of the resonant frequencies of the nineteen single structures, but a common coupling effect of the nineteen structures. The coupling effect between the various structures improved the sound absorption effect of the combination structure and achieved a broadband efficient sound absorption. Because the peak frequencies of the absorption coefficients corresponding to the different depths of neck were close to each other, the curve of the sound absorption coefficient for this acoustic metamaterial of multiple parallel hexagonal Helmholtz resonators was smooth.

**Table 2.** The geometric parameters of the optimized acoustic metamaterial.

| Group | Serial Number | Thickness/mm | Diameter of Micropore/mm | Depth of Neck/mm |
|---|---|---|---|---|
| I | 1 | | 4.47 | 15.7 |
| | 2 | | | 15.2 |
| | 3 | | | 14.4 |
| II | 4 | | 3.54 | 15.9 |
| | 5 | | | 15.2 |
| | 6 | | | 14.7 |
| | 7 | | | 14.3 |
| III | 8 | 40 | 3.75 | 16.7 |
| | 9 | | | 15.3 |
| | 10 | | | 14.1 |
| | 11 | | | 13.1 |
| | 12 | | | 11.8 |
| IV | 13 | | 4.74 | 16.2 |
| | 14 | | | 14.5 |
| | 15 | | | 12.6 |
| | 16 | | | 10.9 |
| V | 17 | | 5.16 | 12.0 |
| | 18 | | | 10.3 |
| | 19 | | | 9.5 |

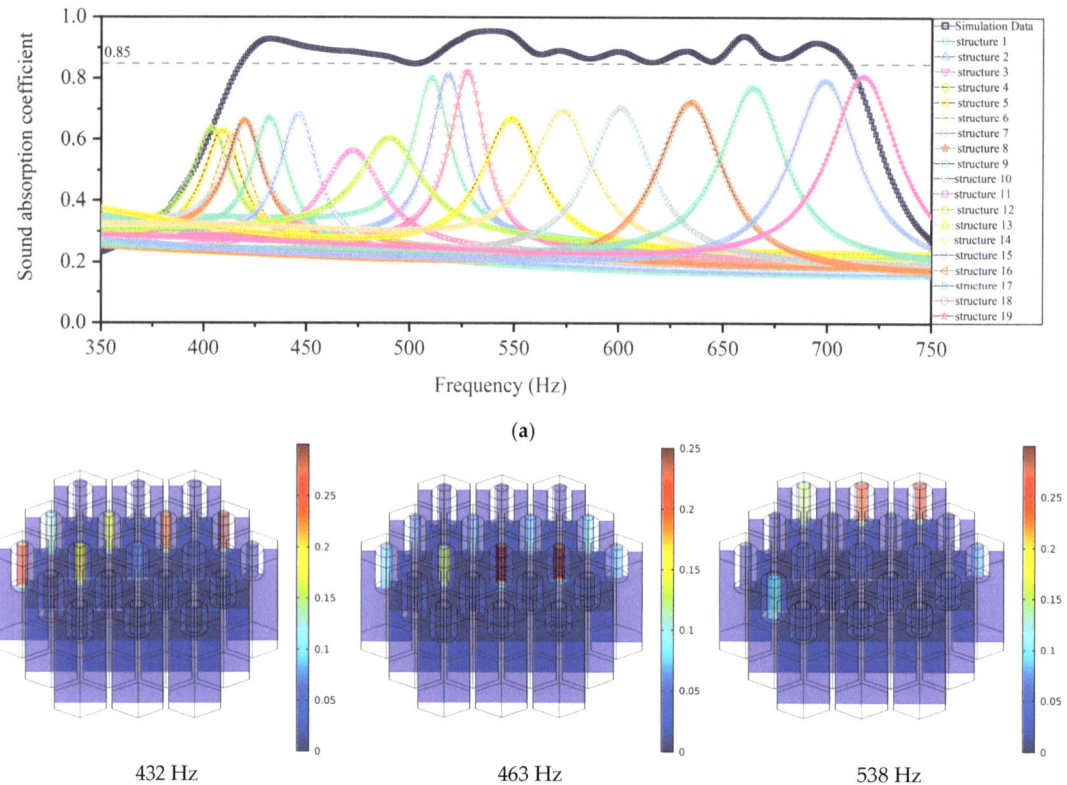

**Figure 7.** Cont.

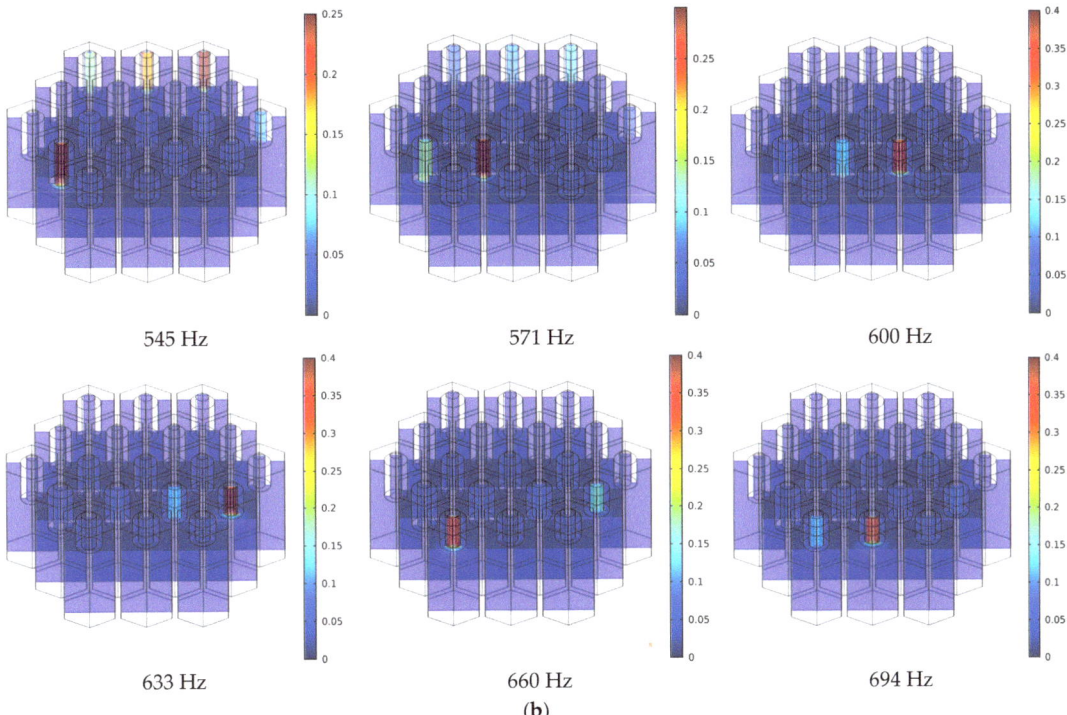

**Figure 7.** The optimization results. (**a**) The optimized sound absorption coefficients obtained in simulation for the acoustic metamaterial of multiple parallel hexagonal Helmholtz resonators with its corresponding nineteen single structures. (**b**) Instantaneous local velocity of corresponding absorption peaks.

In order to reveal the sound absorption mechanism of acoustic metamaterial clearly, the instantaneous local velocity of the corresponding absorption peaks was presented in Figure 7b. As shown in Figure 7b, the highlighted areas were concentrated on the necks. This indicated that the velocity of air particles in the necks was much higher than that of air particles in the rear cavities due to the geometric discontinuity. The thermal viscous effect between the air and the walls of necks converted the acoustic energy into thermal movement and achieved noise attenuation. It could be observed more visually that the acoustic metamaterial achieved the broadband sound absorption effect at the low frequency through the common coupling effect of multiple hexagonal Helmholtz resonators. The closer to the low frequencies, the more hexagonal Helmholtz resonators were involved in sound absorption. Because of the insufficient sound absorption capacity of a single hexagonal Helmholtz resonator within the low frequency range, multiple hexagonal Helmholtz resonators were required to work together to achieve the desired sound absorption effect.

### 4.2. Theoretical Analysis

For further verification of the accuracy and reliability of the optimal design, theoretical modeling of the acoustic metamaterial was also carried out [33–35]. The acoustic impedance $Z$ of the overall structure of acoustic metamaterial of multiple parallel hexagonal Helmholtz resonators with sub-wavelength dimension consists of the acoustic impedance $Z_i$ of all nineteen single Helmholtz resonators connected in parallel, which can be obtained by Equation (4). The acoustic impedance $Z_i$ of single Helmholtz resonator consists of the acoustic impedance of the perforated plate $Z_{im}$ and that of cavity $Z_{ic}$, as shown in Equation (5). The perforated plate impedance $Z_{im}$ can be calculated by Euler's equation, as shown in Equation (6).

$$Z = \frac{1}{\sum_{i=1}^{19}(1/Z_i)} \tag{4}$$

$$Z_i = Z_{im} + Z_{ic} \tag{5}$$

$$Z_{nm} = i\frac{\omega\rho_0 l_i}{\sigma_i}\left[1 - \frac{B_1\left(\frac{d_i}{2}\sqrt{-i\frac{\rho_0\omega}{\mu}}\right)}{\left(d_i\sqrt{-i\frac{\rho_0\omega}{\mu}}\right)\cdot B_0\left(\frac{d_i}{2}\sqrt{-i\frac{\rho_0\omega}{\mu}}\right)}\right]^{-1} + \frac{\sqrt{2\mu\rho_0\omega}}{2\sigma_i} + i\frac{0.85\omega\rho_0 \cdot d_i}{\sigma_i} \tag{6}$$

where $\sigma_i$ is the perforation rate corresponding to a single Helmholtz resonators cavity; $l_i$ is the perforation length; $\rho_0$ is the density of air at room temperature and pressure; $B_1\left(\frac{d_i}{2}\sqrt{-i\frac{\rho_0\omega}{\mu}}\right)$ and $B_0\left(\frac{d_i}{2}\sqrt{-i\frac{\rho_0\omega}{\mu}}\right)$ are the first-order and zero-order first-class Bessel functions, respectively; $\mu$ is the dynamic viscosity coefficient of air; $\omega$ is the acoustic angular frequency and $f$ is the acoustic frequency. The cavity impedance $Z_{ic}$ is gained by the impedance transfer equation, as shown in Equation (7).

$$Z_{ic} = -i\rho_{ce}c_{ce}A/A_c \cot(k_{ce}L) \tag{7}$$

where $A$ is the area of the whole structure; $A_c$ is the cross-sectional area of the cavity; $\rho_{ce}$ is the effective density of air, which can be obtained from Equation (8); $c_{ce}$ is the effective volume compression coefficient of air, which can be gained from Equation (9); $k_{ce}$ is the effective transfer constant of air in the cavity, which can be obtained from Equation (10); $L$ is the length of the rear cavity.

$$\rho_{ce} = \rho_0\left(1 - \frac{2B_1\left(r_i\sqrt{-i\frac{\rho_0\omega}{\mu}}\right)}{\sqrt{-i\frac{\rho_0\omega}{\mu}}B_0\left(r_i\sqrt{-i\frac{\rho_0\omega}{\mu}}\right)}\right)^{-1} \tag{8}$$

$$c_{ce} = \frac{1}{\gamma P_0}\left(\gamma - (\gamma-1)\frac{1 - 2\left(\sqrt{-i\omega\frac{\rho_0 C_p}{\kappa}}\right)^{-1}B_1\left(r_i\sqrt{-i\omega\frac{\rho_0 C_p}{\kappa}}\right)}{B_0\left(r_i\sqrt{-i\omega\frac{\rho_0 C_p}{\kappa}}\right)}\right) \tag{9}$$

$$k_{ce} = \omega\sqrt{\rho_{ce}c_{ce}} \tag{10}$$

where $C_p$ is the specific heat at the constant pressure; $\gamma$ is specific heat ratio, $P_0$ is standard atmospheric pressure at room temperature and pressure.

The total sound absorption coefficient could be determined by Equation (11).

$$\alpha = 1 - \left|\frac{Z - \rho_0 c_0}{Z + \rho_0 c_0}\right|^2 \tag{11}$$

where $c_0$ is the sound speed of air at room temperature and pressure, 343 m/s.

*4.3. Experimental Validation*

4.3.1. Methodology

To verify the sound absorption effect of the acoustic metamaterial, experiments were conducted with the AWA6290T transfer function sound absorption coefficient measurement system (supported by Hangzhou Aihua Instruments Co., Ltd., Hangzhou, China), as shown in Figure 8a, which could detect the sound absorption coefficients of sound absorbing material or structures with normal incidence according to GB/T 18696.2-2002 (ISO 10534-2:1998) "Acoustics—Determination of sound absorption coefficient and impedance in impedance tubes—part 2: Transfer function method", and its schematic diagram is shown in Figure 8b [36–41]. A cylindrical sample with a diameter of 100 mm was manufactured by the low force stereolithography (LFS) 3D printer of Form3 (supported by the Formlabs Inc., Summerville, MA, USA), as shown in Figure 8c, and the prepared sample for the investigated acoustic metamaterial is exhibited in Figure 8d. The proposed acoustic metamaterial of the multiple parallel hexagonal Helmholtz resonators was modeled in the 3D modeling software, and it was further introduced into the Preform software supported by Form3 3D printer. When fabrication of the sample was finished, it was further cleaned by the Formlabsform wash (Formlabs Inc., Boston, MA, USA) to remove residual liquid resin and irradiated for solidification by the FormlabsForm Cure (Formlabs Inc., Boston, MA, USA). The used photosensitive liquid resin in this research was ClearV4, which was purchased from the self-support flagship store of Formlabs 3D printer in JD.com (JD.com Inc., Beijing, China). The acoustic metamaterial made by photosensitive resin 3D printing had a smooth surface and well hardness, which met experimental requirements [36–41]. The AWA6290T detector consisted of the AWA5871 power amplifier, the AWA6290B dynamic signal analyzer, the AWA8551 impedance tube, and corresponding analysis software in the workstation, as shown in Figure 8b. The analysis software could finish the 1/3 OCT analysis and fast Fourier transform (FFT) analysis. Meanwhile, the original incident acoustic wave was also controlled by the signal generation software in the workstation. The detected sample was fixed in the end of the impedance tube, and two microphones were utilized to detect the sound pressure of the incident and reflected acoustic waves, which could derive the sound absorption coefficient at certain frequency according to the transfer function method [36–41]. The distance between the two microphones was set as 70 mm. The detected frequency range was 200–1600 Hz and there were 1502 sampling frequency points in this range. Moreover, for the purpose of elimination of the accidental error, the detection was repeated for 200 times for each sampling frequency point, and the final data were an average of the 200 values obtained in the 200 times of measurement. The measurement process was fully automatic, and took no more than 10 min for one detection [20].

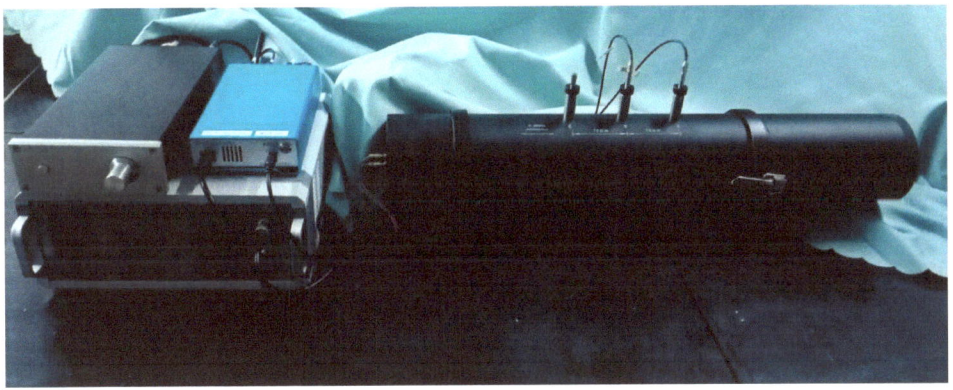

(a)

**Figure 8.** *Cont.*

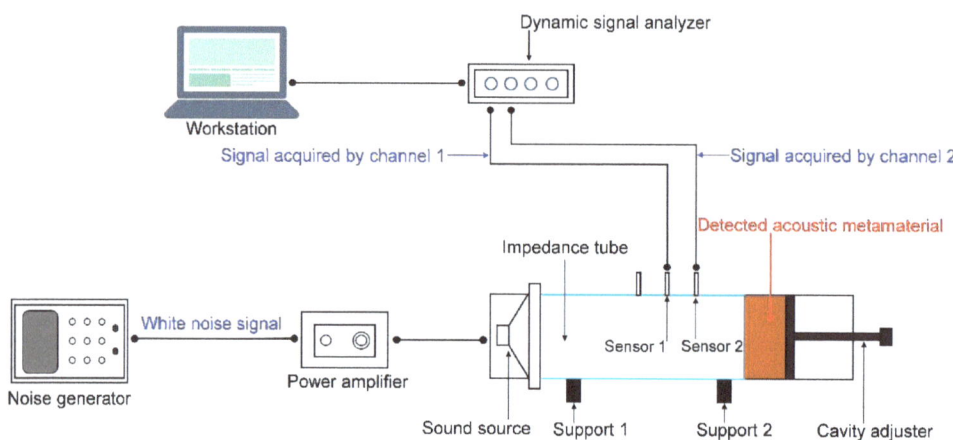

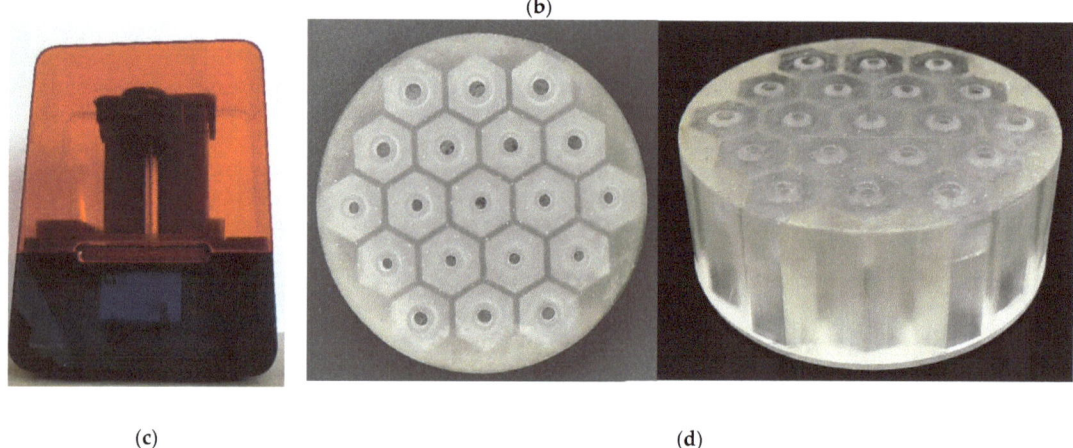

**Figure 8.** Fabrication and detection of the acoustic metamaterial. (**a**) The AWA6290T transfer function sound absorption coefficient measurement system; (**b**) Schematic diagram of the transfer function tube measurement; (**c**) The low force stereolithography (LFS) 3D printer of Form3; (**d**) The 3D printed sample of acoustic metamaterial with the diameter of 100 mm and the thickness of 40 mm.

4.3.2. Experimental Results

Comparisons of the theoretical, simulation and experimental sound absorption coefficients of the proposed acoustic metamaterial of multiple parallel hexagonal Helmholtz resonators are shown in Figure 9. From the curve of experimental data, it was clear that the wave peaks of the absorption coefficients measured experimentally for the acoustic metamaterial were 0.92117, 0.94360, 0.8867, 0.87560, 0.86980, 0.94540, and 0.92770, respectively, which occurred at a frequency of approximately 435.059 Hz, 533.203 Hz, 565.43 Hz, 593.262 Hz, 621.094 Hz, 654.053 Hz, and 687.012 Hz successively. The deviations between the simulation data and experimental data might be due to the imperfection of the sample fabrication process, the inevitable gap between the test sample, and the inner wall of the impedance tube. However, the general view of the two curve trends for simulation data and experimental data exhibited better consistency relative to the theoretical data. This research defined the absorption coefficient above 0.8 as the effective absorption coefficient. Both experimental and simulation data showed that the acoustic metamaterial could achieve effective absorption regarding the noise frequency in the workshop, which verified the accuracy of the optimal design method.

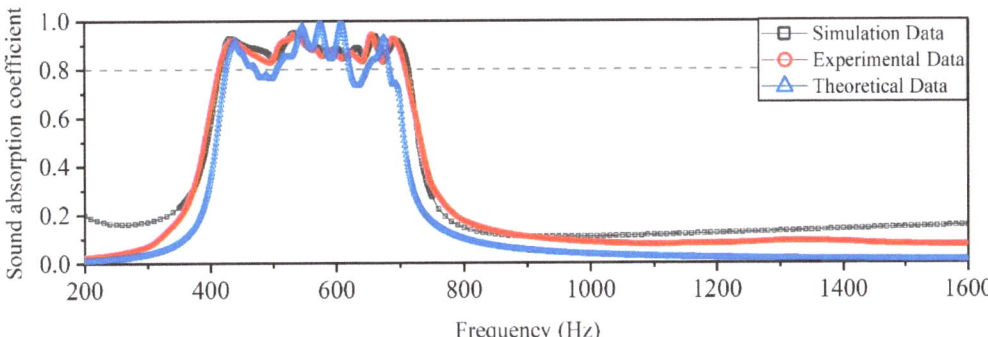

**Figure 9.** Comparisons of theoretical, simulation and experimental sound absorption coefficients of the proposed acoustic metamaterial of multiple parallel hexagonal Helmholtz resonators.

The theoretical data of the optimized structure were obtained according to theoretical model in Section 4.2, as shown in Figure 9. It could be seen from Figure 9 that the deviation of the theoretical data from the experimental data was larger compared with the simulation data. The deviation of the theoretical data and simulation data relative to the experimental data at the target frequency range was quantitatively evaluated by the mean absolute deviation, as shown in Equations (12) and (13), respectively. The deviation of the theoretical data was calculated to be 0.0682, while that of the simulation data was 0.0203. The results showed that the simulation data were better than the theoretical data. This might be because in theoretical calculations, the cavities of the acoustic metamaterial were connected in parallel, and the impedances of each cavity were not affected by each other. However, during the simulation and experiment process, the cavities of the acoustic metamaterial were separated by a certain thickness of wall and affected each other inevitably [12]. Thus, the impedance of the cavities in different regions was different. This also further verified the accuracy and reliability of the optimal design method.

$$Dev1 = average\left(\left|\alpha_{simulation}(f) - \alpha_{experimental}(f)\right|\right) \quad f \in [f_{min}, f_{max}] \quad (12)$$

$$Dev2 = average\left(\left|\alpha_{theoretical}(f) - \alpha_{experimental}(f)\right|\right) \quad f \in [f_{min}, f_{max}] \quad (13)$$

*4.4. Applicability of the Optimal Design*

The accuracy and reliability of the proposed optimal design method were previously verified. Meanwhile, the applicability of the optimal design method was further explored in this research. Two additional application conditions with different noise reduction requirements were selected and samples were prepared for experimental testing based on the optimized parameters. The design requirement of Condition-1 was to attenuate noise at 900–1300 Hz within a space of only 20 mm. The design requirement of Condition-2 was to attenuate the noise at 600–1000 Hz within a space of 30 mm.

The optimized geometric parameters for the design requirement of Condition-1 are shown in Table 3. The corresponding experimental data are shown in Figure 10, which showed that the noise in the range of 842–1355 Hz was attenuated with the sound absorption coefficient above 0.8 in accordance with the requirements of Condition-1.

Table 3. The optimized geometric parameters for design requirements of Condition-1.

| Group | Serial Number | Thickness/mm | Diameter of Micropore/mm | Depth of Neck/mm |
|---|---|---|---|---|
| I | 1 | | | 8.8 |
|   | 2 | | 4.47 | 8.2 |
|   | 3 | | | 7.3 |
| II | 4 | | | 8.1 |
|   | 5 | | 3.54 | 7.6 |
|   | 6 | | | 6.8 |
|   | 7 | | | 6.1 |
| III | 8 | | | 9.1 |
|   | 9 | | | 8.2 |
|   | 10 | 20 | 3.75 | 7.1 |
|   | 11 | | | 6.3 |
|   | 12 | | | 5.1 |
| IV | 13 | | | 9.0 |
|   | 14 | | 4.74 | 7.6 |
|   | 15 | | | 6.4 |
|   | 16 | | | 5.3 |
| V | 17 | | | 6.2 |
|   | 18 | | 5.16 | 5.0 |
|   | 19 | | | 4.5 |

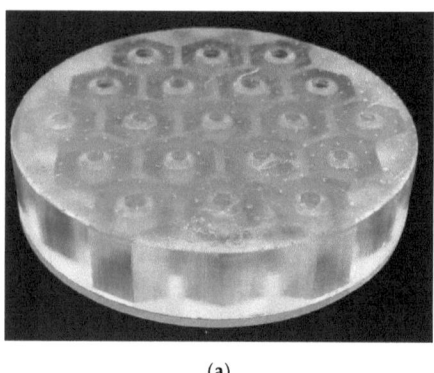

(a)

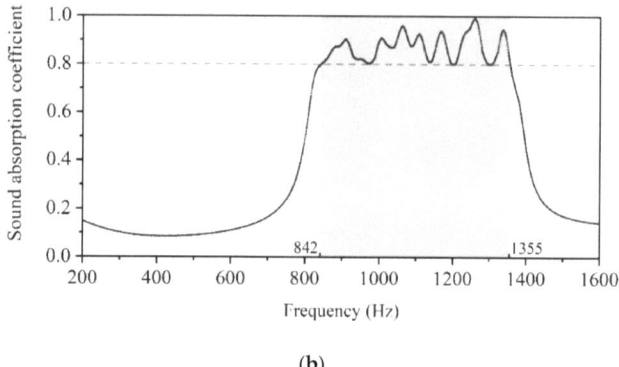

(b)

Figure 10. The optimization results of Condition-1. (a) Schematic diagram of the 3D printed sample with the diameter of 100 mm and the thickness of 20 mm. (b) The corresponding actual sound absorption coefficient.

The geometric parameters optimized for the design requirement of Condition-2 are shown in Table 4. The corresponding experimental data are shown in Figure 11, which showed that the noise in the frequency range 599–1027 Hz was attenuated with the sound absorption coefficient above 0.8, in accordance with the requirements of Condition-2.

Table 4. The optimized geometric parameters for design requirements of Condition-2.

| Group | Serial Number | Thickness/mm | Diameter of Micropore/mm | Depth of Neck/mm |
|---|---|---|---|---|
| I | 1 | | 4.47 | 10.3 |
| | 2 | | | 9.1 |
| | 3 | | | 8.0 |
| II | 4 | | 3.54 | 9.8 |
| | 5 | | | 9.0 |
| | 6 | | | 8.4 |
| | 7 | | | 7.7 |
| III | 8 | 30 | 3.75 | 10.8 |
| | 9 | | | 9.9 |
| | 10 | | | 8.6 |
| | 11 | | | 7.6 |
| | 12 | | | 6.5 |
| IV | 13 | | 4.74 | 9.5 |
| | 14 | | | 8.2 |
| | 15 | | | 7.0 |
| | 16 | | | 5.9 |
| V | 17 | | 5.16 | 6.4 |
| | 18 | | | 5.4 |
| | 19 | | | 4.8 |

(a)

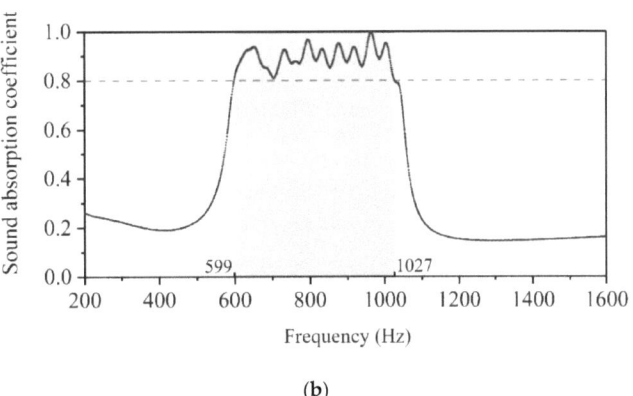
(b)

**Figure 11.** The optimization results of Condition-2. (**a**) Schematic diagram of the 3D printed sample with the diameter of 100 mm and the thickness of 30 mm. (**b**) The corresponding actual sound absorption coefficient.

It demonstrated that optimal design was still effective when the application condition changed. The ability to adjust the sound absorption band of the structure proved the applicability of this proposed optimal design method. It provided a novel method for the design of acoustic metamaterials, with potential application prospects in the fields of noise reduction and sound absorption.

## 5. Conclusions

In summary, an optimal design method to develop the acoustic metamaterials with broadband sound absorption at the low frequency range was proposed. An acoustic metamaterial of multiple parallel hexagonal Helmholtz resonators with the sub-wavelength dimension was employed. Initial parameters of the structure were designed preliminarily, and they were further optimized by the joint simulation method incorporating finite element analysis and cuckoo search algorithm. Subsequently, the theoretical results, simulation results, and experimental results were investigated, respectively.

(1) For the problem of low frequency broadband noise in the workshop and the narrow absorption bandwidth of a single Helmholtz resonator, the acoustic metamaterial of multiple parallel hexagonal Helmholtz resonators was designed and optimized. Its effective absorption capacity was verified by experimental validation. Therefore, the proposed acoustic metamaterial could attenuate the noise with a broad frequency range, which was also beneficial in practical engineering applications.

(2) A joint simulation method incorporating the finite element simulation and cuckoo search algorithm was employed for optimization of the proposed acoustic metamaterial to achieve broadband sound absorption effect at the low frequency range in this research. Compared with the optimization method based on theoretical models, the design method proposed in this research could obtain more accurate optimization results, which satisfied the requirements of certain practical conditions.

(3) The optimal design method proposed in this research could be applied to the absorption needs for different conditions, which also proved its feasibility and practicality. It presented a novel method for the development and application of acoustic metamaterial, which would be favorable to promote its practical applications.

**Author Contributions:** Conceptualization, X.S. and X.Z.; Software, E.W. and C.S.; Validation, F.Y. and E.W.; Investigation, F.Y. and X.Z.; Data curation, F.Y., E.W. and X.S.; Writing—original draft preparation, F.Y., E.W. and X.S.; Writing—review and editing, F.Y., Q.Y., X.Y., W.P. and X.W.; Supervision, X.S. and X.W.; Funding acquisition, X.Y., X.S. and W.P. All authors have read and agreed to the published version of the manuscript.

**Funding:** This research was funded by National Natural Science Foundation of China, grant numbers 12004178 and 52075538; Natural Science Foundation of Jiangsu Province, grant numbers BK20201336 and BK20211356; Natural Science Foundation of Hunan Province, grant number 2020JJ5670; China Postdoctoral Science Foundation, grant number 2021M691579; Natural Science Foundation of Jiangsu Higher Education Institution, grant number 20KJD460003; Scientific Research Foundation for the Introduction of talent of Nanjing Vocational University of Industry Technology, grant number YK20-14-01.

**Institutional Review Board Statement:** Not applicable.

**Informed Consent Statement:** Not applicable.

**Data Availability Statement:** The data that support the findings of this study are available from the corresponding author upon reasonable request.

**Conflicts of Interest:** The authors declare no conflict of interest.

## References

1. Liao, G.X.; Luan, C.C.; Wang, Z.W.; Liu, J.P.; Yao, X.H.; Fu, J.Z. Acoustic wave filtering strategy based on gradient acoustic metamaterial absorbers. *J. Phys. D Appl. Phys.* **2021**, *54*, 335301. [CrossRef]
2. Yang, X.C.; Shen, X.M.; Bai, P.F.; He, X.H.; Zhang, X.N.; Li, Z.Z.; Chen, L.; Yin, Q. Preparation and characterization of gradient compressed porous metal for high-efficiency and thin–thickness acoustic absorber. *Materials* **2019**, *12*, 1413. [CrossRef] [PubMed]
3. Yang, X.C.; Shen, X.M.; Duan, H.Q.; Yang, F.; Zhang, X.N.; Pan, M.; Yin, Q. Improving and Optimizing Sound Absorption Performance of Polyurethane Foam by Prepositive Microperforated Polymethyl Methacrylate Panel. *Appl. Sci.* **2020**, *10*, 2103. [CrossRef]
4. Dong, C.L.; Dong, Y.B.; Wang, Y.; Shi, J.B.; Zhai, S.L.; Zhao, X.P. Acoustic metamaterial absorbers and metasurfaces composed of meta–atoms and meta–molecules. *J. Phys. D Appl. Phys.* **2022**, *55*, 253002. [CrossRef]
5. Xu, Y.C.; Wu, J.H.; Cai, Y.Q.; Ma, F.Y. Acoustic bi–anisotropy in asymmetric acoustic metamaterials. *Appl. Phys. Express* **2020**, *13*, 106503. [CrossRef]
6. Cai, Z.R.; Zhao, S.D.; Huang, Z.D.; Li, Z.; Su, M.; Zhang, Z.Y.; Zhao, Z.P.; Hu, X.T.; Wang, Y.S.; Song, Y.L. Bubble Architectures for Locally Resonant Acoustic Metamaterials. *Adv. Funct. Mater.* **2019**, *29*, 1906984. [CrossRef]
7. Liu, H.X.; Wu, J.H.; Ma, F.Y. High–efficiency sound absorption by a nested and ventilated metasurface based on multi–slit synergetic resonance. *J. Phys. D Appl. Phys.* **2021**, *54*, 205304. [CrossRef]
8. Jimenez, N.; Huang, W.; Romero–Garcia, V.; Pagneux, V.; Groby, J.P. Ultra–thin metamaterial for perfect and quasi–omnidirectional sound absorption. *Appl. Phys. Lett.* **2016**, *109*, 121902. [CrossRef]

9. Zhang, L.; Xin, F.X. Perfect low–frequency sound absorption of rough neck embedded Helmholtz resonators. *J. Acoust. Soc. Am.* **2022**, *151*, 1191–1199. [CrossRef]
10. Sharafkhani, N. A Helmholtz Resonator–Based Acoustic Metamaterial for Power Transformer Noise Control. *Acoust. Aust.* **2021**, *50*, 71–77. [CrossRef]
11. Ismail, A.Y.; Kim, J.; Chang, S.M.; Koo, B. Sound transmission loss of a Helmholtz Resonator-based acoustic metasurface. *Appl. Acoust.* **2022**, *188*, 108569. [CrossRef]
12. Zhu, J.Z.; Qu, Y.G.; Gao, H.; Meng, G. Nonlinear sound absorption of Helmholtz resonators with serrated necks under high-amplitude sound wave excitation. *J. Sound Vib.* **2022**, *537*, 117197. [CrossRef]
13. Gao, Y.X.; Lin, Y.P.; Zhu, Y.F.; Liang, B.; Yang, J.; Yang, J.; Cheng, J.C. Broadband thin sound absorber based on hybrid labyrinthine metastructures with optimally designed parameters. *Sci. Rep.* **2020**, *10*, 10705. [CrossRef] [PubMed]
14. Yan, S.L.; Wu, J.W.; Chen, J.; Xiong, Y.; Mao, Q.B.; Zhang, X. Optimization design and analysis of honeycomb micro–perforated plate broadband sound absorber. *Appl. Acoust.* **2022**, *186*, 108487. [CrossRef]
15. Gao, N.S.; Wang, B.Z.; Lu, K.; Hou, H. Teaching–learning–based optimization of an ultra–broadband parallel sound absorber—ScienceDirect. *Appl. Acoust.* **2021**, *178*, 107969. [CrossRef]
16. Romero-García, V.; Theocharis, G.; Richoux, O.; Merkel, A.; Tournat, V.; Pagneux, V. Perfect and broadband acoustic absorption by critically coupled sub-wavelength resonators. *Sci. Rep.* **2016**, *6*, 19519. [CrossRef]
17. Herrero-Durá, I.; Cebrecos, A.; Picó, R.; Romero-García, V.; García-Raffi, L.M.; Sánchez-Morcillo, V.J. Sound Absorption and Diffusion by 2D Arrays of Helmholtz Resonators. *Appl. Sci.* **2020**, *10*, 1690. [CrossRef]
18. Shen, J.H.; Lee, H.P.; Yan, X. Sound absorption performance and mechanism of flexible PVA microperforated membrane. *Appl. Acoust.* **2021**, *185*, 108402. [CrossRef]
19. Xu, Z.B.; Wang, B.C.; Zhang, S.M.; Chen, R.J. Design and acoustical performance investigation of sound absorption structure based on plastic micro–capillary films. *Appl. Acoust.* **2015**, *89*, 152–158. [CrossRef]
20. Yang, X.C.; Yang, F.; Shen, X.M.; Wang, E.S.; Zhang, X.N.; Shen, C.; Peng, W.Q. Development of Adjustable Parallel Helmholtz Acoustic Metamaterial for Broad Low-Frequency Sound Absorption Band. *Materials* **2022**, *15*, 5938. [CrossRef]
21. Wen, J.; Zhao, H.; Lv, L.M.; Yuan, B.; Wang, G.; Wen, X.S. Effects of locally resonant modes on underwater sound absorption in viscoelastic materials. *J. Acoust. Soc. Am.* **2011**, *130*, 1201. [CrossRef] [PubMed]
22. Ye, C.Z.; Liu, X.W.; Xin, F.X.; Lu, T.J. Influence of hole shape on sound absorption of underwater anechoic layers. *J. Sound Vib.* **2018**, *426*, 54–74. [CrossRef]
23. Zeng, Z.; Zhang, M.; Li, C.; Ren, L.; Wang, P.Y.; Li, J.W.; Yang, W.D.; Pan, Y. Simulation study on characteristics of acoustic metamaterial absorbers based on Mie and Helmholtz resonance for low–frequency acoustic wave control. *J. Phys. D Appl. Phys.* **2021**, *54*, 385501. [CrossRef]
24. Cai, X.J.; Niu, Y.; Geng, S.J.; Zhang, J.J.; Cui, Z.H.; Li, J.W.; Chen, J.J. An under-sampled software defect prediction method based on hybrid multi-objective cuckoo search. *Concurr.-Comp. Pract. Exp.* **2020**, *32*, e5478. [CrossRef]
25. Jain, A.; Chaudhari, N. A novel cuckoo search technique for solving discrete optimization problems. *Int. J. Syst. Assur. Eng.* **2018**, *9*, 972–986. [CrossRef]
26. Yang, X.S.; Deb, S. Cuckoo search: Recent advances and applications. *Neural Comput. Appl.* **2014**, *24*, 169–174. [CrossRef]
27. Qi, X.B.; Yuan, Z.H.; Song, Y. An integrated cuckoo search optimizer for single and multi–objective optimization problems. *PeerJ Comput. Sci.* **2021**, *7*, e370. [CrossRef]
28. Yang, X.C.; Bai, P.F.; Shen, X.M.; To, S.; Chen, L.; Zhang, X.N.; Yin, Q. Optimal design and experimental validation of sound absorbing multilayer microperforated panel with constraint conditions. *Appl. Acoust.* **2019**, *146*, 334–344 [CrossRef]
29. Jha, K.; Gupta, A.; Alabdulatif, A.; Tanwar, S.; Safirescu, C.O.; Mihaltan, T.C. CSVAG: Optimizing Vertical Handoff Using Hybrid Cuckoo Search and Genetic Algorithm-Based Approaches. *Sustainability* **2022**, *14*, 8547. [CrossRef]
30. Nguyen, T.T.; Le, B. Optimization of electric distribution network configuration for power loss reduction based on enhanced binary cuckoo search algorithm. *Comput. Electr. Eng.* **2020**, *90*, 106893. [CrossRef]
31. Yang, F.; Shen, X.M.; Bai, P.F.; Zhang, X.N.; Li, Z.Z.; Yin, Q. Optimization and Validation of Sound Absorption Performance of 10-Layer Gradient Compressed Porous Metal. *Metals* **2019**, *9*, 588. [CrossRef]
32. Sicuaio, T.; Niyomubyeyi, O.; Shyndyapin, A.; Pilesjö, P.; Mansourian, A. Multi-Objective Optimization Using Evolutionary Cuckoo Search Algorithm for Evacuation Planning. *Geomatics* **2022**, *2*, 53–75. [CrossRef]
33. Bergin, M.; Myles, T.A.; Radic, A.; Hatchwell, C.J.; Lambrick, S.M.; Ward, D.J.; Eder, S.D.; Fahy, A.; Barr, M.; Dastoor, P.C. Complex optical elements for scanning helium microscopy through 3D printing. *J. Phys. D Appl. Phys.* **2022**, *55*, 095305. [CrossRef]
34. Zhu, J.; Lao, C.S.; Chen, T.N.; Li, J.S. 3D–printed woodpile structure for integral imaging and invisibility cloaking. *Mater. Des.* **2020**, *191*, 108618. [CrossRef]
35. Nowoświat, A.; Olechowska, M. Experimental Validation of the Model of Reverberation Time Prediction in a Room. *Buildings* **2022**, *12*, 347. [CrossRef]
36. Quan, H.Y.; Zhang, T.; Xu, H.; Luo, S.; Nie, J.; Zhu, X.Q. Photo–curing 3D printing technique and its challenges. *Bioact. Mater.* **2020**, *5*, 110–115. [CrossRef]
37. Grieder, S.; Zhilyaev, I.; Küng, M.; Brauner, C.; Akermann, M.; Bosshard, J.; Inderkum, P.; Francisco, J.; Willemin, Y.; Eichenhofer, M. Consolidation of Additive Manufactured Continuous Carbon Fiber Reinforced Polyamide 12 Composites and the Development of Process-Related Numerical Simulation Methods. *Polymers* **2022**, *14*, 3429. [CrossRef]

38. Shinde, V.V.; Celestine, A.D.; Beckingham, L.E.; Beckingham, B.S. Stereolithography 3D Printing of Microcapsule Catalyst–Based Self–Healing Composites. *ACS Appl. Polym. Mater.* **2020**, *2*, 5048–5057. [CrossRef]
39. Huang, S.B.; Fang, X.S.; Wang, X.; Assouar, B.; Cheng, Q.; Li, Y. Acoustic perfect absorbers via Helmholtz resonators with embedded apertures. *J. Acoust. Soc. Am.* **2019**, *145*, 254–262. [CrossRef]
40. Guo, J.W.; Fang, Y.; Jiang, Z.Y.; Zhang, X. An investigation on noise attenuation by acoustic liner constructed by Helmholtz resonators with extended necks. *J. Acoust. Soc. Am.* **2021**, *149*, 70–81. [CrossRef]
41. Duan, H.Q.; Shen, X.M.; Wang, E.S.; Yang, F.; Zhang, X.N.; Yin, Q. Acoustic multi-layer Helmholtz resonance metamaterials with multiple adjustable absorption peaks. *Appl. Phys. Lett.* **2021**, *118*, 241904. [CrossRef]

Article

# Experimental Studies on Adaptive-Passive Symmetrical Granular Damper Operation

Mateusz Żurawski * and Robert Zalewski

Institute of Machine Design Fundamentals, Warsaw University of Technology, 02-524 Warsaw, Poland
* Correspondence: mateusz.zurawski@pw.edu.pl

**Abstract:** This paper presents experimental studies on a controllable granular damper, whose dissipative properties are provided by the friction phenomenon occuring between loose granular material. In addition, in order to adjust to the current trends in vibration suppression, we built a semi-active device, controlled by a single parameter—underpressure. Such granular structures subjected to underpressure are called Vacuum-Packed Particles. The first section presents the state of the art. A brief description of the most often used intelligent and smart materials for the manufacture of dampers is presented. The main advantages of the proposed device are a simple structure, low construction cost, symmetrical principle of operation, and the ability to change the characteristics of the damper by quickly and suddenly changing the negative pressure inside the granular core. The second section provides a detailed description of the construction and operation principles of the original symmetrical granular damper. A description of its application in the laboratory research test stand is also provided. The third section presents the results of the experimental studies including the recorded damping characteristics of the investigated damper. The effectiveness of the ethylene–propylene–diene grains' application is presented. The two parameters of underpressure and frequency of excitation were considered during the empirical tests. The influence of the system parameters on its global dissipative behavior is discussed in detail. The damper operation characteristics are close to linear, which is positive information from the point of view of the potential adaptive-passive control process. Brief conclusions and the prospective application of vacuum-packed particle dampers are presented in the final section.

**Citation:** Żurawski, M.; Zalewski, R. Experimental Studies on Adaptive-Passive Symmetrical Granular Damper Operation. *Materials* **2022**, *15*, 6170. https://doi.org/10.3390/ma15176170

Academic Editor: Martin Vašina

Received: 29 July 2022
Accepted: 2 September 2022
Published: 5 September 2022

**Publisher's Note:** MDPI stays neutral with regard to jurisdictional claims in published maps and institutional affiliations.

**Copyright:** © 2022 by the authors. Licensee MDPI, Basel, Switzerland. This article is an open access article distributed under the terms and conditions of the Creative Commons Attribution (CC BY) license (https://creativecommons.org/licenses/by/4.0/).

**Keywords:** granular structures; vacuum-packed particles; granular damper; experimental research; semi-active damper

## 1. Introduction

Observing the engineering environment, it can be concluded that vibrations appear almost everywhere. For reasons of safety or comfort, attention should be paid to mechanical vibrations directly affecting devices or structures. The development of materials and manufacturing technology allows for the construction of ever lighter structures [1]. Their low mass very often makes them more susceptible to external factors. That is why the role of vibration eliminators in modern systems is so important [2–4]. New design requirements concerning, inter alia, safety, environmental protection, and economic aspects force changes both in the design of vibration dampers and in the development of new methods of vibration suppression. Currently, many examples can be distinguished in which the phenomenon of resonance is not the only dangerous one. There are several forms of excitation. For example, each vehicle is exposed to varying weather conditions and a variety of road surfaces [5]. It is obvious that the classical damper cannot be used in both cases. The simplest passive damper [6] has permanent dissipation properties, so it is not possible to adjust the damping force to the current external conditions. For such a case, smart and intelligent materials should be taken into account. Applying the previously mentioned group of structures makes it possible to change the damping properties of vibration reduction devices. The advantages of this type of solution are so significant that more often active or semi-active attenuators can be found in practical engineering applications. At the same time complex laboratory tests are constantly being carried out to improve this type of devices.

According to the definitions presented in [7], passive vibration dampers are relatively simple technical devices. They are used to change the stiffness and damping of the structure in the desired way and, at the same time, do not require an external power supply. Characteristic features of this type of absorber include permanent dissipation properties. Due to the previously mentioned simplicity of construction, they are still the most often used vibration reduction systems [8]. Another example is the use of the double-layer permanent-magnet buffer [9]. The authors showed the possibility of damping the vibrations caused by the impact load of the launcher.

Based on [10], active vibration damping consists of the dissipation of vibration energy through the application of appropriate external forms of energy to the investigated dynamical system. Generally, active dampers allow for a dynamic change in the damping characteristics depending on the various conditions. The entire process is managed by a controller which, based on sensor data, sets the appropriate actuator damping values. The active vibration damping system [10] has the greatest possibilities, but at the same time requires the use of additional equipment and the supply of energy from external sources. In the case of such types of system, intelligent materials are often used.

Semi-active vibration damping [7] does not introduce an additional energy into the system. It enables changing the dissipation properties of the actuator responsible for efficient damping. In fact, the system can adapt to the changing exciting load.

Scientists and researchers aim to propose novel designs of devices that could be an alternative to the commonly used controlling absorbers. Several papers [11–14] present an approach in which mechanical vibrations were damped by means of an innovative adaptive impact damper. The main factor reducing vibrations is the possibility of mass movement inside a container of variable volume. This type of damper is effective especially for resonant frequencies or free vibrations. Another adaptive device was described in [15,16]. The main innovation of that approach is the ability to control the size of the valve in the cylinder which directly affects the effectiveness of shock absorption. A further example of devices that can change their damping properties as a result of external factors are Vacuum-Packed Particles. Detailed descriptions of this type of absorber can be found in [17–21]. The discussed damping attenuators are characterized by a granular core with variable pressure inside. By changing the pressure, the contact forces between the grains can be controlled [22–25]. As a result, the characteristics of the hysteresis function, describing the dependence of the forces generated by the devices as the effect of compression and tension are changed. The main disadvantage of the described devices is an effective damping only in the compression stage of the core. Therefore, the authors of this paper propose an approach that allows obtaining the symmetrical characteristics of the force-displacement hysteresis loops. Such a solution allows for the extension of the applicability of this type of device.

To further emphasize the importance of symmetry in mechanical engineering, several works based on solutions that increase the reliability of the structure by ensuring symmetry should be cited [26–30]. A frequent effect of the influence of excitation on a structure is its unforeseen destruction. Therefore, reliability analyses of structures characterized by a symmetrical structure are carried out. These are implemented through the innovative method for symmetry representations and elastic redundancy for members of tensegrity structures [26]. An analogous approach to modeling can be found in systems called origami structures. New types of thin-walled origami tube based on the Kresling origami pattern have been proposed [27]. These studies were extended and improved through the use of Convolutional Neural Networks [28]. Engineering applications where there are vibration and bifurcation problems and the stability of systems exhibiting symmetry have been studied using group theory [29] and layered space grids [30]. An interesting issue is the effect of nanomaterials and research fibers on the mechanical properties of symmetrical polymer pomposites [31]. In this approach, the impact and tensile properties of the polypropylene nanocomposites with graphene nanosheets, nanoclay, and basalt fibers were explored.

The paper outline is as follows. After the introduction to the subject, the construction of a symmetrical granular damper is presented and described in detail. Then, the test stand and experimental results are presented. The experimental analysis was performed for several different parameters of the underpressure and the excitation frequency. Finally, the results are compared, and the general characteristics of the presented novel device are determined. The paper is summarized with synthetic conclusions.

## 2. Prototype of the Symmetrical Granular Damper

### Damper Construction

For many decades, engineers and scientists have been trying to propose newer solutions, whose principles are often very complicated. Therefore, the mechanical systems and the materials have become very advanced in their structure. This offers great opportunities to design adaptive, active, or semi-active devices. As a result, the discussed solutions are very expensive or have a very narrow range of applicability. One of the main assumptions of this paper was to propose an innovative device that would be characterized by a simple structure with low cost, features of adaptability, and a wide range of application possibilities.

The following description presents a vibration damper with a granular core having variable and controllable dissipation properties. The system consists of a rigid cylinder encased by a sealed granular core made of Vacuum-Packed Particles (VPP). The ability to control the pressure in the granular container allows changing the macroscopic damping properties of the device. This solution enables obtaining a symmetrical response characteristic and changable dissipation properties depending on the pressure inside the core and the initial tension of the springs. The design and principle of operation of the presented innovative symmetrical granular damper have been patented [32].

Depicted in Figures 1 and 2, the vibration damper with variable dissipation properties consists of a plastic cylinder (1) with flanges (10, 11) attached at its outer periphery to its opposite edges. The flanges (10, 11) of the cylinder (1) are rigidly connected by three parallel bars (5) to the first and the second retaining discs (13, 14) located at the ends of the damper. In the sealed chamber of the cylinder (1), there is a ring-shaped granular core (4) made of VPP. Inside the granular core (4) there is a rigid cylindrical ring (3). The valve (2) is mounted in the casing of the cylinder (1) to regulate the partial vacuum inside the granular core (4). The sealing of the cylinder chamber (1) consists of an elastic annular carcass (12) made of rubber, fixed on the opposite sides of the cylinder (1), between its edges, and the ends of the cylindrical core (3). The cylindrical core (3) is coaxially connected to a bar (17) slidably mounted on the first retaining disc (13). The granular core (4) transmits the linear load exerted on the cylindrical core (3). In the axis of the cylinder (1), on the opposite sides of the cylindrical core (3), there are two springs (6, 7) provided to compensate the semi-plastic deformation of the granular core (4) in the axial direction under the influence of the linear load. The spring element (6) is mounted between the cylindrical core (3) and the first retaining disk (13). The second spring (7) is located between the cylindrical core (3) and the second retaining disk (14). The ends of the springs (6, 7) are supported in the spring seats (18) setting their axial position. The rod (17) is positioned in the axis of the first spring (6). In the axis of the second spring (7), there is a telescopic guide element, the first member of which (16) is attached to the cylindrical body (3), and the second member (15) is attached to the second guide disc (14). Nuts (8) with washers (9) were used for the bolted connections of the cylinder (1) and the retaining discs (13, 14) with the rods (5).

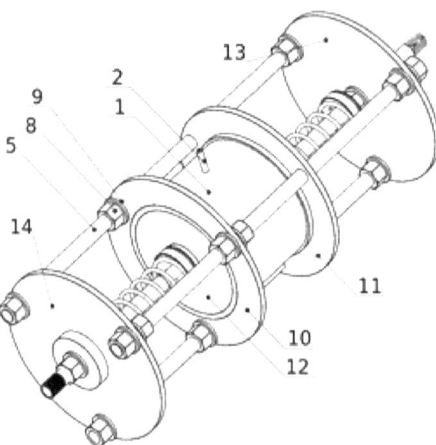

**Figure 1.** Scheme 1 of the symmetrical granular damper.

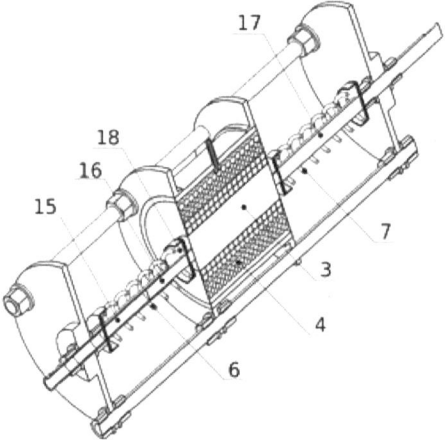

**Figure 2.** Scheme 2 of the symmetrical granular damper.

For a detailed description of the operating principle of the proposed device, a simplified diagram is presented in Figures 3 and 4. The granular material is located between a fixed outer sleeve (2) and a movable inner sleeve (1). The thin silicone sealing plates (3) are glued to the upper and lower parts of the device. Such a solution protects the loose material from falling out, and due to its elastic properties, it allows maintaining full tightness during the movement of the central sleeve.

The stationary outer sleeve (2) and the movable inner sleeve (1) simulate the phenomenon of shearing.

Existing devices [13,14,17,18,21] reflect the effective ability of the mechanical damping. Despite many advantages, the functionality of the aforementioned structures is subject to significant limitations. This causes the limited use of classic granular dampers in technical applications. The main disadvantage of the existing solutions is that they only work effectively in one direction when the granular core is compressed.The proposed design eliminates the disadvantages of classic approaches, thus extending the scope of applicability in mechanical structures subjected to various excitations.

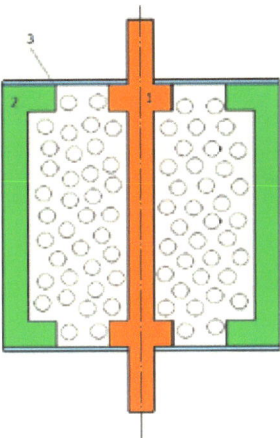

**Figure 3.** Scheme of the principle of the operation—initial stage.

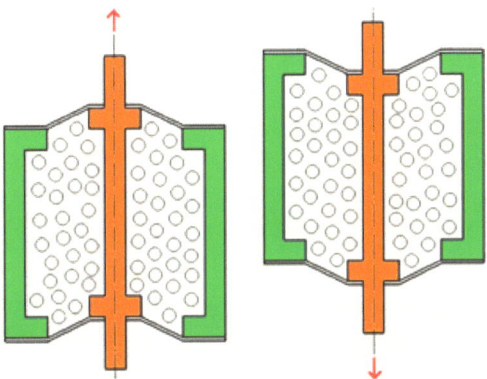

**Figure 4.** Scheme of the principle of the operation—operating stage.

The relatively simple design assumptions of the damper made it possible to manufacture all the elements and test the device on a testing machine in order to confirm the previously mentioned assumptions about the symmetrical damping characteristics. The test stand (Figure 5) consists of an electric motor connected to the disk, to which an eccentrical pusher was attached. Hence, the reciprocating movement of the device was achieved.

With the applied movement mechanism a sine kinematic excitation rule was considered in which both amplitude and frequency could be adjusted. In the conducted research, two quantities were recorded: force and displacement. The force was measured by a piezoelectric sensor installed in the lower part of the test stand, while the displacement measurement was performed by an inductive displacement sensor. In the case of the tests carried out on a granular damper, the two parameters were regulated. The amplitude was constant at 5 mm. The frequency of the motor's rotation ($f$) was set by means of an inverter. On the other hand, the underpressure was generated with the use of a vacuum pump made by AGA Labor (Figure 6).

**Figure 5.** Test stand.

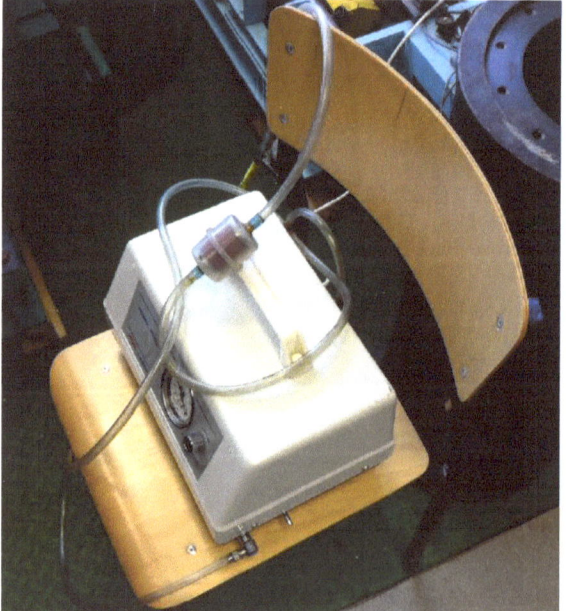

**Figure 6.** AGA Labor vacuum pump.

## 3. Experimental Results

In order to ensure the comprehensiveness of the tests, experiments were carried out for different rotational speeds and for variable values of partial vacuum inside the granular core. For the selected types of grains, measurements were carried out for five rotational frequencies (Table 1).

**Table 1.** Frequency of the excitation.

| $f$ | $f_1$ | $f_2$ | $f_3$ | $f_4$ | $f_5$ |
|---|---|---|---|---|---|
| Frequency [Hz] | 0.4 | 0.9 | 1.5 | 2.1 | 2.8 |

The excitation amplitude was constant $A = 5$ [mm]. For such assumptions, the function describing the harmonic input takes the form:

$$u = A\sin(2\pi ft) \tag{1}$$

During the laboratory tests of the Symmetrical Vacuum-Packed Particle Damper (SVPPD) the most important aspect was investigating the influence of the partial vacuum value, generated inside the granular system—a factor that directly influences the dissipation properties. The internal partial vacuum was changed from 0 to 0.09 [MPa] by means of a vacuum pump (Table 2). In order to investigate the influence of the pressure on the damping characteristics as precisely as possible, the pressure was changed by 0.01 [MPa].

**Table 2.** The symbols and values of the assumed underpressure.

| up | up00 | up01 | up02 | up03 | up04 | up05 | up06 | up07 | up08 | up09 |
|---|---|---|---|---|---|---|---|---|---|---|
| Pressure [MPa] | 0.00 | 0.01 | 0.02 | 0.03 | 0.04 | 0.05 | 0.06 | 0.07 | 0.08 | 0.09 |

The SVPPD was filled with ethylene–propylene–diene grains (EPDM). The low price and the ability to choose the size and color has contributed to the wide use of EPDM granules (Figure 7) in the construction of various types of facilities as a top layer or filling for football fields, running tracks, and playgrounds for children.

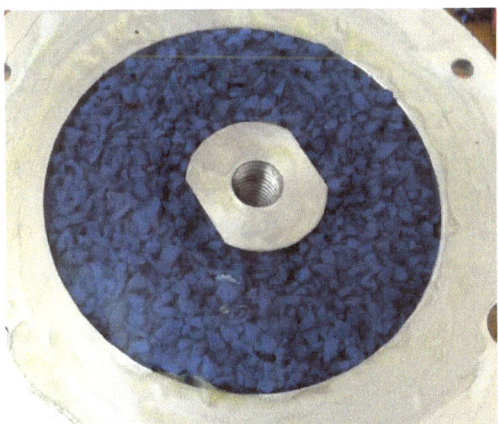

**Figure 7.** SVPP damper with EPDM grains.

The physical and mechanical properties of the applied grains are shown in Table 3.

Table 3. EPDM material properties.

| Material | EPDM |
|---|---|
| Density [g/cm$^3$] | 1.6 |
| Hardness (Shore A) | 60 |
| Young's modulus [MPa] | 2.16 |
| Temperature range of application [C] | $-50 \to 130$ |

From the experimental tests, it was possible to analyze both the influence of the rotational speed on the damping properties of the device and the influence of underpressure on the increase in the damping force. The main goal of the first group of experiments was to determine the characteristics of the damping force as a function of displacement for the selected frequency $f_1$. The results are shown in Figure 8.

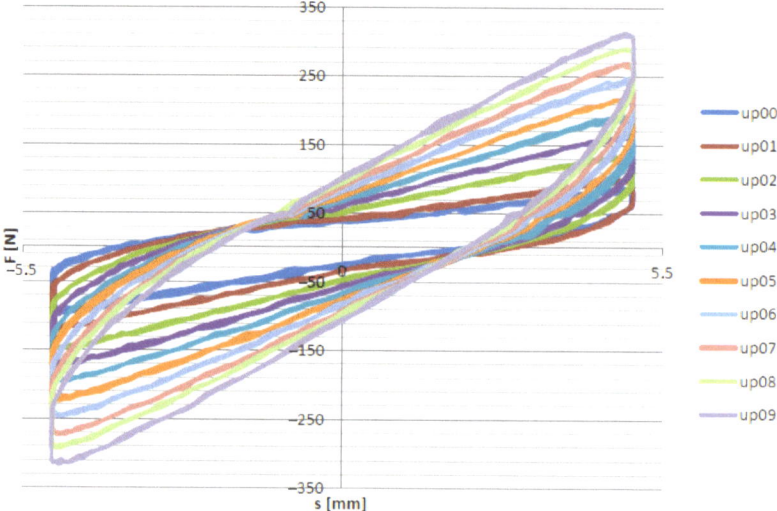

Figure 8. Force/displacement curve for frequency $f_1$.

Figure 8 presents the recorded force-displacement hysteresis loops for the selected loading frequency and underpressure values. The shapes of the recorded data were symmetrical. The maximal value achieved by the SVPPD force was approximatelly 325 N, which was observed for the highest value of the internal underpressure. Decreasing the partial vacuum caused a decrease in the maximum value of the force. The results of the experimental tests for the remaining values of the excitation frequency are presented in Figure 9.

Regardless of the rotational speed, the crucial increase in damping force values was visible during the increase in the applied underpressure. It is also very important that the character was symmetrical during the shear of the granular media. By carrying out tests for various excitation frequencies, it was also possible to investigate the influence of this parameter on the values of the damping force. The figures below show the influence of the rotational speed on the recorded forces for the SVPPD.

When analyzing the above figure, it should be noted that in the case of the zero underpressure value (Figure 10), some differences were noticeable in the values and the characteristics. This type of phenomenon can be caused by the shear of the granulate (as opposed to the compression/stretching) and the gravitational forces that result in the irregular distribution of the grains. The characteristics began to coincide with the higher stiffness of the structure (higher underpressure).

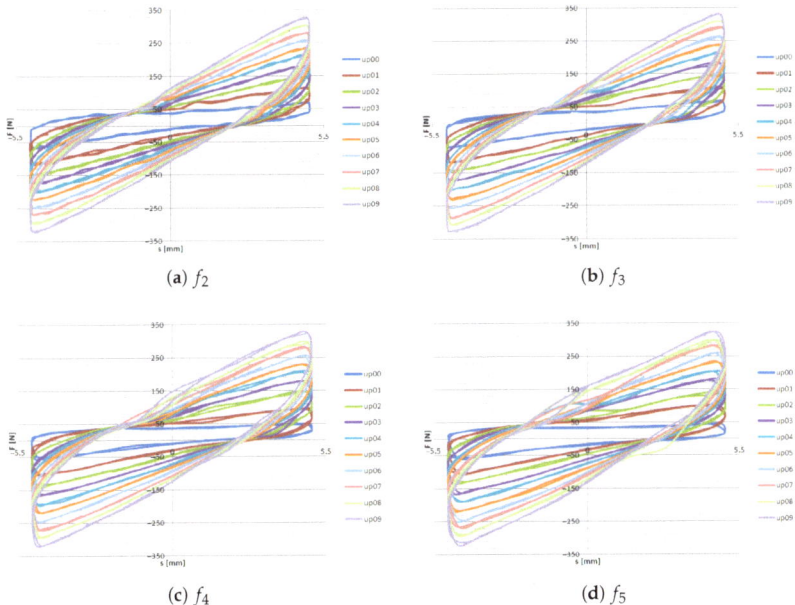

(a) $f_2$

(b) $f_3$

(c) $f_4$

(d) $f_5$

**Figure 9.** Force/displacement curves for various frequencies $f_2 \rightarrow f_5$.

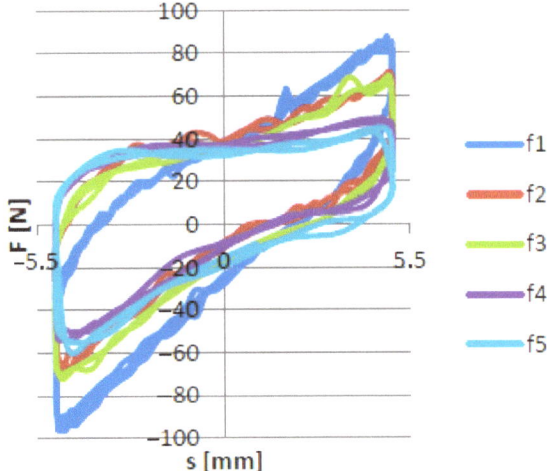

**Figure 10.** Force/displacement characteristics for various rotational speeds for up00 underpressure.

Figures 11 and 12 show that the excitation frequency in the case of the EPDM material had no major influence on the dissipation properties of the damper when the higher values of partial vacuum were generated inside the damper. Based on such an observation it can be concluded that the EPDM granules allow obtaining high damping forces in each case. Moreover, it can be observed that the shapes of the recorded hysteresis loops were similar.

The dependence between underpressure and maximal dissipation force is presented in Figure 13.

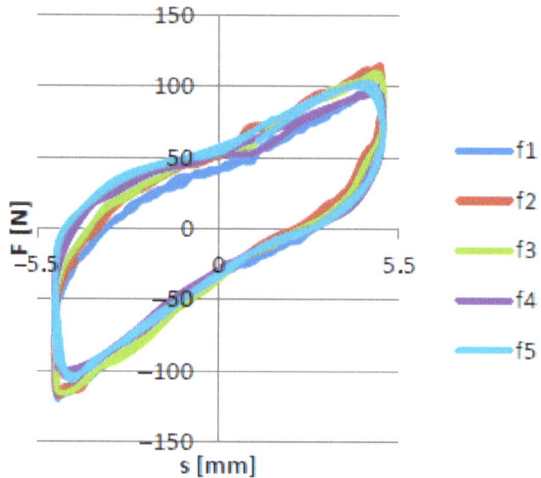

**Figure 11.** Force/displacement characteristics for various rotational speeds for up01 underpressure.

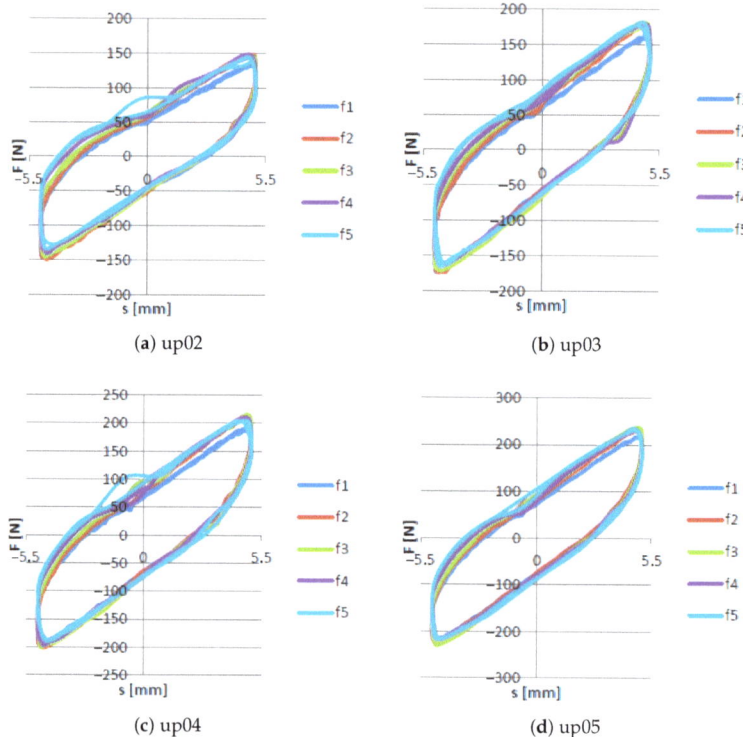

(a) up02

(b) up03

(c) up04

(d) up05

**Figure 12.** *Cont.*

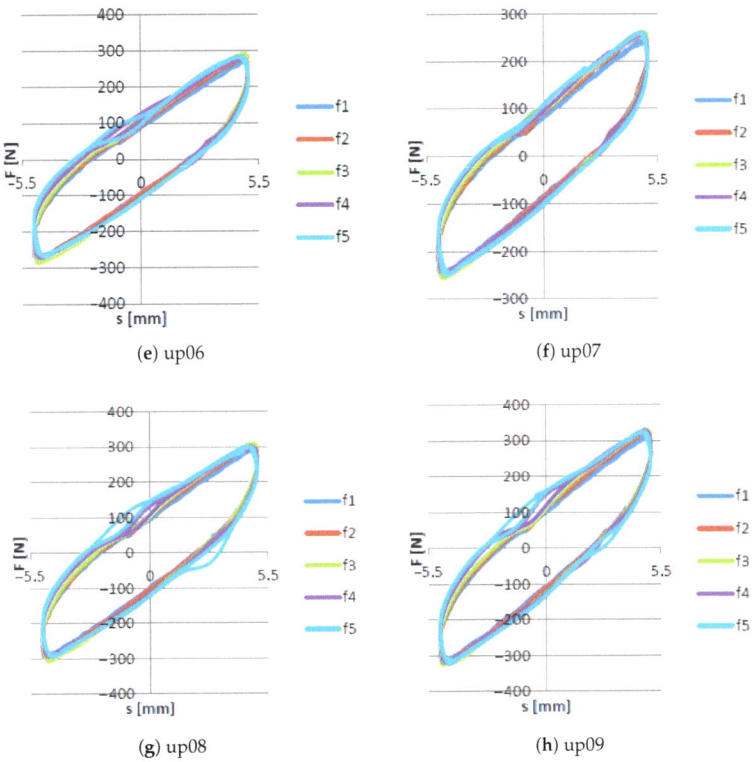

**Figure 12.** Force/displacement characteristics for various rotational speeds for up02 → up09 underpressure.

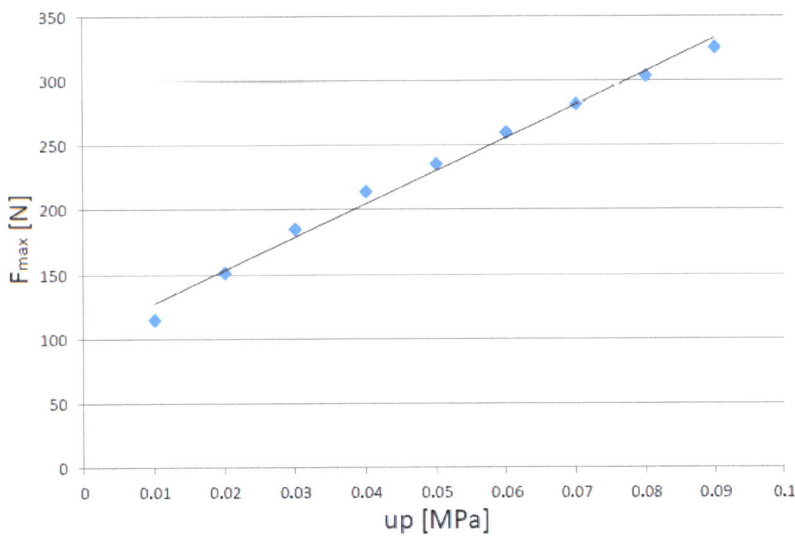

**Figure 13.** Force/underpressure characteristics for $f_2$ rotational speed.

The obtained results were a function close to linear and had values between 115 [N] and 324 [N]. Analogous dependencies for the remaining excitation frequency cases ($f_3$, $f_4$, and $f_5$) were also characterized by similar linearities. This means that the presented structure can be used in a wide range of work. In the case of semi-active dampers, this is a strong advantage, especially in practical engineering applications.

## 4. Summary

The characteristics of vibrations for an innovative symmetrical Vacuum-Packed Particles absorber were experimentally investigated. The process of pumping out the air from the sealed granular core allowed changing the internal grain structure causing jamming of the loose material being placed inside the damper. This is how the range of deformations of the granular core can be controlled. This phenomenon enabled control of the symmetric dissipation forces. Experimental tests were carried out on the physical model of a granular shear damper. The results presented force-displacement hysteresis loops for excitation frequency and underpressure values. Regardless of the rotational speed, the crucial increase in damping forces values was visible for various implemented underpressures. It should be noticed, that for the zero underpressure state, the various frequencies had a nonlinear influence on the damping forces. However, generally, the change in the underpressure in the granular core caused an almost linear change in the forces generated by the symmetrical damper. The novelty of the presented original damper prototype was based on the use of the shear phenomenon of loose materials and thus obtaining a symmetrical dissipation characteristic, different from the case of the compression and tension of granular structures. Since the shear effect between the grains is the same regardless of the direction of the damper movement, the dissipation characteristic should remain symmetrical. Experimental tests carried out on an experimental stand confirmed the theoretical assumptions.

The novel symmetrical granular damper can be treated as an alternative to current semi-active devices and can be implemented at low-cost in an engineering environment. Many mechanical systems are subject to variable excitation. This cause a multiple analysis of the optimal parameters that would allow for the most effective damping of mechanical vibrations. The proposed damper design may constitute the basis for a device included in cyberphysical systems. These are systems based on the measurement of significant dynamic quantities and their use in advanced numerical analysis, often carried out in the data cloud and enriched with artificial intelligence algorithms; then, with appropriate mechatronic controllers, the process of controlling an innovative damper is carried out. Due to the resistance of the granulate to the external factors, the proposed symmetrical damper can be used in difficult environmental conditions. The process of changing the underpressure and the transition state to obtain the final damping properties is very short. Therefore, the proposed structure can be used wherever there are rapid and sudden changes in the parameters of the suppressed system.

**Author Contributions:** Conceptualization, M.Ż. and R.Z.; methodology, M.Ż. and R.Z.; software, M.Ż.; validation, M.Ż. and R.Z.; formal analysis, M.Ż. and R.Z.; investigation, M.Ż. and R.Z.; resources, R.Z.; data curation, M.Ż.; writing—original draft preparation, M.Ż.; writing—review and editing, R.Z.; visualization, M.Ż. and R.Z.; supervision, R.Z.; project administration, M.Ż.; funding acquisition, M.Ż. All authors have read and agreed to the published version of the manuscript.

**Funding:** This research received no external funding. The APC was funded by Grant Wewnętrzny dla Pracowników Politechniki Warszawskiej w Dyscyplinie Inżynieria Mechaniczna, grant number 504/04678/1152/43.090006.

**Conflicts of Interest:** The authors declare no conflict of interest.

**Abbreviations**

The following abbreviations are used in this manuscript:

SVPP    Symmetric Vacuum-Packed Particle
EPDM   ethylene–propylene–diene grains

# References

1. Zhang, X.W.; Yu, T.X. Energy absorption of pressurized thinwalled circular tubes under axial crushing. *Int. J. Mech. Sci.* **2009**, *51*, 335–349. [CrossRef]
2. Ahamed, R.; Choi, S.B.; Ferdaus, M.M. A state of art on magneto-rheological materials and their potential applications. *J. Intell. Mater. Syst. Struct.* **2018**, *29*, 2051–2095. [CrossRef]
3. Mojtaba, A.; Kouchakzadeh, M.A. Aeroelastic characteristics of magneto-rheological fluid sandwich beams in supersonic airflow. *Compos. Struct.* **2016**, *143*, 93–102. [CrossRef]
4. Lu, Z.; Wang, Z.; Zhou, Y.; Lu, X. Nonlinear dissipative devices in structural vibration control: A review. *J. Sound Vib.* **2018**, *18*, 18–49. [CrossRef]
5. Nguyen, Q.-H.; Choi, S.-B. Optimal design of MR shock absorber and application to vehicle suspension. *Smart Mater. Struct.* **2009**, *18*, 035012. [CrossRef]
6. Parulekar, Y.M. Passive response control systems for seismic response reduction: A state-of-the-art review. *Int. J. Struct. Stab. Dyn.* **2009**, *9*, 151–177. [CrossRef]
7. Bankar, V.K.; Aradhye, A. A Review on Active, Semi-active and Passive Vibration Dampin. *Int. J. Curr. Eng. Technol.* **2016**, *6*, 2187–2191.
8. Diez-Jimenez, E.; Rizzo, R. Review of Passive Electromagnetic Devices for Vibration Damping and Isolation. *Shock Vib.* **2019**, *11*, 1250707. [CrossRef]
9. Ge, J.; Xie, X.; Sun, Q.; Yang, G. Design and dynamic characteristics of a double-layer permanent-magnet buffer under intensive impact load. *Eng. Struct.* **2021**, *506*, 116158. [CrossRef]
10. Chuaqui, T.R.; Roque, C.M.; Ribeiro, P. Active vibration control of piezoelectric smart beams with radial basis function generated finite difference collocation method. *Intell. Mater. Syst. Struct.* **2018**, *29*, 2728–2743. [CrossRef]
11. Zurawski, M.; Zalewski, R. Damping of Beam Vibrations Using Tuned Particles Impact Damper. *Appl. Sci.* **2020**, *3*, 6334. [CrossRef]
12. Zurawski, M.; Chilinski, B.; Zalewski, R. Concept of an adaptive-tuned particles impact damper. *J. Theor. Appl. Mech.* **2020**, *58*, 3. [CrossRef]
13. Żurawski, M.; Chiliński, B.; Zalewski, R. A novel method for changing the dynamics of slender elements using sponge particles structures. *Materials* **2020**, *13*, 4874. [CrossRef]
14. Zurawski, M.; Zalewski, R. Adaptive method for tuning dynamic properties of beams by means of particles redistribution. *Acta Phys. Pol. A* **2020**, *138*, 240–244 [CrossRef]
15. Holnicki-Szulc, J.; Pawlowski, P.; Wiklo, M. High-performance impact absorbing materials the concept, design tools and applications. *Smart Mater. Struct.* **2003**, *12*, 461. [CrossRef]
16. Graczykowski, C.; Faraj, R. Development of control systems for fluid-based adaptive impact absorbers. *Mech. Syst. Signal Process.* **2019**, *122*, 622–641. [CrossRef]
17. Szmidt, T.; Zalewski, R. Inertially excited beam vibrations damped by Vacuum Packed Particles. *Smart Mater. Struct.* **2014**, *23*, 105026. [CrossRef]
18. Szmidt, T.; Zalewski, R. Application of Special Granular Structures for semi-active damping of lateral beam vibrations. *Eng. Struct.* **2014**, *65*, 13–20. [CrossRef]
19. Rodak, D.; Żurawski, M.; Gmitrzuk, M.; Starczewski, L. Possibilities of Vacuum Packed Particles application in blast mitigation seat in military armored vehicles. *Bull. Pol. Acad. Sci. Tech. Sci.* **2021**, *15*, e138238. [CrossRef]
20. Zalewski, R. Constitutive model for special granular structures. *Int. J. Nonlinear Mech* **2010**, *45*, 279–285. j.ijnonlinmec.2009.11.011. [CrossRef]
21. Zalewski, R.; Chodkiewicz, P.; Shillor, M. Vibrations of a mass-spring system using a granular-material damper. *Appl. Math. Model.* **2016**, *13*, 8033–8047. [CrossRef]
22. Malone, K.F.; Xu, B.H. Determination of contact parameters for discrete element method simulations of granular systems. *Particuology* **2008**, *8*, 521–528. [CrossRef]
23. Renzo, A.D.; Maio, F.P.D. Comparison of contact-force models for the simulation of collisions in DEM-based granular flow codes. *Chem. Eng. Sci.* **2004**, *20*, 525–541. [CrossRef]
24. Thornton, C.; Cummins, S.J.; Cleary, P.W. An investigation of the comparative behaviour of alternative contact force models during inelastic collisions. *Powder Technol.* **2013**, *233*, 30–46. [CrossRef]
25. Kruggel-Emden, H.; Simsek, E.; Rickelt, S.; Wirtz, S.; Scherer, V. Review and extension of normal force models for the Discrete Element Method. *Powder Technol.* **2007**, *7*, 157–173. [CrossRef]

26. Chen, Y.; Feng, J.; Lv, H.; Sun, Q. Symmetry representations and elastic redundancy for members of tensegrity structures. *Compos. Struct.* **2018**, *203*, 672–680. [CrossRef]
27. Jiaqiang, L.; Chen, Y.; Feng, X.; Feng, J.; Sareh, P. Computational Modeling and Energy Absorption Behavior of Thin-Walled Tubes with the Kresling Origami Pattern. *J. Int. Assoc. Shell Spat. Struct.* **2021**, *62*, 71–81. [CrossRef]
28. Zhang, P.; Fan, W.; Chen, Y.; Feng, J.; Sareh, P. Structural symmetry recognition in planar structures using Convolutional Neural Networks. *Eng. Struct.* **2022**, *260*, 114227. [CrossRef]
29. Zingoni, A. Group-theoretic insights on the vibration of symmetric structures in engineering. *Phil.Trans. R. Soc. A* **2014**, *372*, 20120037. [CrossRef]
30. Zingoni, A. On the symmetries and vibration modes of layered space grids. *Eng. Struct.* **2005**, *27*, 629–638. [CrossRef]
31. Thongchom, C.; Refahati, N.; Saffari, P.R.; Saffari, P.R.; Niyaraki, M.N.; Sirimontree, S.; Keawsawasvong, S. An Experimental Study on the Effect of Nanomaterials and Fibers on the Mechanical Properties of Polymer Composites. *Buildings* **2022**, *12*, 7. [CrossRef]
32. Zalewski, R.; Żurawski, M. Tłumik Drgań o Zmiennych włAściwościach Dyssypacyjnych z Rdzeniem Granulowanym. WUP 02-08-2021, 20 April 2021.

Article

# Addition of Two Substantial Side-Branch Silencers to the Interference Silencer by Incorporating a Zero-Mass Metamaterial

Shuichi Sakamoto [1,*], Juung Shin [2], Shota Abe [2] and Kentaro Toda [2]

1 Department of Engineering, Niigata University, Ikarashi 2-no-cho 8050, Nishi-ku, Niigata 950-2181, Japan
2 Graduate School of Science and Technology, Niigata University, Ikarashi 2-no-cho 8050, Nishi-ku, Niigata 950-2181, Japan; t16m034a@gmail.com (J.S.); t16m902a@gmail.com (S.A.); f21b105g@gmail.com (K.T.)
* Correspondence: sakamoto@eng.niigata-u.ac.jp; Tel.: +81-25-262-7003

**Abstract:** Zero-mass metamaterials comprise an orifice and a thin film. The resonance between the film and the air mass of the orifice hole is caused by sound waves, which significantly decreases the transmission loss at a specific frequency. The study novelly incorporates acoustic metamaterials in the delay tube of an interference silencer. In this case, it is determined that an interference silencer and a "side-branch silencer with two different branch pipe lengths" can be realized in a single silencer. At certain frequencies, the acoustic mass of the acoustic metamaterial approaches zero, which results in an interference silencer with the full length of the delay tube applied. At other frequencies, the acoustic metamaterial acts as a rigid wall with high transmission loss, thereby reflecting sound waves at the zero-mass metamaterial location. In this case, it is a side-branch silencer with two different tube lengths, corresponding to the tube lengths from the entrance and exit of the delay tube to the zero-mass metamaterial, respectively. The incorporation of zero-mass metamaterial into an interference-type silencer can introduce the silencing effect of a side-branch silencer with two different branch tube lengths without increasing the volume of the interference-type silencer. Theoretical values were obtained using the transfer matrix. Consequently, the theoretical and experimental values were close, enabling us to predict the transmission loss of the proposed silencer.

**Keywords:** zero-mass metamaterial; transmission loss; interference silencer; side-branch silencer

**Citation:** Sakamoto, S.; Shin, J.; Abe, S.; Toda, K. Addition of Two Substantial Side-Branch Silencers to the Interference Silencer by Incorporating a Zero-Mass Metamaterial. *Materials* **2022**, *15*, 5140. https://doi.org/10.3390/ma15155140

Academic Editor: Martin Vašina

Received: 22 June 2022
Accepted: 22 July 2022
Published: 24 July 2022

**Publisher's Note:** MDPI stays neutral with regard to jurisdictional claims in published maps and institutional affiliations.

**Copyright:** © 2022 by the authors. Licensee MDPI, Basel, Switzerland. This article is an open access article distributed under the terms and conditions of the Creative Commons Attribution (CC BY) license (https:// creativecommons.org/licenses/by/ 4.0/).

## 1. Introduction

In an interference silencer, the peak frequency at which the sound reduction effect can be obtained is uniquely determined by the delay tube length. Thus, the use of multiple delay tubes, depending on the required sound reduction frequency, increases the silencer volume. Therefore, we focused on zero-mass metamaterials [1], in which the acoustic mass approaches zero at a specific frequency.

Acoustic metamaterials are new artificial materials that have been developed in the last few decades. In addition, theoretical analysis development [2] has shown the potential for further new performance and various applications in practical engineering fields [3,4]. For example, it can provide a noise-free environment in certain bandwidths [5] or allow near-zero density operation by connecting tubular waveguides with ultranarrow tubes [6]. Zero-mass metamaterials comprise an orifice and a thin film. The resonance between the film and the air mass of the orifice hole is caused by sound waves, which significantly decreases the transmission loss at a specific frequency [7]. Recently, more complex designs of acoustic metamaterials with membranes have been reported to improve their acoustic performance. For example, holes can be perforated in the membrane [8] to introduce the Helmholtz resonator effect, or constrained structures [9] can be used to suppress certain eigenmodes of the membrane. Moreover, active control can be introduced to tune the acoustic performance [10].

The study novelly incorporates acoustic metamaterials in the delay tube of an interference silencer. In this case, it is determined that an interference silencer and a "side-branch silencer with two different branch pipe lengths" can be realized in a single silencer.

Herein, an overview of the interference silencer with built-in acoustic metamaterials is proposed. At certain frequencies, the acoustic mass of the acoustic metamaterial approaches zero, which results in an interference silencer with the full length of the delay tube applied. At other frequencies, the acoustic metamaterial acts as a rigid wall with high transmission loss, thereby reflecting sound waves at the zero-mass metamaterial location. In this case, it is a side-branch silencer with two different tube lengths, corresponding to the tube lengths from the entrance and exit of the delay tube to the zero-mass metamaterial, respectively.

In the theoretical analysis, the transfer matrix method [11,12] was employed for the interference-type silencer ducts. For the zero-mass metamaterial, the acoustic impedance was calculated using the Equation of motion for a spring-mass (damper) system in which the thin film and air in the orifice hole have the same mass and behave similarly. The acoustic impedance was integrated into the transfer matrix of the pipeline system of the silencer. The transmission loss was calculated for the transfer matrix of the silencer and compared with the experimental results to verify the sound reduction characteristics.

## 2. Measuring Equipment and Samples

### 2.1. Transmission Loss Measurement

The transmission loss was measured using a Brüel & Kjær Type 4206T 4-microphone impedance-measuring tube (Nærum, Denmark) and an Ono Sokki 4-channel-fast Fourier transform (FFT) analyzer DS-3000 (Yokohama, Japan). A schematic of the measurement apparatus is presented in Figure 1. The silencer is placed in a predetermined position, and two microphones are mounted each in front of and behind the sample. A signal generator and amplifier generate sound waves with a reference signal (linear multisine waves). Furthermore, the microphones detect the sound pressure before and after transmission through the sample. In addition, a 4-channel FFT analyzer measures the cross-spectrum via a signal amplifier for the microphones. The transmission loss was calculated in accordance with the American Society for Testing and Materials (ASTM) E2611-09.

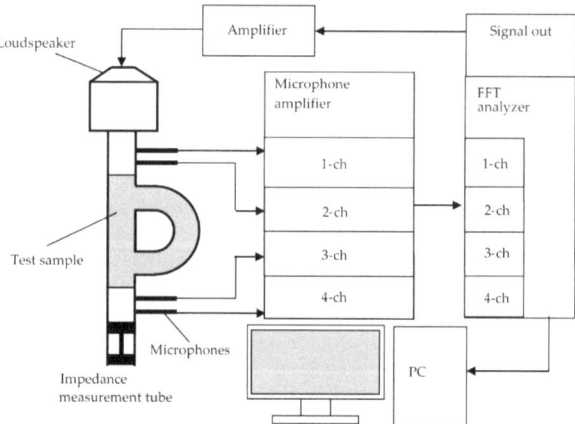

**Figure 1.** Schematic of the impedance measurement tube and measurement equipment.

### 2.2. Zero-Mass Metamaterial and Delay Tube of Interferometric Silencer

Figure 2a presents the delay tube section of the interference silencer with zero-mass metamaterials used herein. Figure 2b presents the dimensions of the delay tube. Tables 1 and 2 present the specifications of the thin films and orifice plates used in the experiments, respectively.

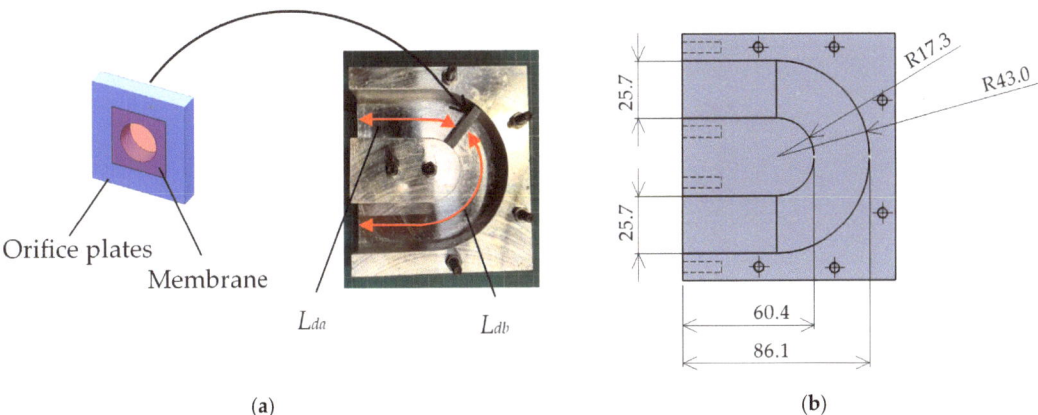

**Figure 2.** (a) Delay tube section of the interference silencer with incorporated zero-mass metamaterial; (b) dimensions of the delay tube.

**Table 1.** Specifications of thin film material.

| Material of Membrane | Nominal Density (kg/m$^3$) | Measured Membrane Thickness (μm) |
|---|---|---|
| Low-density polyethylene | 920 | 11 |

**Table 2.** Specifications of orifice plate.

| Material of Orifice | Thickness of Plate (mm) | Diameter of Hole (mm) |
|---|---|---|
| Photocurable resin | 5 | 10 |

Herein, a 25.7 mm$^2$ orifice plate fabricated by a photo fabrication type 3D printer, Formlabs Form2 (Somerville, MA, USA), was used, and photocuring resins were used. The membrane material was attached and fixed to the orifice plate coated with Vaseline without tension.

The delay tube of the interference silencer used herein was fabricated from an A5052 aluminum alloy. Table 3 shows the chemical composition of the delay tube of the interference silencer used in the experiments. The inner cross-section of the delay tube was a square of 25.7 mm side, and the length of the centerline of the delay tube was 180.9 mm. The delay tube length ($L_d$) was divided into the upstream ($L_{da}$) and downstream ($L_{db}$) lengths, respectively, at the position where the zero-mass metamaterial was installed. The main tube length ($L_m$) was 60.4 mm.

**Table 3.** Chemical composition of A5052 aluminum alloy.

| Element | Ratio (%) | Element | Ratio (%) |
|---|---|---|---|
| Aluminum | Balance | Copper | 0.10 max |
| Magnesium | 2.2–2.8 | Manganese | 0.10 max |
| Chromium | 0.15–0.35 | Zinc | 0.10 max |
| Silicon | 0.25 max | Others, each | 0.05 max |
| Iron | 0.40 max | Others, total | 0.15 max |

The flow of the subsequent experimental and theoretical analysis is shown in the flowchart in Figure 3. For comparison, the sound reduction characteristics of a conventional interference silencer are described in Section 3.

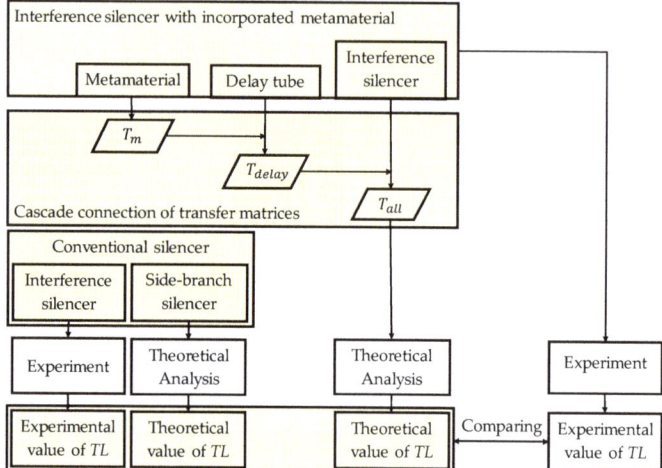

**Figure 3.** Flowchart of experimental and theoretical analysis process.

## 3. Theoretical Analysis

### 3.1. Analysis of Zero-Mass Metamaterial

The zero-mass metamaterial comprises a thin film attached to an orifice plate. Figure 4 presents an analytical model of the zero-mass metamaterial. Herein, the calculation is performed as a one-degree-of-freedom vibration system in which the film and the air in the orifice hole vibrate identically. $M_{mem}$ denotes the mass of the membrane attached to the orifice hole; $M_{air}$, the mass of the air in the orifice hole; $k_{mem}$, the spring constant of the membrane; and $b_{eff}$, the damping coefficient of the zero-mass metamaterial. Here, the spring constant of the thin film $k_{mem}$ was measured using the method of Lee et al. [13].

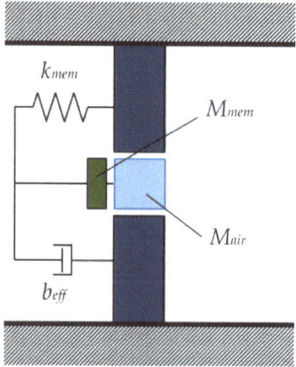

**Figure 4.** Analytical model for zero-mass metamaterials.

In addition, $t$ is defined as the orifice plate thickness; $r$, the orifice hole radius; and $\lambda$, the wavelength of the incident sound wave. Since $r \ll \lambda$, the zero-mass metamaterial is assumed to be a concentrated element model.

The names of the variables near the zero-mass metamaterial are presented in Figure 5. Notably, $p_i$ is the incident sound pressure; $p_r$, the reflected sound pressure; $p_t$, the transmitted sound pressure; $u_i$, the incident particle velocity; $u_r$, the reflected particle velocity; and $u_t$, the transmitted particle velocity. Furthermore, $S$ is the cross-sectional area of the acoustic tube. Note that $p_1$ and $u_2$ are given by Equations (1)–(4) expressing $p_2$ and $u_2$, respectively.

$$p_1 = p_i + p_r \tag{1}$$

$$u_2 = u_i + u_r \tag{2}$$

$$p_2 = p_t \tag{3}$$

$$u_2 = u_t \tag{4}$$

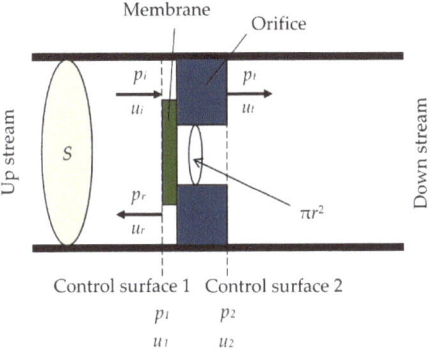

**Figure 5.** Variable name near zero-mass metamaterial.

If $t$ and $r$ are sufficiently smaller than $\lambda$, the membrane and air at the orifice hole are considered to oscillate in unison as described above [1,13], and the particle velocities before and after the zero-mass metamaterial are equal. This scenario is expressed in Equation (5).

$$u_1 = u_2 \tag{5}$$

Here, using the characteristic impedance ($Z_0$) of air, the sound pressure can be expressed as Equations (6)–(8), respectively.

$$p_i = Z_0 u_i \tag{6}$$

$$p_r = -Z_0 u_r \tag{7}$$

$$p_t = Z_0 u_t \tag{8}$$

The relationship between the sound pressure, particle velocity, and transfer matrix is expressed by Equation (9).

$$\begin{bmatrix} p_1 \\ S u_1 \end{bmatrix} = \begin{bmatrix} T_{11} & T_{12} \\ T_{21} & T_{22} \end{bmatrix} \begin{bmatrix} p_2 \\ S u_2 \end{bmatrix} \tag{9}$$

From a previous study on zero-mass metamaterials [1], the Equation of motion for this structure is given by Equation (10), where $\xi$ is the displacement of the membrane and air at the orifice hole, and $\xi(t) = \xi_0 exp(-i\omega t)$.

$$(p_1 - p_2)\pi r^2 = M_{eff}\ddot{\xi} + b_{eff}\dot{\xi} \tag{10}$$

Here, $M_{eff}$ denotes the effective mass defined by Equation (11), and $b_{eff}$ is the effective damping coefficient of the zero-mass metamaterial. To simplify the calculations, $b_{eff}$ is set to zero because its effect on the theoretical value of transmission loss is negligible.

$$M_{eff} = M_{air} + M_{mem} - \frac{k_{mem}}{\omega^2} \tag{11}$$

Reorganizing Equation (10) using Equations (2) and (4), Equation (12) is obtained [1].

$$p_i + p_r - p_t = \left(\frac{1}{\pi r^2}\right)\left(-i\omega M_{eff} + b_{eff}\right)\ddot{\xi} \tag{12}$$

Reorganizing the above Equations using Equations (6) and (7), Equation (13) is obtained.

$$2p_i - 2p_t = p_t\left(\frac{S}{Z_0 \pi^2 r^4}\right)\left(b_{eff} - i\omega M_{eff}\right) \tag{13}$$

Substituting Equation (13) into Equation (9) using Equations (1) and (4), Equation (14) is obtained.

$$\begin{bmatrix} p_1 \\ Su_1 \end{bmatrix} = \begin{bmatrix} 1 & \frac{S}{\pi^2 r^4}\left(b_{eff} - i\omega M_{eff}\right) \\ 0 & 1 \end{bmatrix} \begin{bmatrix} p_2 \\ Su_2 \end{bmatrix} \tag{14}$$

Thus, the transfer matrix $T_m$ of the zero-mass metamaterial is given by Equation (15).

$$T_m = \begin{bmatrix} 1 & \frac{S}{\pi^2 r^4}\left(b_{eff} - i\omega M_{eff}\right) \\ 0 & 1 \end{bmatrix} \tag{15}$$

### 3.2. Theoretical Analysis of Delay Tube with Built-In Zero-Mass Metamaterials

Figure 6 presents the acoustic elements of the delay tube section and equivalent circuit corresponding to the acoustic system. In Figure 6, the tube elements are upstream and downstream of the zero-mass metamaterial and are connected by a cascade connection. The cascade-connected transfer matrix $T$ can be expressed as in Equation (16) using $T_m$ in Equation (15).

$$T_{delay} = T_f T_m T_b = \begin{bmatrix} A & B \\ C & D \end{bmatrix} \tag{16}$$

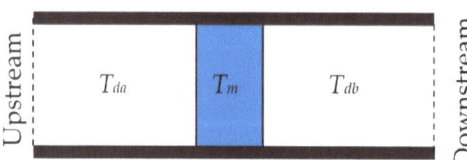
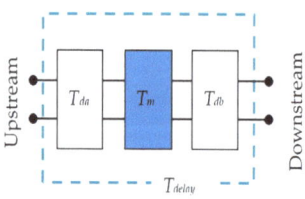

(a) (b)

**Figure 6.** Acoustic elements of the delay tube: (a) acoustic elements of the delay tube and (b) equivalent circuit of the delay tube.

Here, the transfer matrix for the upstream tube element of the metamaterial is presented in Equation (17), using air density $\rho_0$ and speed of sound $c_0$ in the air.

$$T_f = \begin{bmatrix} \cos kl_f & i\frac{\rho_0 c_0}{S}\sin kl_f \\ i\frac{S}{\rho_0 c_0}\sin kl_f & \cos kl_f \end{bmatrix} \tag{17}$$

Similarly, the transfer matrix for the downstream tube element is presented in Equation (18).

$$T_b = \begin{bmatrix} \cos kl_b & i\frac{\rho_0 c_0}{S} \sin kl_b \\ i\frac{S}{\rho_0 c_0} \sin kl_b & \cos kl_b \end{bmatrix} \quad (18)$$

### 3.3. Transfer Matrix for Entire Interference Silencer

Here, $k$ denotes the wavenumber; $l_f$, the length of the upstream delay tube; $l_b$, the length of the downstream delay tube; and $S$, the cross-sectional area of the main and delay tubes. If the transfer matrices $T_{delay}$ and $T_2$ are expressed by Equation (19), $T$ after a parallel connection is expressed by Equation (20).

$$T_{delay} = \begin{bmatrix} A_1 & B_1 \\ C_1 & D_1 \end{bmatrix}, T_2 = \begin{bmatrix} A_2 & B_2 \\ C_2 & D_2 \end{bmatrix} \quad (19)$$

$$T = \begin{bmatrix} \frac{A_1 B_2 + A_2 B_1}{B_1 + B_2} & \frac{B_1 B_2}{B_1 + B_2} \\ C_1 + C_2 + \frac{(A_2 - A_1)(D_1 - D_2)}{B_1 + B_2} & \frac{D_1 B_2 + D_2 B_1}{B_1 + B_2} \end{bmatrix} \quad (20)$$

The cross-sectional views of the entire interference silencer and the equivalent circuit corresponding to its acoustic system are presented in Figure 7. In addition, $T_1$-, $T$-, and $T_1$-connected cascades are $T_{all}$, and are expressed by Equation (21).

$$T_{all} = T_1 T T_1 = \begin{bmatrix} A_{all} & B_{all} \\ C_{all} & D_{all} \end{bmatrix} \quad (21)$$

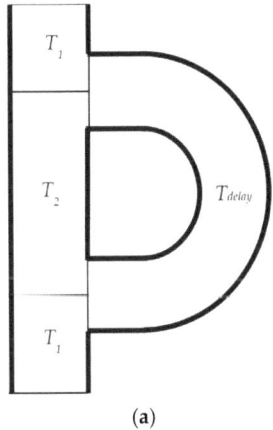
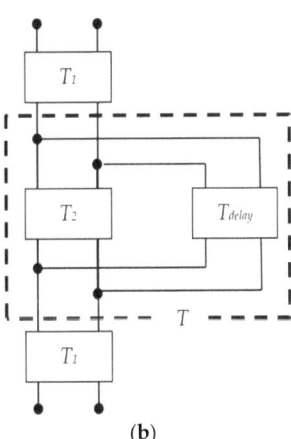

**Figure 7.** Acoustic elements and equivalent circuit of an interference silencer: (**a**) division into acoustic elements and (**b**) equivalent circuit.

Using the four-terminal constants $A_{all}$–$D_{all}$ in Equation (21), the transmission loss (TL) of the interference silencer can be expressed as in Equation (22).

$$TL = 10 \log_{10} \frac{\left| A_{all} + \frac{S}{\rho_0 c_0} B_{all} + \frac{\rho_0 c_0}{S} C_{all} + D_{all} \right|^2}{4} \quad (22)$$

The delay tube is connected perpendicular to the side of the main tube. An opening end correction must be added to the tube end of the delay tube with respect to the geometric tube length. Many studies have been conducted on aperture end correction [14,15]. Because these tubes are equal in the cross-sectional area, a value of 0.8 was used for the aperture end correction value [15]. Therefore, the aperture end correction length was 0.8 times the equivalent circular radius of the 25.7 mm² aperture.

### 3.4. Peak Frequency of Sound Attenuation for Ordinary Interference-Type Silencers and Side-Branch Silencers

For comparison with the silencer proposed herein, the sound reduction characteristics of the conventional interference and side-branch silencers are described.

The phase change amount in the main and delay tubes of the interference-type silencer can be expressed as Equations (23) and (24), respectively.

$$\theta_m = \frac{2\pi f L_m}{c_0} \tag{23}$$

$$\theta_d = \frac{2\pi f L_d}{c_0} \tag{24}$$

When the difference between these two-phase changes satisfies Equation (25), the sound waves propagating in the main and delay tubes merge in the opposite phase at the confluence and are canceled.

$$|\theta_m - \theta_d| = (2n-1)\pi \ (n = 1, 2, 3 \ldots) \tag{25}$$

The above Equation can be rearranged so that the peak frequency of sound attenuation of an interference-type silencer can be expressed by Equation (26).

$$f_{peak} = \left| \frac{(2n-1)c_0}{2(L_m - L_d)} \right| \ (n = 1, 2, 3 \ldots) \tag{26}$$

The peak frequency of the side-branch silencer can be expressed as in Equation (27), where $L_{side}$ denotes the length of the branch tube of the side-branch silencer.

$$f_{side} = \left| \frac{(2n-1)c_0}{4L_{side}} \right| \ (n = 1, 2, 3 \ldots) \tag{27}$$

In an interference-type silencer, the peak attenuation culminates at the frequency $f_{peak}$ satisfying Equations (28)–(31) and its odd-numbered multiples [16]. Therefore, it is sufficient to determine the delay tube length of the interference-type silencer so that the "dip frequency of the TL of the zero-mass metamaterial in the interference-type silencer" and the "peak frequency of the sound reduction of the interference-type silencer" coincide.

The dip frequency of the TL of the zero-mass metamaterial used herein is 1425 Hz; thus, the acoustic mass of the acoustic metamaterial approaches zero at ~1425 Hz in the proposed silencer. Therefore, it is necessary to select the total delay tube length such that the peak frequency of sound reduction is ~1425 Hz.

$$L_d - L_m = \frac{\lambda}{2} \tag{28}$$

$$L_d + L_m = \lambda \tag{29}$$

$$L_d = \frac{3c}{4f} \tag{30}$$

$$L_m = \frac{c}{4f} \tag{31}$$

The peak sound attenuation frequencies of the side-branch silencers corresponding to the two branch tube lengths divided by the zero-mass metamaterial are presented in Table 4.

Table 4. Peak frequency of the side-branch silencer sound attenuation at each branch tube length.

| | | Length of Side-Branch Tube $L_{side}$ (mm) | | | |
|---|---|---|---|---|---|
| | | 50.0 | 60.0 | 125.9 | 115.9 |
| Peak Frequency $f_{side}$ (Hz) | $n = 1$ | 1701.5 | 1418.0 | 675.7 | 734.0 |
| | $n = 2$ | 5104.5 | 4253.8 | 2027.2 | 2202.1 |

## 4. Experimental and Theoretical Values for Transmission Loss

### 4.1. Transmission Loss of Single Zero-Mass Metamaterial

The zero-mass metamaterial comprises an orifice and a thin film. The orifice plate has a hole diameter $d$ of 10 mm and a thickness $t$ of 5 mm, and the film is made of low-density polyethylene with a thickness $t_m$ of 11 μm.

Figure 8 presents the experimental and theoretical values for the zero-mass metamaterial. Evidently, sound waves penetrate the zero-mass metamaterial at a dip frequency of 1425 Hz.

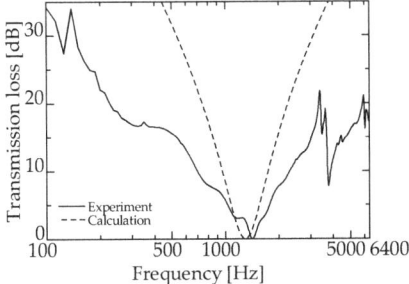

Figure 8. Transmission loss of single zero-mass metamaterial (experimental and theoretical).

In addition, sound waves are considered to be reflected in the frequency band where the *TL* is significant.

### 4.2. Comparison of Experimental and Theoretical Values

A comparison of the theoretical and experimental values based on the theoretical analysis described in the previous section is shown.

Figures 9 and 10 present the experimental and theoretical values for zero-mass metamaterials installed at $L_{da}$ = 50 and 60 mm, respectively.

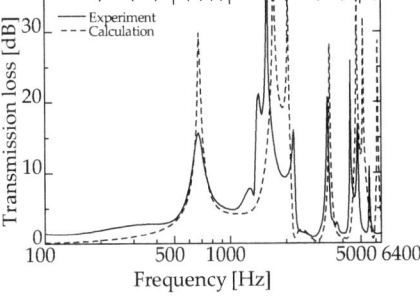

Figure 9. Experimental and theoretical values of the interference silencer with incorporated zero-mass metamaterial (zero-mass metamaterials installed at $L_{da}$ = 50 mm).

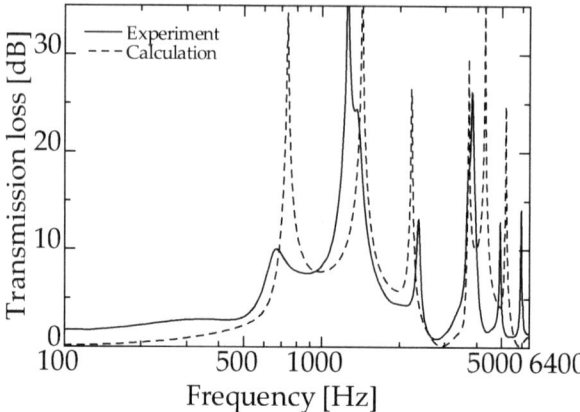

**Figure 10.** Experimental and theoretical values of the interference silencer with incorporated zero-mass metamaterial (zero-mass metamaterials installed at $L_{da}$ = 60 mm).

As can be seen from Figure 8, the dip frequency of the *TL* of the integrated zero-mass metamaterial is 1425 Hz, indicating that for the proposed silencer, the acoustic mass of the acoustic metamaterial approaches zero at a specific frequency, i.e., around 1425 Hz. Therefore, the total length of the delay tube of the interference silencer was set so that the peak frequency of sound reduction is ~1425 Hz.

Figures 9 and 10 present the experimental and theoretical sound attenuation peaks that are roughly consistent with the designed attenuation frequency of 1425 Hz for an interference silencer with $L_d$ = 180.9 mm (Figures 9 and 10). Therefore, it can be assumed that the acoustic mass of the zero-mass metamaterial approaches zero as expected.

The difference in the peak sound attenuation magnitude and frequency between the experimental and theoretical values is probably because the sound waves incident on the zero-mass metamaterial is incompletely transmitted at the dip frequency of the *TL* and are incompletely reflected at the frequency of the high *TL*.

Figures 11 and 12 present the experimental values of an ordinary interference silencer and the experimental values of a zero-mass metamaterial at $L_{da}$ = 50 and 60 mm, respectively (redrawn from Figures 9 and 10).

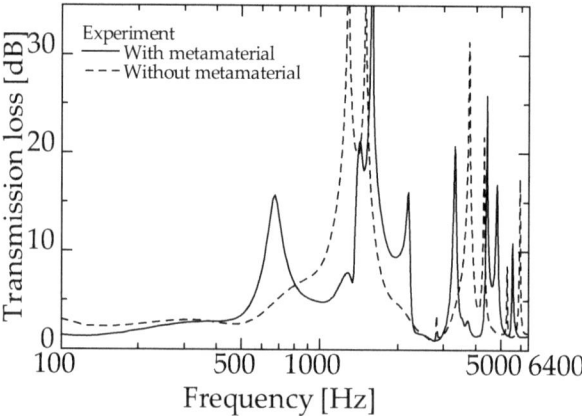

**Figure 11.** Comparison of the experimental values with and without zero-mass metamaterial (zero-mass metamaterials at $L_{da}$ = 50 mm).

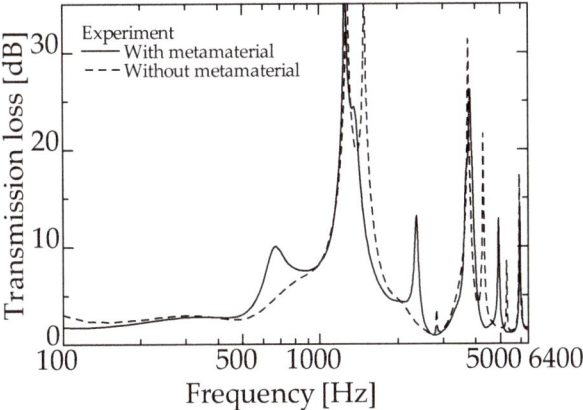

**Figure 12.** Comparison of the experimental values with and without zero-mass metamaterial (zero-mass metamaterials at $L_{da}$ = 60 mm).

For both experimental values in Figure 11, a sound reduction peak is observed near 1425 Hz. This frequency band is the dip frequency band of the zero-mass metamaterial (Figure 7). The same is true for Figure 12.

These facts suggest that the proposed silencer worked similarly to an ordinary interference silencer. Thus, the delay tube with zero-mass metamaterials acted like the delay tube of an ordinary interference-type silencer. The above results confirm that the acoustic mass of the zero-mass metamaterial is near zero around the dip frequency presented in Figure 8.

Figure 13 shows the theoretical values for two different conventional side-branch silencers with $L_d$ = 50 and 125.9 mm, respectively, and the experimental values for a zero-mass metamaterial installed at $L_{da}$ = 50 mm (redrawn from Figure 13).

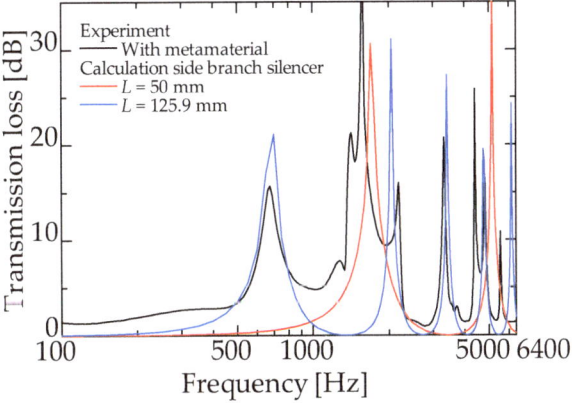

**Figure 13.** Comparison with simple side-branch silencer (experimental: interference silencer with zero-mass metamaterial; theoretical: side-branch silencer with 50 mm and 125.9 mm branch tube length).

In Figure 13, the experimental values of the proposed silencer exhibit sound attenuation peaks at 675, 1700, and 2000 Hz, in addition to the attenuation peak of the interference silencer at 1425 Hz. The sound reduction peaks at 675 and 2000 Hz correspond to the first-order, third-order, and subsequent sound reduction peak frequencies (blue line in Figure 13) of the side-branch silencer with a 125.9 mm branch tube length. The experimental peak of the proposed silencer at 1700 Hz corresponds to the 1700 Hz peak of the side-branch silencer with a 50 mm branch tube length (red line in Figure 13).

Figure 14 presents the theoretical values for two types of conventional side-branch silencers, with $L_d$ = 60 and 115.9 mm, respectively, and the experimental values for a zero-mass metamaterial installed at $L_{da}$ = 60 mm (redrawn from Figure 14). Exactly as in Figure 13, the experimental values of the proposed silencer exhibit sound reduction peaks at 735, 1420, and 2200 Hz, in addition to the sound reduction peak near 1425 Hz of the interference silencer. Similarly, the 735 and 2200 Hz sound reduction peaks correspond to the respective peak frequencies of the side-branch silencer with a 115.9 mm branch tube length (blue line in Figure 14). The 1420 Hz peak of the proposed silencer corresponds to the 1420 Hz peak of the side-branch silencer with a 60 mm branch tube length (red line in Figure 14).

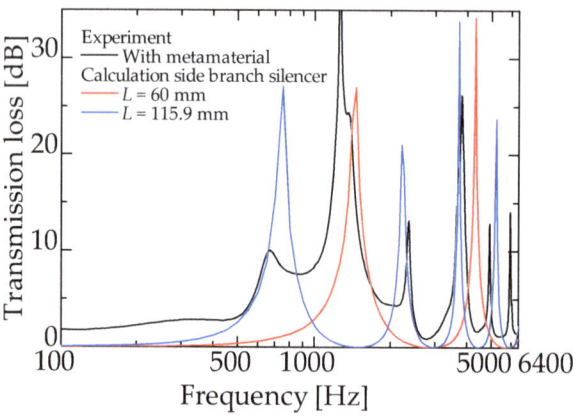

**Figure 14.** Comparison with simple side-branch silencer (experimental: interference silencer with zero-mass metamaterial; theoretical: side-branch silencer with 60 mm and 115.9 mm branch tube length).

The above results indicate that the zero-mass metamaterial installed in the delay tube acts as a rigid wall with high *TL* at frequencies other than the specified frequency. Furthermore, the delay tube is divided into two closed-end tubes with two different tube lengths, acting as two side-branch silencers. These results confirm that the zero-mass metamaterial has an acoustically significant mass in the bands other than the dip frequency of the zero-mass metamaterial.

## 5. Conclusions

Herein, the performance of a silencer with zero-mass metamaterial built into the delay tube of an interference-type silencer was evaluated via theoretical analysis of TL and experiments, and the following conclusions were obtained:

1. The incorporation of zero-mass metamaterial into an interference-type silencer can introduce the silencing effect of a side-branch silencer with two different branch tube lengths without increasing the volume of the interference-type silencer. Consequently, the total volume of the interference and two side-branch silencers could be reduced by half.
2. The incorporation of zero-mass metamaterials in the interference silencer increased the sound reduction peaks.
3. Theoretical analysis of the interference-type silencer with built-in zero-mass metamaterial was conducted using the Equations of motion, and theoretical values were obtained using the transfer matrix. Consequently, the theoretical and experimental values were close, enabling us to predict the *TL* of the proposed silencer.

**Author Contributions:** Conceptualization, S.S.; Data curation, J.S. and K.T.; Formal analysis, S.A.; Project administration, S.S.; Software, S.A.; Supervision, S.S. All authors have read and agreed to the published version of the manuscript.

**Funding:** This research received no external funding.

**Institutional Review Board Statement:** Not applicable.

**Conflicts of Interest:** The authors declare no conflict of interest.

## Nomenclature

| | |
|---|---|
| ASTM | American Society for Testing and Materials |
| $A$–$D$, $A_{all}$–$D_{all}$ | Four-terminal constants |
| $b_{eff}$ | Damping coefficient |
| $c_0$ | Speed of sound in the air (m/s) |
| $d$ | Orifice plate diameter (mm) |
| $f$, $f_{peak}$, $f_{side}$ | Peak frequency (Hz) |
| FFT | Fast Fourier transform |
| $I$ | Imaginary unit |
| $k$ | Wavenumber |
| $k_{mem}$ | Spring constant of the thin film (N/m) |
| $L_d$, $L_{da}$, $L_{db}$, $L_m$, $L_{side}$ | Length of the tube (mm) |
| $M_{air}$ | Mass of the air in the orifice hole |
| $M_{eff}$ | Effective mass |
| $M_{mem}$ | Mass of the membrane attached to the orifice hole |
| $p_1$, $p_2$, $p_i$, $p_r$, $p_t$ | Sound pressure (Pa) |
| $r$ | Orifice hole radius (mm) |
| $S$ | Cross-sectional area of the acoustic tube (mm$^2$) |
| $t$ | Orifice plate thickness (mm) |
| $t_m$ | Film thickness (mm) |
| $T$, $T_1$, $T_2$, $T_{11}$, $T_{12}$, $T_{21}$, $T_{22}$, $T_{all}$, $T_b$, $T_{delay}$, $T_f$, $T_m$ | Transfer matrix |
| TL | Transmission loss (dB) |
| $u_1$, $u_2$, $u_i$, $u_r$, $u_t$ | Particle velocity (m/s) |

## References

1. Park, J.J.; Lee, K.J.B.; Wright, O.B.; Jung, M.K.; Lee, S.H. Giant Acoustic Concentration by Extraordinary Transmission in Zero-Mass Metamaterials. *Phys. Rev. Lett.* **2013**, *110*, 244302. [CrossRef] [PubMed]
2. Gric, T.; Eldlio, M.; Cada, M.; Pistora, J. Analytic solution to field distribution in two-dimensional inhomogeneous waveguides. *J. Electromagn. Waves Appl.* **2015**, *29*, 1068–1081. [CrossRef]
3. Liu, J.; Guo, H.; Wang, T. A Review of Acoustic Metamaterials and Phononic Crystals. *Crystals* **2020**, *10*, 305. [CrossRef]
4. Zangeneh-Nejad, F.; Fleury, R. Active times for acoustic metamaterials. *Rev. Physics.* **2019**, *4*, 100031. [CrossRef]
5. Yang, Z.; Mei, J.; Yang, M.; Chan, N.; Sheng, P. Membrane-type acoustic metamaterial with negative dynamic mass. *Phys. Rev. Lett.* **2008**, *101*, 204301. [CrossRef] [PubMed]
6. Fleury, R.; Alù, A. Extraordinary sound transmission through density-near-zero ultranarrow channel. *Phys. Rev. Lett.* **2013**, *111*, 055501. [CrossRef] [PubMed]
7. Cummer, S.A.; Christensen, J.; Alù, A. Controlling sound with acoustic metamaterials. *Nat. Rev. Mater.* **2016**, *1*, 16001. [CrossRef]
8. Langfeldt, F.; Kemsies, H.; Gleine, W.; von Estorff, O. Perforated membrane-type acoustic metamaterials. *Phys. Lett. A* **2017**, *381*, 1457–1462. [CrossRef]
9. Wang, X.; Zhao, H.; Luo, X.; Huang, Z. Membrane-constrained acoustic metamaterials for low frequency sound insulation. *Appl. Phys. Lett.* **2016**, *108*, 041905. [CrossRef]
10. Xiao, S.; Ma, G.; Li, Y.; Yang, Z.; Sheng, P. Active control of membrane-type acoustic metamaterial by electric field. *Appl. Phys. Lett.* **2015**, *106*, 091904. [CrossRef]
11. Sasao, H. Introduction to Acoustical Analysis Using Excel-Analysis of Acoustical Structure Characteristics-(4) Analysis of Duct Silencer Using Excel. *Air-Cond. Sanit. Eng.* **2007**, *81*, 51–58. (In Japanese)
12. Sasao, H. Introduction to Acoustic Analysis by Excel-Analysis of Acoustic Structure Characteristics-(2) Analysis of Sound Absorbing Structure by Excel. *Air-Cond. Sanit. Eng.* **2006**, *80*, 99–105. (In Japanese)
13. Lee, S.H.; Park, C.M.; Seo, Y.M.; Wang, Z.G.; Kim, C.K. Acoustic metamaterial with negative density. *Phys. Lett. A* **2009**, *373*, 4464–4469. [CrossRef]
14. Ingård, U. On the theory and design of acoustic resonators. *J. Acoust. Soc. Am.* **1953**, *25*, 1037–1061. [CrossRef]

15. Benade, H.A. Measured end corrections for woodwind toneholes. *J. Acoust. Soc. Am.* **1967**, *41*, 1609. [CrossRef]
16. Nakano, A. Noise control technology (3). *J. Environ. Conserv. Eng.* **1988**, *17*, 745–752. (In Japanese) [CrossRef]

Article

# Effect of Sheet Vibration on the Theoretical Analysis and Experimentation of Nonwoven Fabric Sheet with Back Air Space

Shuichi Sakamoto [1,*], Ryo Iizuka [2] and Takumi Nozawa [2]

1. Department of Engineering, Niigata University, Ikarashi 2-no-cho 8050, Nishi-ku, Niigata City 950-2181, Japan
2. Graduate School of Science and Technology, Niigata University, Ikarashi 2-no-cho 8050, Nishi-ku, Niigata City 950-2181, Japan; t16m007d@gmail.com (R.I.); t18a093a@gmail.com (T.N.)
* Correspondence: sakamoto@eng.niigata-u.ac.jp; Tel.: +81-25-262-7003

**Abstract:** The purpose of this study was to improve the accuracy of the theoretical analysis of sound absorption mechanisms when a back air space is used in nonwoven fabrics. In the case of a nonwoven sheet with a back air space, it can be shown that there is a difference between the experimental results and theoretical analysis results obtained using the Miki model when the area of the nonwoven sheet is large. Therefore, in this study, the accuracy of the theoretical values was improved using the plate vibration model in conjunction with the Miki model. The experimental results showed that when the vibration of the nonwoven sheet was suppressed, the sound absorption coefficient was higher than that of the vibration-prone nonwoven sheet alone. The sound absorption coefficient at the peak frequency was increased by >0.2, especially for 3501BD. Using the support frame, the sound absorption coefficient at the peak frequencies of 3A01A and 3701B was increased to 0.99. In the theoretical analysis of a large-area, vibration-prone nonwoven fabric, in which the vibration of the nonwoven fabric was taken into account, the theoretical values were in agreement with the experimental values, and the accuracy of the theoretical values was improved. Comparing the theoretical values for nonwoven fabrics without high ventilation resistance, the sound absorption coefficient was greater when vibration was not considered. Therefore, it was suggested that the vibration of the nonwoven fabric hinders sound absorption.

**Keywords:** nonwoven fabric sheet; back air space; vibration; absorption coefficient

**Citation:** Sakamoto, S.; Iizuka, R.; Nozawa, T. Effect of Sheet Vibration on the Theoretical Analysis and Experimentation of Nonwoven Fabric Sheet with Back Air Space. *Materials* **2022**, *15*, 3840. https://doi.org/10.3390/ma15113840

Academic Editor: Martin Vašina

Received: 24 March 2022
Accepted: 25 May 2022
Published: 27 May 2022

**Publisher's Note:** MDPI stays neutral with regard to jurisdictional claims in published maps and institutional affiliations.

**Copyright:** © 2022 by the authors. Licensee MDPI, Basel, Switzerland. This article is an open access article distributed under the terms and conditions of the Creative Commons Attribution (CC BY) license (https:// creativecommons.org/licenses/by/ 4.0/).

## 1. Introduction

The acoustic properties of nonwovens vary along with changes in the fiber diameter, density, ventilation resistance, etc. [1–3]. It is also known that thin nonwoven fabric sheets exhibit broad sound absorption curves corresponding to the thickness of the air space behind them. An accurate prediction of the acoustic properties of such sound-absorbing mechanisms is useful in noise engineering applications.

The acoustic properties of nanofiber nonwoven fabrics have recently been studied [4,5], and nonwoven fabrics have attracted attention as sound-absorbing materials. In addition, the acoustic properties of porous materials, such as nonwoven fibers, and sound-absorbing materials, such as permeable membranes, have been studied [6–8]. Furthermore, it was observed that sound wave vibration affects the acoustic properties of these thin membranes or plates [9–11].

The purpose of this study was to improve the accuracy of the theoretical analysis of sound absorption mechanisms when a back air space is used in nonwoven fabrics. In the case of a nonwoven sheet with a back air space, it can be shown that there is a difference between the experimental results and theoretical analysis results obtained using the Miki model [2] when the area of the nonwoven sheet is large. Therefore, the accuracy of the theoretical values was improved using the plate vibration model [12] in conjunction with the Miki model [2]. The aim was to keep the difference between the experimental and theoretical values of the sound absorption coefficient within 0.1.

In these experiments, the sound absorption coefficient was measured under the following conditions: (1) a sample with a nonwoven fabric sheet and a back air space and (2) a sample with a vibration-suppressed nonwoven fabric sheet and a back air space. A support frame with honeycomb-shaped openings was utilized to suppress the vibration of the nonwoven fabric sheet so as to experimentally reproduce a theoretical model that does not account for vibration.

Experiments were conducted using nonwoven fabrics with different ventilation resistance. Theoretical analysis of the presence or absence of vibration of the nonwoven fabrics was then performed for each of these experimental values.

## 2. Measurement Equipment and Samples

### 2.1. Measurement Equipment

Sound absorption coefficients were measured using a Brüel & Kjær Type 4206 two-microphone impedance measuring tube, as shown in the schematic diagram in Figure 1. A sample was attached to the impedance measurement tube, a sinusoidal signal was generated by the signal generator in the fast Fourier transform (FFT) analyzer, and sound waves radiated into the tube by means of a loudspeaker. The FFT analyzer then measured the transfer function between the sound pressure signals from the two microphones attached to the measurement tube. The measured transfer function was used to calculate the normal incident sound absorption coefficient in accordance with ISO 10534-2. The frequency range over which the measurement device can be used is 50–1600 Hz.

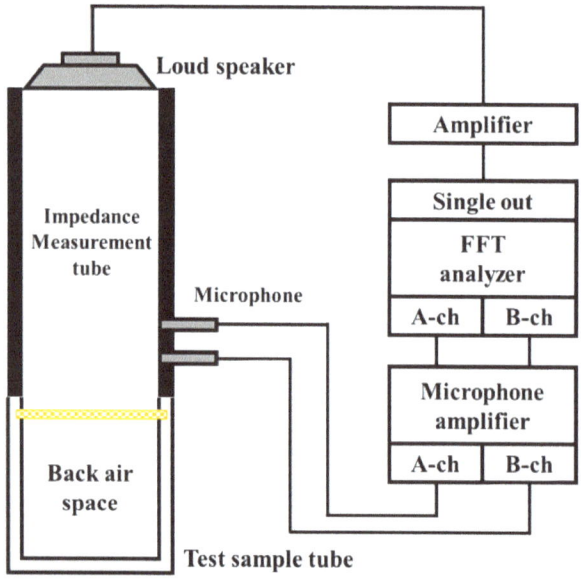

**Figure 1.** Schematic diagram of a two-microphone impedance tube for absorption coefficient measurements.

The ventilation resistance $R_n$ of nonwoven fabrics was measured using a Kato Tech KES-F8-AP1 permeability tester. This is an air permeability tester that injects a constant flow of air into a sample via a plunger and cylinder. A precise pressure gage was used to quantify the pressure loss caused by the sample at a constant flow rate of 4 cc/cm$^2$/s ($4 \times 10^{-2}$ m/s). The measurement result allowed the ventilation resistance $R_n$ (kPa s/m) to be calculated directly. Flow resistivity $\sigma_n$ was calculated by dividing the ventilation resistance $R_n$ by the thickness of the nonwoven fabric.

## 2.2. Measurement Samples

Figure 2 shows microphotographs of the six nonwoven fabric sheet samples used in this study. The nonwoven fabric sheet samples were cut into a circular shape with a diameter of 110 mm. The nonwoven fabrics used in the experiment were spun-bound fabrics, which were manufactured in thin sheets with fibers that were almost perpendicular to the thickness direction.

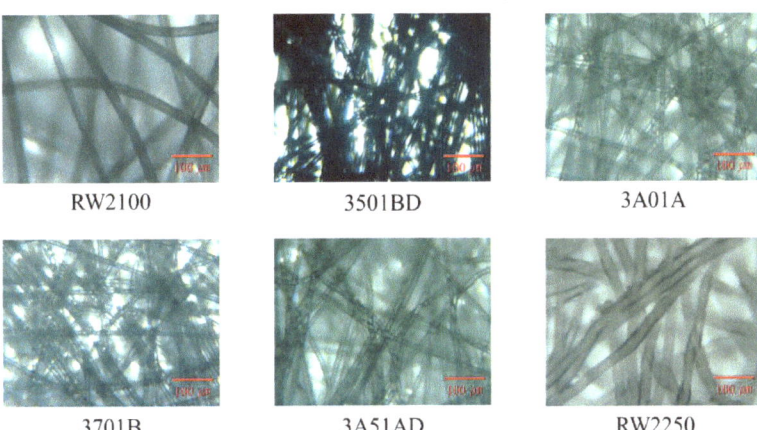

**Figure 2.** Microphotographs of nonwoven fabric sheets.

Table 1 shows the specifications of each nonwoven fabric with respect to ventilation resistance. These nonwovens differed in terms of their ventilation resistance, thickness, and area density, where flow resistivity $\sigma_n$ is the ventilation resistance $R_n$ divided by the thickness $t_n$ of the nonwoven sheet.

**Table 1.** Specifications of nonwoven fabrics.

| Name | Ventilation Resistance $R_n$ (kPa × s/m) | Thickness $t_n$ (mm) | Flow Resistivity $\sigma_n = R_n/t_n$ (kPa × s/m²) | Area Density $\rho_A$ (g/m²) | Material |
|---|---|---|---|---|---|
| RW2100 | 0.1767 | 0.58 | 303.6 | 100 | Polypropylene |
| 3501BD | 0.2021 | 0.19 | 1064 | 50 | Polyester |
| 3A01A | 0.3363 | 0.39 | 862.2 | 100 | Polyester |
| 3701B | 0.3763 | 0.24 | 1568 | 70 | Polyester |
| 3A51AD | 0.5283 | 0.45 | 1171 | 152 | Polyester |
| RW2250 | 1.350 | 0.82 | 1646 | 250 | Polypropylene |

Figure 3a,b show a photograph of a nonwoven fabric with a support frame and a schematic diagram of the support frame, respectively. The nonwoven fabric with a support frame shown is a sample in which the vibration of the nonwoven fabric was suppressed by pasting the nonwoven fabric onto a support frame, which was made out of SUS304. The outer diameter $D_f$ = 109.90 mm, thickness $t_f$ = 1.00 mm, the width of the frame was 1.0 mm, and aperture was a regular hexagonal honeycomb shape, where each face was 3.44 mm long. The aperture ratio was as large as 0.73 to ensure that the mesh was acoustically negligible.

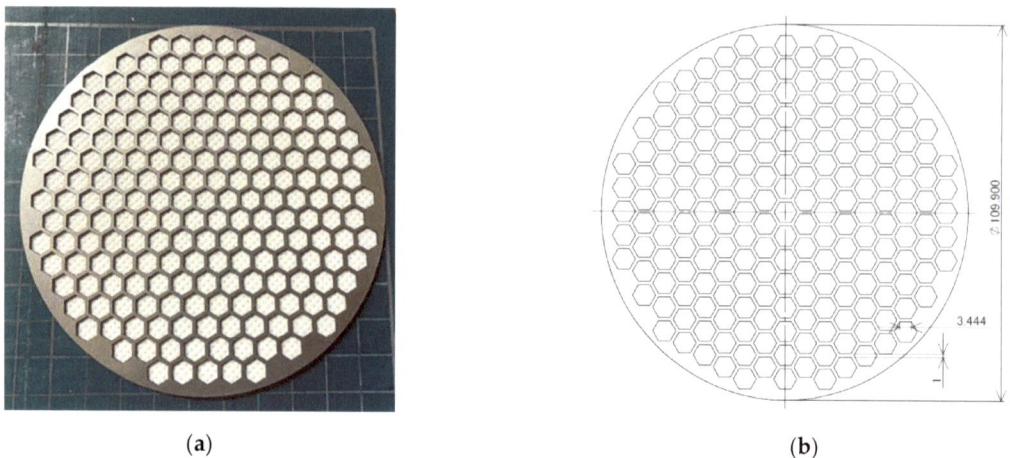

**Figure 3.** Frame for supporting the nonwoven fabric: (**a**) photograph; (**b**) dimensions of the frame.

Figure 4a,b present a photograph and schematic diagrams of the sample tubes, respectively. The material of the sample tube was aluminum alloy, and the dimensions were: inner diameter $D_i$ = 100 mm, outer diameter 120 mm, and length of the back air space $L$ = 100 mm.

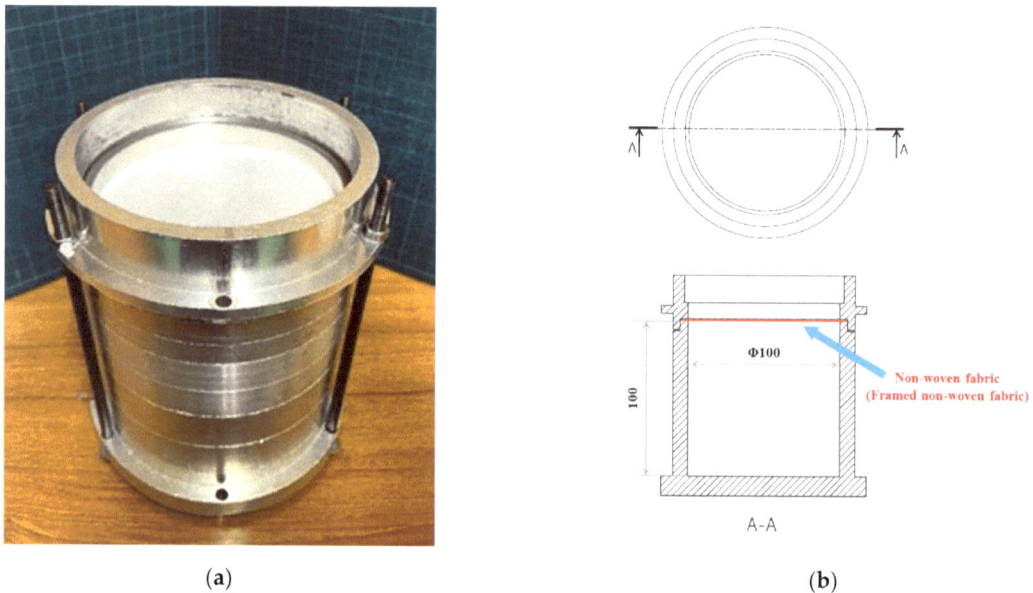

**Figure 4.** Sample tube: (**a**) photograph; (**b**) dimensions.

As shown in Figure 4b, a sample was placed at the top end of the sample tube, and the sound absorption coefficient was measured for a nonwoven fabric sheet with a back air space. In this paper, the vibration of the nonwoven fabric was created via self-excited vibration caused by sound waves incident on the nonwoven fabric. In this experiment, the vibration of the nonwoven fabric was excited via the sound wave in order to measure the sound absorption coefficient.

## 3. Theoretical Analyses

### 3.1. Analysis Model Corresponding to the Measured Sample

In this section, the sound absorption coefficient of nonwoven fabric with a back air space was derived through theoretical analysis using a transfer matrix.

Figure 5a,b show the equivalent circuits corresponding to the measured samples. Figure 5a is a schematic of a nonwoven fabric sample with a support frame, and Figure 5b is a schematic of a nonwoven sheet sample. The right-hand sides of Figure 5a,b show equivalent circuits corresponding to the analytical model described in Section 3.6, respectively.

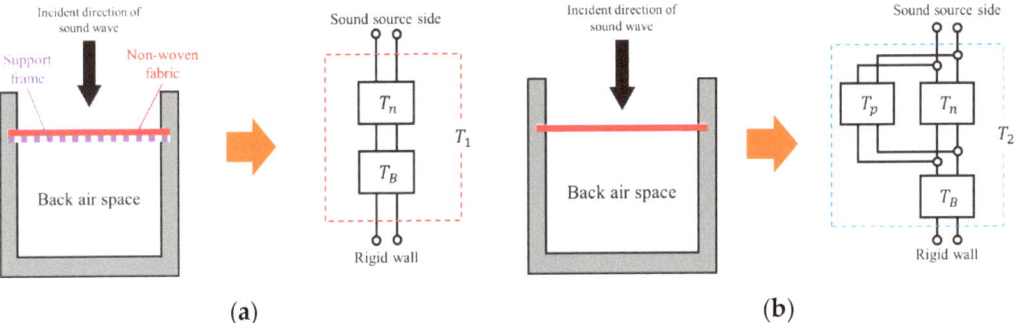

**Figure 5.** Equivalent circuits corresponding to test samples: (**a**) framed nonwoven fabric; (**b**) nonwoven fabric sheet.

First, the analytical model shown in Figure 5a did not consider the vibration of the nonwoven fabric because the vibration of the nonwoven fabric was suppressed by the support frame. However, the analytical model in Figure 5b considered the vibration of the nonwoven fabric. Therefore, it can be seen that there were three types of elements that composed these analytical models: the porous characteristics of the nonwoven fabric, the air layer behind the fabric, and the vibration of the nonwoven fabric.

The mathematical expressions of the equivalent circuits on the right side of Figure 5a,b are explained in Section 3.6.

### 3.2. Transfer Matrix Based on a One-Dimensional Wave Equation

Assuming that the sound pressure and particle velocity at the incident surface of the sound wave are $p_1$ and $u_1$, respectively, the sound pressure and particle velocity at the end of the measurement tube are $p_2$ and $u_2$, respectively, and the cross-sectional area of the measurement tube is $S$. The transfer matrix $T$ between the incident and end surfaces can be expressed using the one-dimensional wave equation as follows [13]:

$$\begin{bmatrix} p_1 \\ Su_1 \end{bmatrix} = T \begin{bmatrix} p_2 \\ Su_2 \end{bmatrix} = \begin{bmatrix} A & B \\ C & D \end{bmatrix} \begin{bmatrix} p_2 \\ Su_2 \end{bmatrix} \quad (1)$$

where the four-terminal constants $A$, $B$, $C$, and $D$ in the transfer matrix $T$ are expressed as follows:

$$T = \begin{bmatrix} A & B \\ C & D \end{bmatrix} = \begin{bmatrix} \cos h(\gamma l) & \frac{Z_c}{S}\sin h(\gamma l) \\ \frac{S}{Z_c}\sin h(\gamma l) & \cos h(\gamma l) \end{bmatrix} \quad (2)$$

where $\gamma$ is the propagation constant, $Z_c$ is the characteristic impedance, and $l$ is the length of the measurement tube. When Equation (2) is substituted for $A$, $B$, $C$, and $D$ in Equation (1) and expressed as a series of equations for $p_1$ and $u_1$, they are one-dimensional wave equations for sound pressure and particle velocity.

### 3.3. Transfer Matrix for Nonwoven Fabrics as Porous Materials

The propagation constant $\gamma_n$ and characteristic impedance $Z_n$ for a nonwoven fabric can be expressed as in Equations (3) and (4) using the Miki model [2]. Note that $j$ is an imaginary unit.

$$\gamma_n = \rho c \left\{ 1 + 0.0699 \left( \frac{f}{\sigma_n} \right)^{-0.632} \right\} - j\rho c \left\{ 0.107 \left( \frac{f}{\sigma_n} \right)^{-0.632} \right\} \tag{3}$$

$$Z_n = 0.160 \frac{\omega}{c} \left( \frac{f}{\sigma_n} \right)^{-0.618} + j\frac{\omega}{c} \left\{ 1 + 0.109 \left( \frac{f}{\sigma_n} \right)^{-0.618} \right\} \tag{4}$$

The flow resistivity $\sigma_n$ of the nonwoven fabric in Equations (3) and (4) can be expressed as Equation (5).

$$\sigma_n = \frac{R_n}{t_n} \tag{5}$$

where $R_n$ is the ventilation resistance of the nonwoven fabric and $t_n$ is the thickness of the nonwoven fabric.

The transfer matrix $T_n$ of the nonwoven fabric is obtained by substituting Equations (3) and (4) into the transfer matrix in Equation (2) as follows to yield:

$$T_n = \begin{bmatrix} \cosh(\gamma_n t_n) & \frac{Z_n}{S_n}\sinh(\gamma_n t_n) \\ \frac{S_n}{Z_n}\sinh(\gamma_n t_n) & \cosh(\gamma_n t_n) \end{bmatrix} = \begin{bmatrix} A_n & B_n \\ C_n & D_n \end{bmatrix} \tag{6}$$

### 3.4. Transfer Matrix for the Vibration of Nonwoven Sheet

The acoustic impedance $Z_p$ for a vibrating sheet is expressed using Equation (7) [12].

$$Z_p = \frac{b}{S_n^2} + j\frac{\omega m}{S_n^2} = \frac{2\xi\sqrt{mk_a}}{S_n^2} + j\frac{\omega \rho_A S_n}{S_n^2} = \frac{2\xi k_a}{\omega_0 S_n^2} + j\frac{\omega \rho_A}{S_n} \tag{7}$$

where $S_n$ is the area of the nonwoven fabric, $\rho_A$ is the area density of the nonwoven fabric, $\xi$ is the damping ratio, and $\omega$ is the angular frequency of the sound wave ($\omega = 2\pi f$, $f$: frequency). The attenuation constant $b$ of the nonwoven fabric, mass $m$ of the nonwoven fabric, natural angular frequency $\omega_0$, and spring constant $k_a$ of the air layer can be expressed using Equations (8)–(11), respectively, as follows

$$b = 2\xi\sqrt{mk_a} \tag{8}$$

$$m = \rho_A S_n \tag{9}$$

$$\omega_0 = \sqrt{\frac{k_a}{m}} \tag{10}$$

$$k_a = \frac{\Gamma p_0 S_t}{L} \tag{11}$$

where $\Gamma$ is the specific heat ratio, $p_0$ is the atmospheric pressure, $S_t$ is the cross-sectional area of the measurement tube ($S_t = S_n$), and $L$ is the length of the back air layer. The damping ratio $\xi$ is set to 0.1.

Substituting Equation (7) for $Z_p$ in the following equation, the transfer matrix $T_p$ of the vibrating sheet is expressed using the right side of the following equation:

$$T_p = \begin{bmatrix} 1 & Z_p \\ 0 & 1 \end{bmatrix} = \begin{bmatrix} 1 & \frac{2\xi k_a}{\omega_0 S_n^2} + j\frac{\omega \rho_A}{S_n} \\ 0 & 1 \end{bmatrix} = \begin{bmatrix} A_p & B_p \\ C_p & D_p \end{bmatrix} \tag{12}$$

## 3.5. Transfer Matrix for Back Air Space

The transfer matrix $T_B$ for the back air space can be expressed as follows, assuming that damping in the transfer matrix based on the one-dimensional wave equation in Equation (2) is negligible:

$$T_B = \begin{bmatrix} \cos kL & j\frac{\rho c}{S_t}\sin kL \\ j\frac{S_t}{\rho c}\sin kL & \cos kL \end{bmatrix} \qquad (13)$$

where $k$ is the wavenumber, $\rho$ is the density of air, and $c$ is the speed of sound in air.

## 3.6. Equivalent Circuit and Transfer Matrix Corresponding to the Analytical Model

For the nonwoven fabric with a support frame in Figure 5a, the transfer matrix $T_1$ can be obtained via the cascade connecting the transfer matrix $T_n$ using the Miki model in Equation (6) and the transfer matrix $T_B$ for the back air space in Equation (13), as follows:

$$\begin{aligned} T_1 &= T_n \times T_B \\ &= \begin{bmatrix} \cosh(\gamma_n t_n) & \frac{Z_n}{S_n}\sinh(\gamma_n t_n) \\ \frac{S_n}{Z_n}\sinh(\gamma_n t_n) & \cosh(\gamma_n t_n) \end{bmatrix} \times \begin{bmatrix} \cos kL & j\frac{\rho c}{S_t}\sin kL \\ j\frac{S_t}{\rho c}\sin kL & \cos kL \end{bmatrix} \\ &= \begin{bmatrix} A_1 & B_1 \\ C_1 & D_1 \end{bmatrix} \end{aligned} \qquad (14)$$

In addition, when the acoustic elements are connected sequentially, such as in silencer design, the transfer matrix cascades [13].

For the nonwoven sheet shown in Figure 5b alone, the transfer matrix $T_n$ using the Miki model in Equation (6) and the transfer matrix $T_p$ for the vibration of the nonwoven sheet in Equation (12) are connected in parallel [12,14,15]. The parallel connection occurs because there are two sound-absorbing principles for one incident surface [12,14]. Owing to the parallel connection, sound pressure acts equally on both sound-absorbing principles and particle velocity is diverted more toward the sound-absorbing principle with lower impedance. Next, the transfer matrix $T_2$ is obtained via a cascade connecting the transfer matrix $T_B$ of the back air layer in Equation (13) to them, as shown in the following Equation [13].

$$\begin{aligned} T_2 &= \begin{bmatrix} \frac{A_n B_p + A_p B_n}{B_n + B_p} & \frac{B_n B_p}{B_n + B_p} \\ C_n + C_p + \frac{(A_p - A_n)(D_n - D_p)}{B_n + B_p} & \frac{D_n B_p + D_p B_n}{B_n + B_p} \end{bmatrix} \begin{bmatrix} \cos kL & j\frac{\rho c}{S_t}\sin kL \\ j\frac{S_t}{\rho c}\sin kL & \cos kL \end{bmatrix} \\ &= \begin{bmatrix} A_2 & B_2 \\ C_2 & D_2 \end{bmatrix} \end{aligned} \qquad (15)$$

## 3.7. Derivation of the Sound Absorption Coefficient

The sound absorption coefficient for the transfer matrices $T_1$ and $T_2$ obtained in Section 3.6 was calculated. The four-terminal constants of the transfer matrices $T_1$ and $T_2$ correspond to the four-terminal constants $A$, $B$, $C$, and $D$ of the transfer matrix $T$ in Equation (1).

Since the ends of $T_1$ and $T_2$ are rigid walls, $u_2 = 0$; therefore, Equation (1) can be expressed as:

$$\begin{bmatrix} p_1 \\ Su_1 \end{bmatrix} = \begin{bmatrix} Ap_2 \\ Cp_2 \end{bmatrix} \qquad (16)$$

The specific acoustic impedance $Z$ of this acoustic system, as seen from the plane of incidence, is shown in Equation (17):

$$Z = \frac{p}{u} \qquad (17)$$

where, from $p = p_1$ and $Su_1 = S_t u$, the specific acoustic impedance $Z$ is expressed as in Equation (18).

$$Z = \frac{p}{u} = \frac{p}{S_t u} S_t = \frac{p_1}{Su_1} S_t = \frac{A}{C} S_t \qquad (18)$$

Here, the relationship between specific acoustic impedance $Z$ and reflectance $R$ is expressed using the following equation

$$R = \frac{Z - \rho c}{Z + \rho c} \qquad (19)$$

The sound absorption coefficient $\alpha$ is expressed using the reflection coefficient $R$ as follows

$$\alpha = 1 - |R|^2 \qquad (20)$$

## 4. Comparisons of Experimental and Theoretical Values

A comparison between the sound absorption coefficient based on the theoretical analysis and the experimental results is shown in Figure 6a–f.

In Figure 6a–d, for both nonwoven sheets, the experimental values for the nonwoven fabric with a support frame (black line) were consistent with the theoretical values (red line), which did not account for the vibration of the nonwoven fabric. In Figure 6e,f, the trends of the experimental (black line) and theoretical (red line) values were in agreement at low frequencies, but less so at high frequencies.

In Figure 6a–d, the experimental values for the nonwoven sheet alone (black dashed line) were consistent with the theoretical values (blue line), which considered the vibration of the nonwoven fabric. Similarly, in the case of the nonwoven sheet alone, the agreement between the experimental (black dashed line) and theoretical (blue line) values was poor for Figure 6e,f, but much better in the low-frequency range.

As described above, the relatively good agreement between the experimental and theoretical trends suggests that the vibration of the nonwoven fabric with the support frame was nearly suppressed, and that of the nonwovens alone suggested that the nonwovens were vibrating. This is because the support frame reduced the mass of the nonwoven fabric per aperture to less than 1/100 and the frequency band in which the nonwoven fabric's spring-mass system vibrates is considered to be approximately one order of magnitude higher. Therefore, the vibration of the nonwoven fabric in the frequency band of interest is considered to be nearly completely suppressed in terms of the sound absorption coefficient.

Meanwhile, the vibration of the support frame itself, independently of the type of nonwoven fabric, was observed as a spike-like decrease in sound absorption coefficient near 650 Hz.

The effect of the vibration component in the analytical model is also discussed. In the theoretical values shown in Figure 6a–e, the sound absorption coefficient is generally higher over the entire frequency range when vibration is not considered. The simulation results indicate that the vibration of the nonwoven fabric hindered sound absorption for nonwoven fabrics that do not have high ventilation resistance.

For RW2100, shown in Figure 6a, the difference in experimental values with and without the support frame was the smallest among the nonwovens used in this study. The theoretical values also showed the smallest difference depending on whether vibration was taken into account. This suggests that when the ventilation resistance was low, as in the case of RW2100, the excitation force due to sound pressure was low, and thus the nonwoven fabric was less likely to vibrate.

RW2250, shown in Figure 6f, was the only nonwoven fabric in this study to show a higher sound absorption coefficient experimentally with a single nonwoven fabric than with a support frame (except in the low-frequency range). Additionally, the theoretical sound absorption coefficient was higher when vibration was taken into account than when it was not. This suggests that RW2250 had a significantly higher ventilation resistance than the other nonwovens used in this study, and its effectiveness as a plate vibration type sound absorber was better than its functionality as a porous (permeable) material, and thus it showed a different trend compared to the other nonwovens sheets.

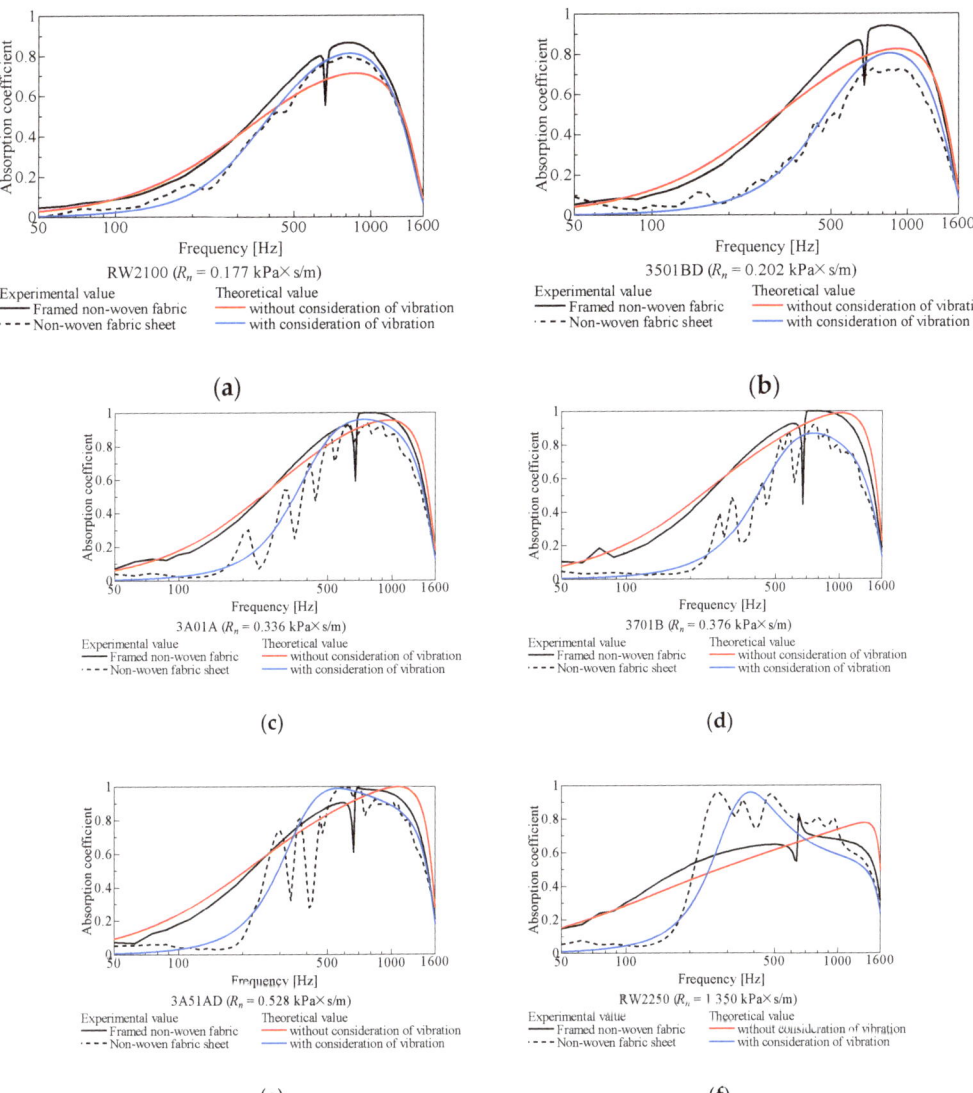

**Figure 6.** Comparison of experimental and theoretical values (length of back air space L = 100 mm): (**a**) RW2100; (**b**) 3501BD; (**c**) 3A01A; (**d**) 3701B; (**e**) 3A51AD; (**f**) RW2250.

## 5. Conclusions

The following conclusions were obtained from the experimental results and the theoretical analysis of a nonwoven fabric sheet with a back air space.

The vibration of the nonwoven fabric was suppressed by attaching a support frame with honeycomb-shaped apertures to the large-area, vibration-prone nonwoven fabric. This led to an experiment corresponding to a theoretical analytical model that did not account for the vibration of the nonwoven fabric. The experimental results showed that when the vibration of the nonwoven sheet was suppressed, the sound absorption coefficient was higher than that of the vibration-prone nonwoven sheet alone. The sound absorption coefficient at the peak frequency was increased by >0.2, especially for 3501BD. Using the support frame, the sound absorption coefficient at the peak frequencies of 3A01A and 3701B was increased to 0.99. However, the opposite trend was observed for RW2250, which showed the highest ventilation resistance.

In the theoretical analysis of a large-area, vibration-prone nonwoven fabric, in which the vibration of the nonwoven fabric was taken into account, the theoretical values agreed with the experimental values, and the accuracy of the theoretical values was improved.

By comparing the theoretical values for nonwoven fabrics without high ventilation resistance, the sound absorption coefficient was greater when vibration was not considered. Therefore, the results suggested that the vibration of the nonwoven fabric hindered sound absorption.

The following secondary points were also found: A comparison of the theoretical values for nonwoven fabrics with high ventilation resistance showed that the sound absorption coefficient was higher when vibration was considered. However, this was not the case in the low-frequency range.

The trends observed for the theoretical values were generally consistent with those of the experimental values.

**Author Contributions:** Conceptualization, S.S. and R.I.; methodology, S.S.; software, R.I.; validation, T.N.; formal analysis, R.I.; investigation, T.N.; data curation, T.N.; writing—original draft preparation, S.S. and R.I.; writing—review and editing, S.S.; visualization, R.I.; project administration, S.S. All authors have read and agreed to the published version of the manuscript.

**Funding:** This research received no external funding.

**Acknowledgments:** Idemitsu Kosan Co., Ltd. and Toyobo Co., Ltd. provided the nonwoven fabric sheets used in this study. The authors appreciate their support.

**Conflicts of Interest:** The authors declare no conflict of interest.

## References

1. Delany, M.E.; Bazley, E.N. Acoustic Properties of Fibrous Absorbent Materials. *Appl. Acoust.* **1970**, *3*, 105–116. [CrossRef]
2. Miki, Y. Acoustical properties of porous materials-Modifications of Delany-Bazley models. *Acoust. Soc. Jpn.* **1990**, *11*, 19–24. [CrossRef]
3. Komatsu, T. Improvement of the Delany-Bazley and Miki models for fibrous sound-absorbing materials. *Acoust. Sci. Technol.* **2008**, *29*, 121–129. [CrossRef]
4. Kurosawa, Y. Development of sound absorption coefficient prediction technique of ultrafine fiber. *Trans. JSME* **2016**, *82*, 837. (In Japanese). [CrossRef]
5. Panneton, R. Comments on the limp frame equivalent fluid model for porous media. *J. Acoust. Soc. Am.* **2007**, *122*, EL217. [CrossRef] [PubMed]
6. Biot, M.A. Theory of propagation of elastic waves in a fluid-saturated porous solid. I. Low-frequency range. *J. Acoust. Soc. Am.* **1956**, *28*, 168–178. [CrossRef]
7. Biot, M.A. Theory of propagation of elastic waves in a fluid-saturated porous solid. II. Higher frequency range. *J. Acoust. Soc. Am.* **1956**, *28*, 179–191. [CrossRef]
8. Allard, J.F.; Atalla, N. *Propagation of Sound in Porous Media*; John Wiley & Sons, Ltd.: Hoboken, NJ, USA, 2009. [CrossRef]
9. Sakagami, K.; Kiyama, M.; Morimoto, M.; Takahashi, D. Sound Absorption of a Cavity-Backed Membrane: A Step Towards Design Method for Membrane-Type Absorbers. *Appl. Acoust.* **1996**, *49*, 237–247. [CrossRef]
10. Sakagami, K.; Kiyama, M.; Morimoto, M.; Takahashi, D. Detailed Analysis of the Acoustic Properties of a Permeable Membrane. *Appl. Acoust.* **1998**, *54*, 93–111. [CrossRef]
11. Sakagami, K.; Kiyama, M.; Morimoto, M.; Yairi, M. A note on the effect of vibration of a microperforated panel on its sound absorption characteristics. *Acoust. Sci. Technol.* **2005**, *26*, 204–207. [CrossRef]
12. Sakamoto, S.; Fujisawa, K.; Watanabe, S. Small plate vibration sound-absorbing device with a clearance and without surrounding restriction: Theoretical analysis and experiment. *Noise Control. Eng. J.* **2021**, *69*, 30–38. [CrossRef]
13. Suyama, E.; Hirata, M. The Four Terminal Matrices of Tube System Based on Assuming of Plane Wave Propagation with Frictional Dissipation: Acoustic Characteristic Analysis of Silencing Systems Based on Assuming of Plane Wave Propagation with frictional dissipation part 2. *J. Acoust. Soc. Jpn.* **1979**, *35*, 165–170. (In Japanese). [CrossRef]

14. Sakamoto, S.; Takakura, R.; Suzuki, R.; Katayama, I.; Saito, R.; Suzuki, K. Theoretical and Experimental Analyses of Acoustic Characteristics of Fine-grain Powder Considering Longitudinal Vibration and Boundary Layer Viscosity. *J. Acoust. Soc. Am.* **2021**, *149*, 1030–1040. [CrossRef] [PubMed]
15. Li, C.; Cazzolato, B.; Anthony, Z. Acoustic impedance of micro perforated membranes: Velocity continuity condition at the perforation boundary. *J. Acoust. Soc. Am.* **2016**, *139*, 93–103. [CrossRef] [PubMed]

Article

# A New Efficient Approach to Simulate Material Damping in Metals by Modeling Thermoelastic Coupling

Christin Zacharias *, Carsten Könke and Christian Guist

Institute of Structural Mechanics, Bauhaus-University Weimar, 99423 Weimar, Germany; carsten.koenke@uni-weimar.de (C.K.); christian.guist@uni-weimar.de (C.G.)
* Correspondence: christin.zacharias@uni-weimar.de

**Abstract:** The realistic prediction of material damping is crucial in the design and dynamic simulation of many components in mechanical engineering. Material damping in metals occurs mainly due to the thermoelastic effect. This paper presents a new approach for implementing thermoelastic damping into finite element simulations, which provides an alternative to computationally intensive, fully coupled thermoelastic simulations. A significantly better agreement between simulation results and experimental data was achieved, when compared with the empirical damping values found in the literature. The method is based on the calculation of the generated heat within a vibration cycle. The temperature distribution is determined by the mechanical eigenmodes and the energy converted into heat, and thus dissipated, is calculated. This algorithm leads to modal damping coefficients that can then be used in subsequent analyses of dynamically excited oscillations. The results were validated with experimental data obtained from vibration tests. In order to measure material damping only, a test setup excluding friction and environmental influences was developed. Furthermore, comparisons with fully coupled thermoelastic simulations were performed. It was clear that the new approach achieved results comparable to those of a computationally expensive, coupled simulation with regard to the loss factors and frequency response analyses.

**Keywords:** thermoelastic damping; finite element simulation; modal damping; dynamic simulation; experimental damping measurement

Citation: Zacharias, C.; Könke, C.; Guist, C. A New Efficient Approach to Simulate Material Damping in Metals by Modeling Thermoelastic Coupling. Materials 2022, 15, 1706. https://doi.org/10.3390/ma15051706

Academic Editor: Martin Vašina

Received: 20 January 2022
Accepted: 18 February 2022
Published: 24 February 2022

**Publisher's Note:** MDPI stays neutral with regard to jurisdictional claims in published maps and institutional affiliations.

**Copyright:** © 2022 by the authors. Licensee MDPI, Basel, Switzerland. This article is an open access article distributed under the terms and conditions of the Creative Commons Attribution (CC BY) license (https://creativecommons.org/licenses/by/4.0/).

## 1. Introduction

The simulation of dynamic processes is essential in almost every engineering discipline, including structural and mechanical engineering, aerospace engineering and micro- and nanotechnology. Besides the stiffness and mass of a structure, the accurate modeling of damping behavior is highly important to predict amplitudes and frequencies correctly, to avoid resonance phenomena, etc. Nevertheless, the estimation of damping parameters is challenging. Several mathematical models have been established to consider energy dissipation, e.g., viscous damping or Rayleigh damping. Although these assumptions are sufficient in many applications, they can lead to inaccurate predictions. Therefore, in practice, it is often necessary to perform experimental vibration analyses in addition to the computational development process. To avoid this effort, there is a need for precise and fast usable damping models.

One reason behind the difficulty in modeling energy dissipation realistically lies in the diversity of sources. In the considered structures, three main causes can be distinguished. Within the material, energy is dissipated through physical processes (material damping). At the levels of bearings, suspensions and connecting joints, dissipating friction occurs (joint damping). Furthermore, mechanical vibrations are attenuated by the interaction with the environment (air or radiation damping).

This paper focuses on material damping. In recent years, material damping has become increasingly important in developing and improving advanced materials and structures. High intrinsic energy dissipation guarantees favorable vibration and acoustic behavior.

Examples are carbon-fiber-reinforced composites, e.g., those presented by Xu et al. [1] or Attard et al. [2], and sandwich panels, as described by Ledi [3]. Special honeycomb sandwich panels are outlined, e.g., by Arunkumar et al. [4] and Sadiq et al. [5]. There are several factors influencing the material damping discussed in the literature, for example the orientation of the fibers in fiber-reinforced composites (see [6–8]) or the manufacturing. Wesolowski et al. studied the material damping in laminated composite structures in comparison with conventional materials [9]. We present a simplified and convenient method for the simulation of material damping of homogeneous materials in the framework of finite element modeling. The main physical effect causing energy dissipation within the material is thermoelastic relaxation. Depending on the microstructure and the environment, other sources of damping can be dislocations, friction on grain boundaries, electromagnetic effects, etc.

In a thermoelastic solid there is a coupling effect between the elastic strain field and the temperature field. Therefore, mechanical vibrations cause temperature variations. In compressed areas, the temperature increases and, for areas in tension, it decreases. The resulting gradient affects irreversible heat fluxes followed by an increase in entropy and a transformation of mechanical energy into heat.

The temperature gradient equalizes within a characteristic relaxation time, $\tau_i$. It is the reciprocal of the thermal peak frequency, i.e., the eigensolution of the heat conduction equation.

$$\tau_i = \frac{1}{\omega_i} \quad (1)$$

The amount of thermoelastic damping corresponds to the mechanical eigenfrequency of the structure. The highest loss factor is achieved if the relaxation time approximately coincides with the reciprocal of the natural frequency (see Figure 1). In this case, the period length is exactly sufficient to bring the system back to equilibrium by heat flows.

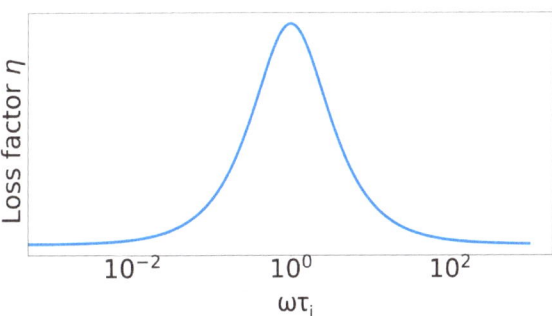

**Figure 1.** Dependency of the loss factor on the excitation cycle frequency $\omega$.

Research on thermoelastic coupling has a long history and goes back to the work of Duhamel in 1837 [10], as described in [11]. In 1956, Biot provided a mathematical formulation of thermoelastic materials and solution techniques [12]. These theories were extended and refined in the following years and established in standard books, e.g., by Parkus [13] or Nowacki [14]. Over several years, intensive research has emphasized the development of simplified analytical calculations of thermoelastic damping.

In 1937, Zener [15–17] published a series of seminal articles concerning energy dissipation in thin bending beams with one-dimensional heat conduction. The author developed a well-established formula to calculate the loss factor $\eta$.

$$\eta = \frac{\alpha^2 E T_0}{\rho C_p} \frac{\omega \omega_i}{\omega^2 \omega_i^2} \quad (2)$$

The first fraction describes the material-dependent damping potential, the so-called relaxation strength, where $\alpha$ = the thermal expansion coefficient, $E$ = Young's modulus, $T_0$ = the temperature of the solid, $C_p$ = the heat capacity under constant pressure and $\rho$ = density. The second fraction of the formula is frequency-dependent with the current circular frequency, $\omega$. Zener developed the expression for thermoelastic damping containing an infinite series of the thermal frequencies, $\omega_i$, and the corresponding thermal modes. He showed that, in thin beams, only the first thermal frequency is relevant and simplified the model to the above-mentioned formula. The thermal peak frequency $\omega_i$ is calculated using $C_p$, the thermal conductivity, $\lambda$, the density, $\rho$, and the thickness of the beam, $h$.

$$\frac{1}{\omega_i} = \tau_i = \left(\frac{h}{\pi}\right)^2 \frac{C_p \rho}{\lambda} \tag{3}$$

Zener showed that the restriction to the first thermal mode is sufficient in the case of thin bending beams.

Some years later, Chadwick [18,19] and Alblas [20,21] extended Zener's approach and applied it to general three-dimensional solids. In a more recent publication, Lifshitz et al. [22] revised the thermoelastic beam theory and refined the solution by also using higher thermal eigensolutions. Their formula has been used as basis in several further studies (e.g., [23–25]).

In general, two approaches for calculating thermoelastic damping can be distinguished, namely, dissipated mechanical energy as the imaginary solution of thermoelastically coupled elastic equation or as generated entropy due to temperature gradients (see Section 2.1). In this paper, we focus on the entropy method. Concerning this topic, Kinra et al. [26] and Bishop et al. [27,28] derived a calculation based on the second law of thermodynamics. Duwel et al. [29] formulated a strongly (two-way) coupled and a weakly (one-way) coupled approach to calculate thermoelastic dissipation in microresonators and presented experimental results. Chandorkar et al. [30] emphasized the superposition of mechanical and thermal modes to consider thermoelastic relaxation. Hao et al. [31] extended the approach to anisotropic materials and embedded the calculation in a finite element algorithm. Tai et al. [32,33] developed simplified concepts for beams and plates based on the entropy theory.

Most of the cited literature refers to micro- and nanoresonators, since thermoelastic damping is often the dominant loss mechanism in microstructures [34]. In macrostructures, thermoelastic damping can be a significant loss mechanism, especially in thin-walled structures. For example, Cagnoli [35] showed this in experiments and calculations on thin circular disks. Concerning nonlinear damping models, the recent publications by Huang et al. [36], as well as Amabili [37], should be mentioned.

Even if other damping mechanisms, especially joint damping, take a decisive role in complex structures, the calculation of material damping is an important component for the realistic prediction of the dynamical behavior.

The described analytical approaches allow exact calculations of thermoelastic damping to be conducted, but are not suitable for complex geometries. Serra et al. [38] developed a finite element formulation for thermoelastically coupled problems. The authors applied the underlying theory on shell and solid elements that are characterized by coupled damping and stiffness matrices and considered thermoelastic damping. These elements showed a good agreement with the experimental data in statistical analyses and dynamic simulations requiring direct time-integration methods. The significant disadvantage, in this case, is the enormous computational effort for fully coupled thermoelastic simulations.

We present an approach that allows the calculation of thermoelastic damping to be performed based on the stress and strain distribution in the mode shapes. Using an entropy method, modal damping coefficients were determined and applied in finite element simulations of dynamically excited components in the time and frequency domain. This allowed us to consider material damping in time-efficient simulations. Furthermore, the damping coefficients of specific mode shapes were measured in physical experiments to verify the simulated results.

## 2. Materials and Methods

### 2.1. Simulation

Since Zener's first publications in the 1930s [15,17], there have been two main approaches to interpreting thermoelastic damping, explained as follows:

- Dissipated mechanical energy: The calculation of thermoelastic damping is based on the phase lag between stresses and corresponding strains. The thermoelastically coupled differential equations of elasticity and heat conduction have to be solved in the complex domain. The loss factor is equal to the ratio of the imaginary and real parts of the solution.
- Entropy approach: The amount of dissipated energy is equal to the heat generated during the elastic vibration. The energy transferred into heat can be calculated by analyzing the heat flows in the structure causing an increase in entropy. The loss factor is obtained from the quotient of dissipated energy to total strain energy. This approach is used in the present paper. The general procedure of the calculation is shown in the flowchart in Figure 2.

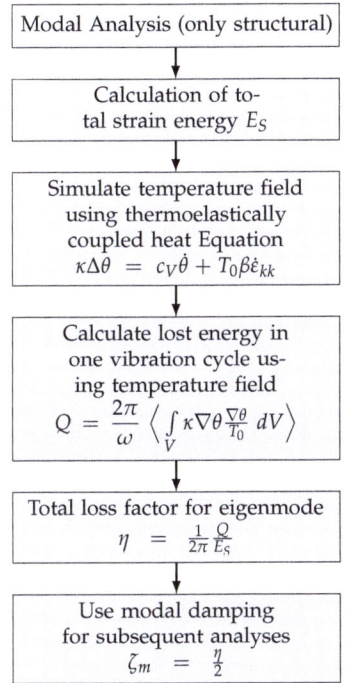

**Figure 2.** Flow chart for numerical solution procedure.

The amount of generated heat during one vibration cycle is determined starting from the rate of entropy of a solid with volume, $V$.

$$\dot{S} = \int_V \frac{-\nabla \mathbf{q}}{T} \, dV = -\int_V \nabla\left(\frac{\mathbf{q}}{T}\right) dV + \int_V \mathbf{q} \nabla \frac{1}{T} \, dV \tag{4}$$

where $S$ denotes the entropy, $\mathbf{q}$ is the vector of heat flux and $T$ is the absolute temperature of the solid. Following the derivation in [39], the equation can be transformed using the divergence theorem. If the terms are converted into a surface integral, the first integral becomes zero. The vibration is defined to be adiabatic, i.e., there is no heat flux over the boundaries of the solid.

$$\dot{S} = \int_V \mathbf{q} \nabla \frac{1}{T} \, dV \tag{5}$$

The absolute temperature, $T$, is defined as the sum of the reference temperature, $T_0$, and the temperature increment, $\theta$. The expression $1/T$ is replaced by its Taylor expansion in the vicinity of the reference temperature. Under the assumption of very small temperature variations in the present thermoelastic problems, the first term produces a zero integral.

$$\dot{S} = \int_V \mathbf{q} \nabla \left( \frac{1}{T_0} - \frac{\theta}{T_0^2} \right) dV \approx \int_V -\mathbf{q} \nabla \frac{\theta}{T_0} \, dV \tag{6}$$

The same assumption allows us to calculate the rate of produced heat using the reference temperature. Using Fourier's law

$$\mathbf{q} = -\kappa \nabla \theta \tag{7}$$

to substitute $\mathbf{q}$, the following formula is obtained:

$$\dot{Q} = \int_V \kappa \nabla \theta \frac{\nabla \theta}{T_0} \, dV \tag{8}$$

with the thermal conductivity $\kappa$. Considering one vibration cycle from $t = 0$ to $t = T$,

$$Q = \frac{2\pi}{\omega} \left\langle \int_V \kappa \nabla \theta \frac{\nabla \theta}{T_0} \, dV \right\rangle \tag{9}$$

where $\langle f(x) \rangle = \frac{1}{T} \int_0^T f(t) \, dt$ denotes the time average of the function. The thermoelastic loss factor, $\eta$, is defined as the ratio of generated thermal energy, $Q$, to total strain energy, $E_S$.

$$\eta = \frac{1}{2\pi} \frac{Q}{E_S} \tag{10}$$

The loss factor of a component modeled with three-dimensional solid elements is obtained in a procedure combining several numerical solution steps. A loss factor is calculated for each eigenmode in order to apply modal damping in subsequent analyses. For this purpose, a purely structural natural frequency analysis is run first. From the eigensolutions, the (normalized) strain distributions are calculated.

The coupling of elastic strain field and temperature field is mathematically described by the thermoelastically coupled heat equation, e.g., as derived by Biot in 1956 [12].

$$\kappa \Delta \theta = c_V \dot{\theta} + T_0 \beta \dot{\varepsilon}_{kk} \tag{11}$$

where $\beta = 3\frac{\alpha K}{c_V}$ is defined as the thermoelastic coupling constant with the elastic bulk modulus $K$ and the thermal expansion coefficient $\alpha$. It describes the relation between an adiabatic volume change and the temperature variation in a solid.

In the coupled heat equation, the mechanical volumetric strains, $\varepsilon_{kk}$, act as a source term. Equation (11) is solved numerically in the time domain to obtain the temperature variations in the solid over one vibration cycle.

For the solution, the software package FEniCS was used. Therefore, the problem was implemented in a finite element environment, based on the weak form of the coupled differential problem. The time derivatives were discretized by a finite difference algorithm [40].

$$\dot{\theta} = \frac{\theta^{n+1} - \theta^n}{\Delta t} \tag{12}$$

$$\dot{\varepsilon}_{kk} = \frac{\varepsilon_{kk}^{n+1} - \varepsilon_{kk}^n}{\Delta t} \tag{13}$$

The temperature is calculated using the thermoelastically coupled heat equation. Once the temperature has been determined, the loss factor can be calculated using Equations (9) and (10). In the approach described, a weak thermoelastic coupling is used, i.e., the elastic strain field influences the temperature distribution, but not vice versa.

The calculated loss factor, $\eta$, is specific for an eigenmode. In the mode-based subsequent analyses, they can be used as modal damping values, $\zeta_m$.

$$\zeta_m = \frac{\eta}{2} \tag{14}$$

## 2.2. Material

In the experimental studies, two types of specimens were considered. First, a simple plate geometry was investigated to validate the general procedure and verify the simulation results. In a second step, the method was extended to a more complex three-dimensional geometry. This was designed in the style of a simplified gearbox case. Both components are shown in Figure 3.

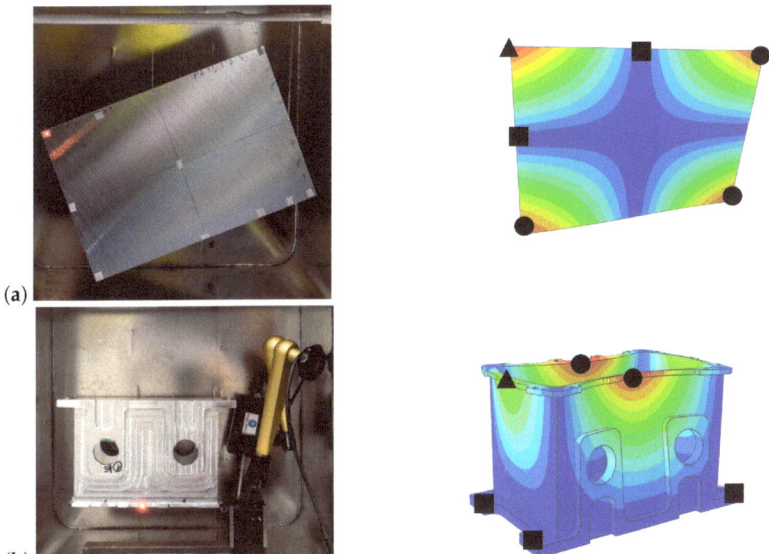

**Figure 3.** Experimental setup and positions of suspension (■), measurement points (●) and excitation points (▲—measurement and excitation) on the first three eigenmodes of the rectangular aluminum plate (**a**) and the box component (**b**).

In order to ensure comparability, both bodies were made of the same aluminum alloy (AlMg4.5Mn0.7). The material parameters are listed in Table 1.

Table 1. Material parameters of the used aluminum alloy.

| Parameter | Value | Unit |
|---|---|---|
| Young's modulus $E$ | $7 \times 10^{10}$ | N/m$^2$ |
| Poisson's ratio $\nu$ | 0.34 | |
| density $\rho$ | 2660 | kg/m$^3$ |
| thermal conductivity $\lambda$ | 140 | W/mK |
| heat capacity $C_p$ | 904 | J/K |
| thermal expansion coefficient $\alpha$ | $2.31 \times 10^{-5}$ | K$^{-1}$ |

*2.3. Experimental Studies*

The experiments aimed to measure the material damping. Therefore, other sources of energy dissipation had to be eliminated. To avoid air damping effects, the experimental setup was located within a vacuum chamber. With the available device, a pressure of approximately 5 mbar was achieved. All the processes during the experiments had to be controllable from the outside.

Furthermore, the impact of joint damping, which occurs as friction on bearings, suspensions and joints, had to be eliminated. Friction arises due to relative movements; therefore, the displacement at bearings should be inhibited. In the experiment, the specimens were suspended on very thin elastic strings that were placed in the zero lines of the mode shapes to simulate a free bearing. Therefore, a new experimental setup had to be taken for each eigenmode. Two examples for bearing and excitation points are shown in Figure 3. For a better understanding of the shown eigenmodes, video simulations are included in the Supplementary Material Videos S1 and S2.

The excitation was also designed in such way to avoid additional damping sources. Therefore, the specimens were excited with an automatic impulse hammer, which guaranteed a minimum of contact. The hammer was installed within the vacuum chamber and controlled from the outside.

As a result, the velocity of a single point was measured. The recording of the data was also performed in a contactless manner using a laser Doppler vibrometer.

The velocity over time was measured and analyzed with an algorithm based on LabView software. By applying an exponential curve fit on the decay function, the damping coefficient, $\zeta$, was determined. This measure could be transformed easily into other damping characteristics, e.g., the loss factor, $\eta$.

## 3. Results

*3.1. Rectangular Plate*

Figure 4 provides the correlation among the magnitude of displacement of the plate structure $u$, the hydrostatic stress distribution $\sigma_{kk}$ and the generated heat $Q$ in a specific mode shape. Since the eigenmodes are normalized vectors, the amplitudes were not significant and all quantities were scaled to 1 in the figure. The vibration of the plate follows a sine curve over time and the results are shown at time $t = T/4$, i.e., at the time step of maximum deflection, maximum stress level and maximum thermoelastic heat generation. Furthermore, Figure 4g–i display a cut on the mid-plane level of the plate because the heat flowed in the thickness direction. Therefore, the highest amount of energy was dissipated in the mid-level. The heat flow at the surface was negligible.

From the figure, it is apparent that the generated heat (and therefore the dissipated energy) correlated directly with the spatial distribution of the hydrostatic stress. The thermoelastic coupling only affected those parts of the stress or strain tensor that were associated with a volume change. Locations with a high hydrostatic stress level show large heat production.

Figure 5 presents the experimental data and the simulation of the loss factors. For comparison, the analytical calculation according to Zener for a specimen thickness of 3 mm is shown. The experiments were conducted on four individual plates that were identical in construction and four measurement points on each plate were used. The results present an average of the results per eigenfrequency. The standard deviation is shown as bars; however, in the case of the simple plate, the value was too small to be visible.

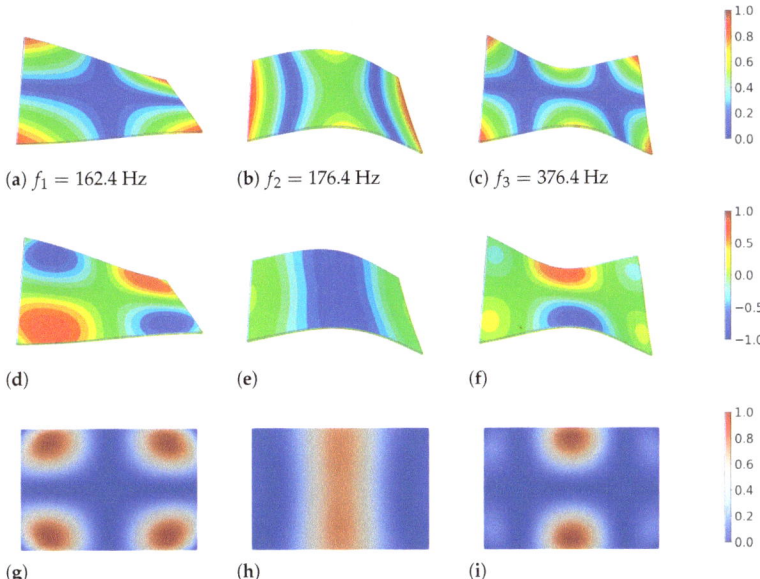

**Figure 4.** (**a**–**c**) Mode shapes of the rectangular aluminum plate 300 mm × 200 mm × 3 mm, deflection normalized to 1. (**d**–**f**) Distribution of hydrostatic pressure at the surface of the plate, $\sigma_{kk} = \sigma_{xx} + \sigma_{yy} + \sigma_{zz}$. (**g**–**i**) Spatial distribution of the dissipated energy, $Q$, in the first three eigenmodes of the plate at time $T/4$. A cut at mid-plane level of the plate is displayed. The values were normalized to 1 with respect to the maximum in each mode since they were based on relative displacements and temperature fields.

Comparing the two data series, it can be seen that the loss factors measured experimentally were slightly higher than the simulated data. This discrepancy may be explained by the experimental challenges. Every disruptive factor in the experimental setup caused an increase in damping. Due to the very low magnitude of the loss factors measured in this study, the experiments were very sensitive to inaccuracies. The available vacuum chamber reached only a rough vacuum of ≈5 mbar. The remaining air resistance led to a small increase in the damping coefficient. Furthermore, the suspension was due to practical reasons realized at the edge of the plate. Therefore, it did not always matched exactly the zero line of the mode shape. This also affected the measured damping due to joint friction. Overall, the results of the simulations match those of the experiments well. Especially, the relative differences between the mode shapes could be represented very well.

As mentioned above, Serra et al. [38] developed a fully coupled finite element formulation considering thermoelastic loss. This theory was implemented in the FE-Software package ANSYS as SOLID226, SOLID227 or PLANE223 elements. These element types are available for full transient (time–domain) and full harmonic (frequency–domain) analyses.

Both components studied in the present contribution were simulated with these elements to compare the fully coupled approach with the simplified method presented here. The plate was discretized with SOLID226 elements (20-node hexahedrons) with an element size of 2 mm and 4 layers of elements in the thickness direction. A full harmonic analysis was performed with a point load applied sinusoidally at the corner point of the plate in the out-of-plane direction. No further boundary conditions were defined to simulate a free suspension.

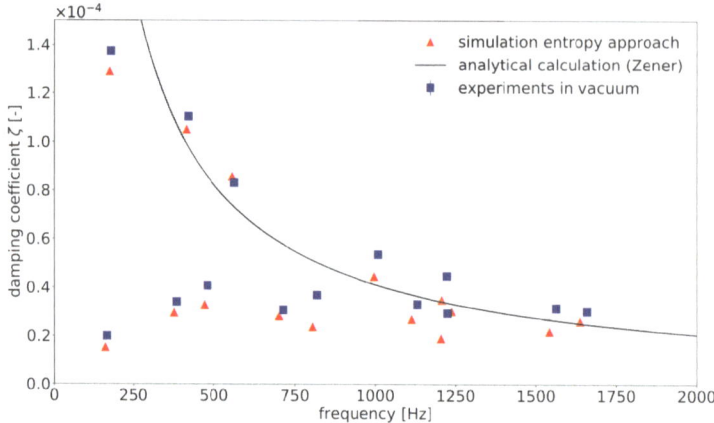

**Figure 5.** Comparison of experimental data, simulated damping coefficients and analytically calculated values with equation according to Zener [15].

The results of each frequency step include a real and an imaginary part of the solution. In order to calculate the loss factor, the complex data for the total strain energy were extracted per element and summed up over the whole structure. The loss factor is defined as the quotient of imaginary and real strain energy.

$$\eta = \frac{|\text{Im}(E_S)|}{|\text{Re}(E_S)|} \tag{15}$$

Figure 6 provides the results obtained from the ANSYS analysis in comparison to the loss factors calculated with the simplified approach. It is apparent from the diagram that the simplified energy approach produced only sample points at the resonant frequencies, whereas the outcome of the fully coupled harmonic response analysis was one loss factor per frequency step. If the excitation frequency in the simulation matches exactly the eigenfrequency of the plate, there is a pole in the curve. Therefore, the absolute height of the peaks depends on the frequency discretization and is not representative. The calculated loss factors followed the course of the graph satisfactorily. Both the magnitude and the differences in the mode-dependent damping factors could be represented accurately.

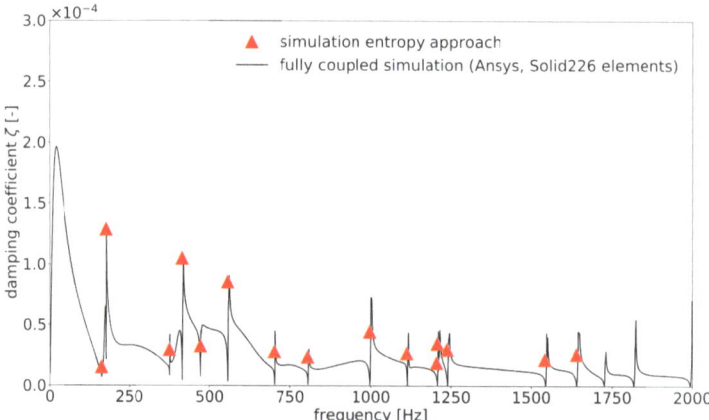

**Figure 6.** Comparison of the damping coefficients of the modal-based entropy approach with the fully coupled thermoelastic ANSYS simulation for the rectangular plate.

## 3.2. Complex Three-Dimensional Geometry

Figure 7 shows the correlation of the magnitude of displacement, $u$ (a–c), the hydrostatic stress, $\sigma_{kk}$ (d–f), and the generated heat, $Q$ (g–i), in a specific mode shape for the complex geometry. The eigenmodes 2, 5 and 7 were chosen as examples. Just as in Figure 4, all data were scaled to 1. The heat conduction, $Q$ (or, equivalently, the dissipated energy), was plotted at $t = T/4$. The points of maximal heat generation did not occur on the faces of the box but at the corners and edges, where the stress peaks were located. This is shown graphically in the detail plot under (j). If high stresses and strains occurred locally due to geometry reasons, there was a large difference in stress in respect to neighboring areas. This was followed by a temperature gradient with short paths of heat conduction, which led to locally high energy dissipation.

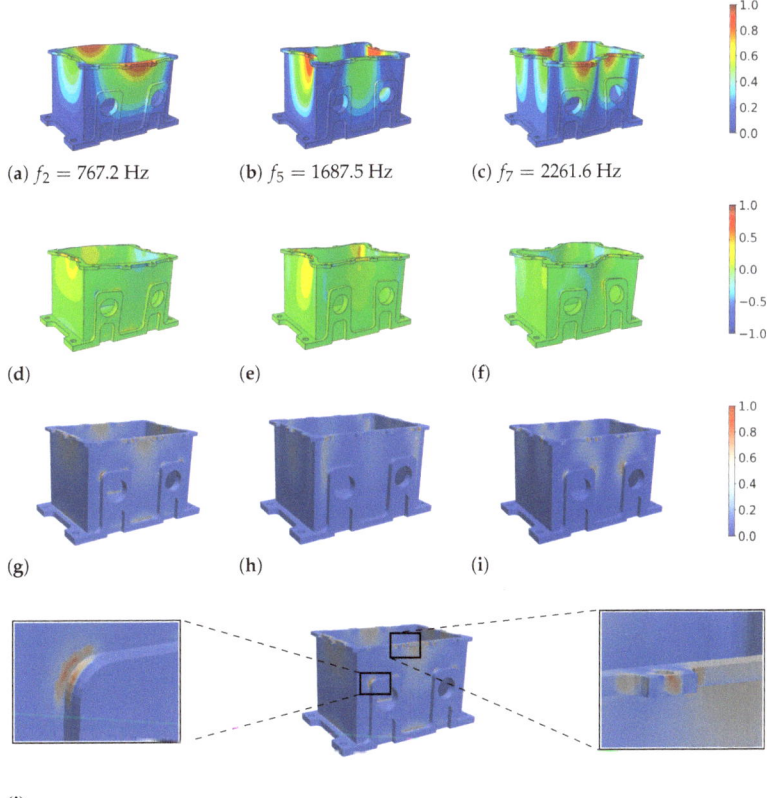

**Figure 7.** (**a**–**c**) Mode shapes 2, 5 and 7 of the box component; displacement magnitude normalized to 1. (**d**–**f**) Distribution of hydrostatic pressure $\sigma_{kk} = \sigma_{xx} + \sigma_{yy} + \sigma_{zz}$. (**g**–**i**) Spatial distribution of the dissipated energy, $Q$, at time $T/4$. The values were normalized to 1 with respect to the maximum in each mode since they were based on relative displacements and temperature fields. (**j**) Detailed view of locations with high energy dissipation.

Figure 8 compares the experimental values of the damping coefficients $\zeta$ with the simulated data obtained using the entropy method. The experiments were performed on two identical components and two measurement points were used for each eigenmode. The values shown are averaged measurement results. The standard deviation is represented by bars. The damping coefficient of the first eigenmode had a higher value than the following one. The other coefficients remained on the same level, with a slightly decreasing trend. This course was consistent with both calculated and measured values. Normally, the experimentally measured values are always slightly higher than the calculated ones (except in the second eigenmode). This is mainly due to the systematic problems in the experimental implementation, as already explained in Section 3.1. For practical reasons, all samples were suspended in the holes at the bottom of the case. These locations were not deformed in all the eigenmodes considered. However, they were not always the exact zero lines of the system, so there may have been additional effects due to frictional damping. In addition, high vacuum could not be achieved with the available vacuum chamber. During the experiments, air pressure of about 5 mbar was always maintained, which led to a slight air resistance during the oscillation. Taking these facts into account, there was overall good agreement between simulation and experiment.

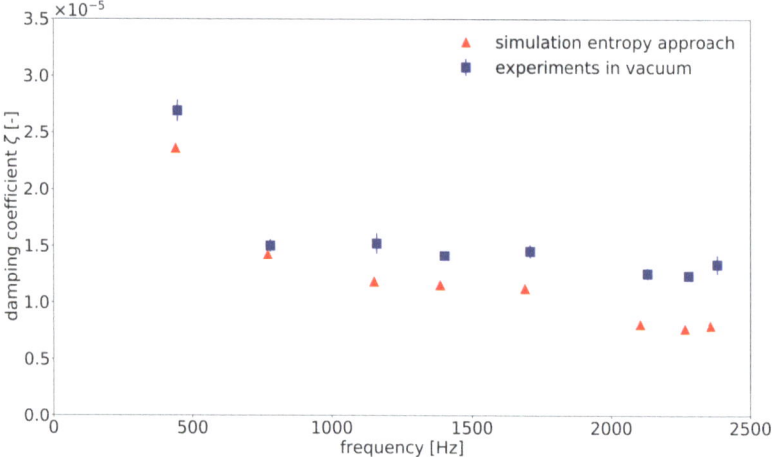

**Figure 8.** Comparison of the experimental data and the damping coefficient determined by the entropy approach for the box component.

Figure 9 exhibits the damping coefficients calculated by the entropy approach with the damping coefficient curve extruded from an ANSYS harmonic simulation with discretization by SOLID227 elements (fully coupled thermoelastic tetrahedron elements). In the full harmonic simulation, the system response was calculated for each frequency step. Afterwards, the loss factor was obtained by dividing the imaginary part of the strain energy by the real part of the strain energy. Therefore, one data point was calculated for each frequency step. The results of the entropy simulation were determined as discrete damping coefficients at the resonant frequencies. The curve of the fully coupled simulation had poles at the resonant frequencies; therefore, an exact comparison is difficult. The damping values calculated by the entropy approach followed the course of the curve satisfactorily.

In the next step, the calculated damping coefficients were applied in a frequency–domain simulation based on modal superposition. For this purpose, the discretely determined damping coefficients were used as modal damping ratios. Figure 10 shows a comparison of a single-point displacement in the modal-based harmonic simulation with the same displacement component in the full harmonic simulation. For this analysis, a single load F in the y-direction was applied harmonically at a point in the middle of the upper edge of the component, as shown in Figure 11. The displacement in the y-direction was determined at a node denoted by P. The course of the curves and the width of the peaks agreed satisfactorily. The absolute height of the peaks was not significant because it depended on the frequency discretization. Only at eigenfrequency 5 at approximately 1687 Hz were there inaccuracies. In this eigenmode, the considered node was close to a zero line (see Figure 7b), so that the oscillation mode could not be reproduced well in both simulations. This led to distortions of the results.

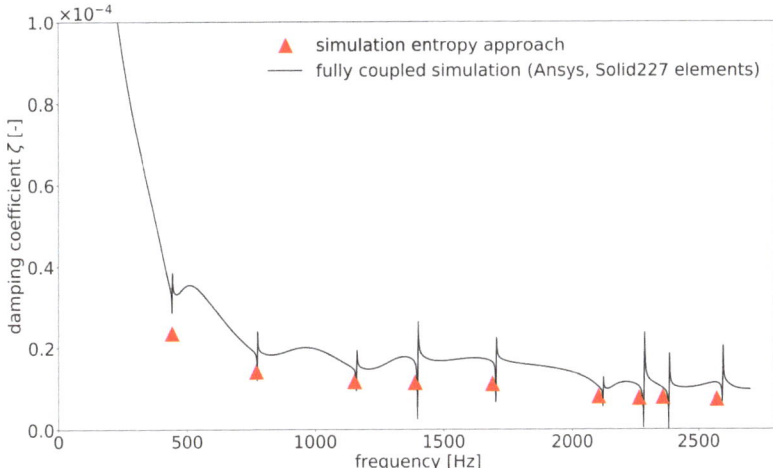

**Figure 9.** Comparison of the damping coefficients of the modal based entropy approach with the fully coupled thermoelastic ANSYS simulation for the box component.

Note that only frequencies f > 200 Hz are displayed here. The first natural frequency occurred at 437 Hz, i.e., the relevant frequency domain was covered. No boundary conditions were applied to the model to ensure comparability with the experiment. Therefore, a large influence of the rigid body modes occurred at low frequencies in the full harmonic simulation.

Overall, the excellent agreement of the curves shows that, with the approach presented here, results equivalent to those of a fully coupled calculation could be achieved with a significant reduction in computing time.

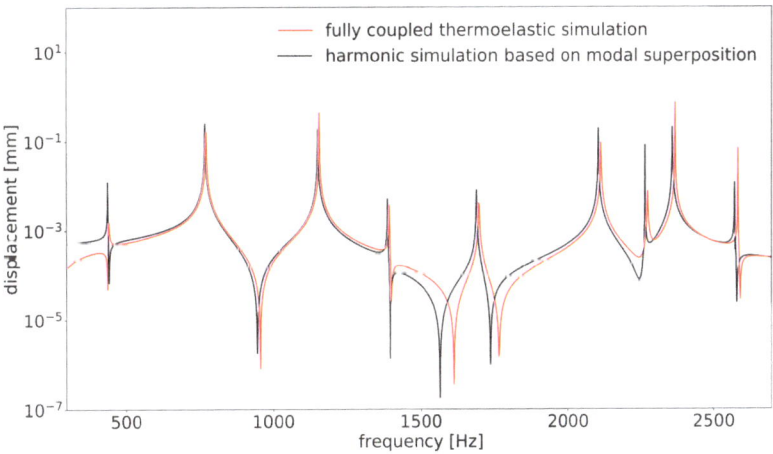

**Figure 10.** Comparison of displacement in y-direction of single node. Red curve: fully coupled simulation with thermoelastic elements SOLID226. Black curve: harmonic simulation based on modal superposition, use of modal damping coefficients that were calculated with the entropy approach.

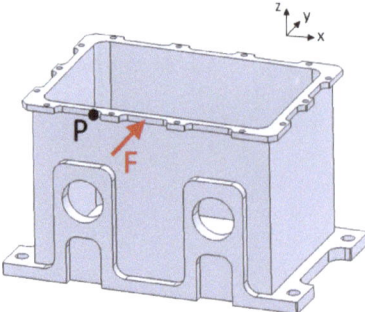

**Figure 11.** Excitation point of force F and measurement point P of displacement in frequency–domain analysis.

## 4. Discussion

First, the results demonstrate that thermoelastic damping is clearly mode-dependent. The damping coefficient of the aluminum alloy considered was in the range from $1 \times 10^{-5}$ to $1 \times 10^{-4}$, but the values of the individual mode shapes of a component differed from each other by a factor of 4 or more. In the literature, one general loss factor or damping coefficient is often given for a material (see, for example, [41,42]). Usually, it is not sufficient to use a global damping value for the simulation, as illustrated by Figures 12 and 13. In these diagrams, the system response of the rectangular aluminum plate in the first eigenmode to an impulse excitation is shown. Both time and frequency domain data were taken at a corner point of the plate. Figure 12 shows the time series of the velocity, filtered to the first natural frequency (band pass filter 160 Hz–170 Hz). The experimental data were compared with the finite element simulation using a global damping factor of 0.05% (following [41], where 0.04% was proposed). Furthermore, the results were set against the finite element simulation using the modal damping determined by the entropy approach presented here. It was obvious that the assumption of a global damping coefficient led to a decay curve that deviated strongly from the experimentally measured time course. Figure 13 shows the same system response in the frequency domain. The amount of damping is represented by the width of the peaks. The assumed global damping of 0.05% agreed well with the calculated damping coefficient of the second eigenmode, which is why the curves overlapped well in this range. A possible explanation for this might be that the second eigenmode was close to a pure beam bending; therefore, it could be calculated with Zener's theory. In the other eigenmodes, especially at the third natural frequency at approximately 376 Hz, the spectrum showed the deviation in the width of the peaks.

In comparison with the fully coupled simulations, good accuracy was achieved in the calculation of the loss factors, as could be shown with both components. The comparisons shown in Figures 6, 9 and 10 demonstrate good agreement. At the moment, it is not possible to apply the fully coupled element formulations for finite element analyses based on modal superposition. Therefore, especially for large systems, enormous computation times is required, since the thermoelastically coupled system of equations must be solved in every time or frequency step. The calculation of the system response in the frequency range under harmonic excitation took almost 6 days for the box component (2700 frequency steps). This procedure is usually not suitable for practical use. The frequency response analysis based on modal position using modal damping coefficients offers a great advantage in terms of computing time. Taking all the analytical steps into account, the system response for the box component could be solved in about 2 h. In addition, for the simpler system of the plate, the computation time could be reduced by more than half (5000 frequency steps). The computing times are displayed in Table 2. The bottleneck in the analysis remained the determination of the modal damping coefficients by the method presented here. This calculation step was the only one that was performed on a local computer and was not parallelized, so there is still potential for improvement in this respect. Furthermore, the disadvantages of a mode superposition method in finite element simulation must also be taken into account; the coupling effects of the mode shapes could not be represented and the accuracy of the solution depended on the number of considered modes.

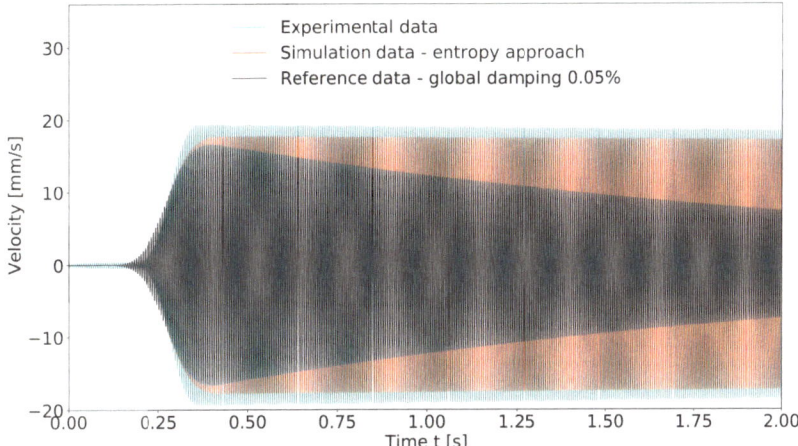

**Figure 12.** Comparison of velocity time series of impulse-excited plate structure: experiment, simulation with global structural damping and simulation with modal damping. The velocity was measured/analyzed on a corner point of the plate.

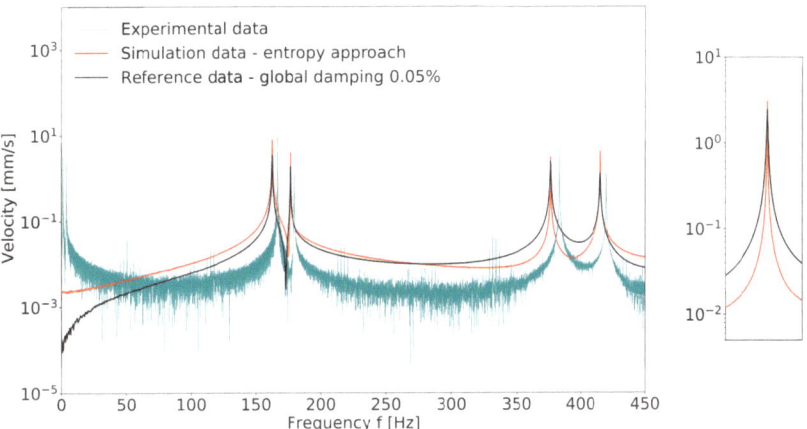

**Figure 13.** Comparison of velocity frequency spectrum of impulse-excited plate structure: experiment, simulation with global structural damping and simulation with modal damping. The velocity was measured/analyzed on a corner point of the plate.

In comparison with the simulated values, the experimental studies deliver always slightly higher damping coefficients. A possible explanation could be that, in the thermoelastic simulation, not all physical effects in the material are captured; therefore, lower values are calculated. However, it is more likely that the deviations are caused by the limitations of the experimental setup. In particular, the box component is difficult to install in the vacuum chamber due to its dimensions and weight. Therefore, for example, in our study, compromises had to be made in the design of the bearings. Since the suspension could not always be realized exactly in the zero line of the eigenmode, slight damping effects due to friction occurred. In addition, a complete vacuum was not achieved in the experimental setup, so that a small influence due to air damping could be recorded. The experimental setup should be improved so that measured values can better compare with numerically simulated ones.

**Table 2.** Computing time in CPU hours for a frequency–domain analysis using a direct solution procedure with fully coupled thermoelastic elements and a mode superposition solution procedure with modal damping. The modal damping coefficients were determined with the entropy approach provided in the paper. For the rectangular plate, 5000 frequency steps were analyzed and, for the box component, 2700 frequency steps were analyzed. All simulations (except for the calculation of the modal damping coefficients) were run on 16 cores of a compute node with 2 Intel Xeon Processor E5-2650 v4 CPUs with a base clock of 2.2 GHz and 128 GB RAM.

|  | Plate | Box Component |
| --- | --- | --- |
| number of elements | 9600 el. | 291,320 el. |
| fully coupled simulation SOLID226/227 | 34.8 | 2234.6 ($\approx$6 days) |
| entropy approach SOLID186/187 | 16.5 | 35.1 |
| eigenfrequency analysis | 1.2 | 7.5 |
| modal damping coeff. | 9.3 | 17.9 |
| harmonic analysis | 6.0 | 9.7 |

## 5. Conclusions

We here present a method to calculate modal thermoelastic damping coefficients for arbitrary components discretized with solid elements. The analysis is based on an increase in entropy during mechanical vibrations that leads to energy dissipation. Modal damping ratios were calculated and used in subsequent finite element simulations based on modal superposition.

The results were compared with simulations using a thermoelastically fully coupled element approach. It is shown that similar results were obtained for both the loss factors and the system response in the frequency spectrum in significantly lower computation times.

The comparison of the modal damping coefficients with the experimental data showed good agreement. The physically measured damping values are always slightly higher than the calculated ones.

In conclusion, the method seems suitable for calculating material damping. However, in order to make realistic predictions of the damping behavior of structures, other causes of energy dissipation must also be taken into account. Suitable models for friction and air damping must be used in the simulations.

In addition, the extension of the method to other materials might prove an important area for future research. The method was developed and tested on aluminum. As a next step, anisotropic materials or laminated components could be investigated.

Furthermore, interesting research questions can be derived from the possibility to locate the areas of high energy dissipation in the simulation. This provides a good starting point for the optimization of geometries or surfaces.

**Supplementary Materials:** The following supporting information can be downloaded at https://www.mdpi.com/article/10.3390/ma15051706/s1, Video S1: Simulation of mode 1 of the rectangular aluminum plate. The points of suspension, excitation and measurement are shown. Video S2: Simulation of mode 2 of the aluminum gearbox case. The points of suspension, excitation and measurement are shown.

**Author Contributions:** Conceptualization, C.Z., C.K. and C.G.; methodology, C.Z.; software, C.Z.; investigation, C.Z.; resources, C.Z. and C.K.; data curation, C.Z.; writing—original draft preparation,

C.Z.; writing—review and editing, C.K. and C.G.; visualization, C.Z.; supervision, C.K.; project administration, C.G. All authors have read and agreed to the published version of the manuscript.

**Funding:** This research study received no external funding.

**Institutional Review Board Statement:** Not applicable.

**Informed Consent Statement:** Not applicable.

**Data Availability Statement:** Not applicable.

**Conflicts of Interest:** The authors declare no conflict of interest.

## References

1. Xu, Z.; Ha, C.S.; Kadam, R.; Lindahl, J.; Kim, S.; Wu, H.F.; Kunc, V.; Zheng, X. Additive manufacturing of two-phase lightweight, stiff and high damping carbon fiber reinforced polymer microlattices. *Addit. Manuf.* **2020**, *32*, 101106. [CrossRef]
2. Attard, T.L.; He, L.; Zhou, H. Improving damping property of carbon-fiber reinforced epoxy composite through novel hybrid epoxy-polyurea interfacial reaction. *Compos. Part B Eng.* **2019**, *164*, 720–731. [CrossRef]
3. Ledi, K.S.; Hamdaoui, M.; Robin, G.; Daya, E.M. An identification method for frequency dependent material properties of viscoelastic sandwich structures. *J. Sound Vib.* **2018**, *428*, 13–25. [CrossRef]
4. Arunkumar, M.P.; Jagadeesh, M.; Pitchaimani, J.; Gangadharan, K.V.; Babu, M.C. Sound radiation and transmission loss characteristics of a honeycomb sandwich panel with composite facings: Effect of inherent material damping. *J. Sound Vib.* **2016**, *383*, 221–232. [CrossRef]
5. Sadiq, S.E.; Bakhy, S.H.; Jweeg, M.J. Optimum vibration characteristics for honey comb sandwich panel used in aircraft structure. *J. Eng. Sci. Technol.* **2021**, *16*, 1463–1479.
6. Le Guen, M.J.; Newman, R.H.; Fernyhough, A.; Emms, G.W.; Staiger, M.P. The damping-modulus relationship in flax-carbon fibre hybrid composites. *Compos. Part B Eng.* **2016**, *89*, 27–33. [CrossRef]
7. Ma, L.; Chen, Y.L.; Yang, J.S.; Wang, X.T.; Ma, G.L.; Schmidt, R.; Schröder, K.U. Modal characteristics and damping enhancement of carbon fiber composite auxetic double-arrow corrugated sandwich panels. *Compos. Struct.* **2018**, *203*, 539–550. [CrossRef]
8. Li, H.; Niu, Y.; Li, Z.; Xu, Z.; Han, Q. Modeling of amplitude-dependent damping characteristics of fiber reinforced composite thin plate. *Appl. Math. Model.* **2020**, *80*, 394–407. [CrossRef]
9. Wesolowski, M.; Barkanov, E. Improving material damping characterization of a laminated plate. *J. Sound Vib.* **2019**, *462*, 114928. [CrossRef]
10. Duhamel, J.M.C. Second memoire sur les phenomenes thermo-mecaniques. *J. l'Ecole Polytech.* **1837**, *15*, 1–57.
11. Maugin, G. *Duhamel's Pioneering Work in Thermo-Elasticity and Its Legacy*; Springer International Publishing: Berlin/Heidelberg, Germany, 2014.
12. Biot, M.A. Thermoelasticity and Irreversible Thermodynamics. *J. Appl. Phys.* **1956**, *27*, 240–253. [CrossRef]
13. Parkus, H. *Thermoelasticity*; Springer: Wien, Austria, 1976.
14. Nowacki, W. Thermoelasticity. In *International Series of Monographs in Aeronautics and Astronautics. Division 1: Solid and Structural Mechanics*; Addison-Wesley Publishing Company: Boston, MA, USA, 1962; Volume 3.
15. Zener, C. Internal Friction in Solids. I. Theory of Internal Friction in Reeds. *Phys. Rev.* **1937**, *52*, 230–235. [CrossRef]
16. Zener, C.; Otis, W.; Nuckolls, R. Internal Friction in Solids III. Experimental Demonstration of Thermoelastic Internal Friction. *Phys. Rev.* **1938**, *53*, 100–101. [CrossRef]
17. Zener, C. Internal Friction in Solids II. General Theory of Thermoelastic Internal Friction. *Phys. Rev.* **1938**, *53*, 90–99. [CrossRef]
18. Chadwick, P. On the propagation of thermoelastic disturbances in thin plates and rods. *J. Mech. Phys. Solids* **1962**, *10*, 99–109. [CrossRef]
19. Chadwick, P. Thermal damping of a vibrating elastic body. *Mathematika* **1962**, *9*, 38–48. [CrossRef]
20. Alblas, J.B. On the general theory of thermo-elastic friction. *Appl. Sci. Res.* **1961**, *10*, 349–362. [CrossRef]
21. Alblas, J.B. A note on the theory of thermoelastic damping. *J. Therm. Stress.* **1981**, *4*, 333–355. [CrossRef]
22. Lifshitz, R.; Roukes, M.L. Thermoelastic damping in micro- and nanomechanical systems. *Phys. Rev. B* **2000**, *61*, 5600–5609. [CrossRef]
23. Wong, S.; Fox, C.; McWilliam, S. Thermoelastic damping of the in-plane vibration of thin silicon rings. *J. Sound Vib.* **2006**, *293*, 266–285. [CrossRef]
24. De, S.; Aluru, N. Theory of thermoelastic damping in electrostatically actuated microstructures. *Phys. Rev. B* **2006**, *74*, 144305. [CrossRef]
25. Sun, Y.; Tohmyoh, H. Thermoelastic damping of the axisymmetric vibration of circular plate resonators. *J. Sound Vib.* **2009**, *319*, 392–405. [CrossRef]
26. Kinra, V.K.; Milligan, K.B. A Second-Law Analysis of Thermoelastic Damping. *J. Appl. Mech.* **1994**, *61*, 71–76. [CrossRef]
27. Bishop, J.; Kinra, V. *Some Improvements in the Flexural Damping Measurement Technique*; ASTM Special Technical Publication: Philadelphia, PA, USA, 1992; pp. 457–470.
28. Bishop, J.; Kinra, V. Elastothermodynamic damping in laminated composites. *Int. J. Solids Struct.* **1997**, *34*, 1075–1092. [CrossRef]

29. Duwel, A.; Candler, R.; Kenny, T.; Varghese, M. Engineering MEMS resonators with low thermoelastic damping. *J. Microelectromech. Syst.* **2006**, *15*, 1437–1445. [CrossRef]
30. Chandorkar, S.; Candler, R.; Duwel, A.; Melamud, R.; Agarwal, M.; Goodson, K.; Kenny, T. Multimode thermoelastic dissipation. *J. Appl. Phys.* **2009**, *105*, 043505. [CrossRef]
31. Hao, Z.; Xu, Y.; Durgam, S. A thermal-energy method for calculating thermoelastic damping in micromechanical resonators. *J. Sound Vib.* **2009**, *322*, 870–882. [CrossRef]
32. Tai, Y.; Li, P.; Zuo, W. *An Entropy Based Analytical Model for Thermoelastic Damping in Micromechanical Resonators*; Applied Mechanics and Materials; Trans Tech Publications Ltd.: Baech, Switzerland, 2012; Volume 159, pp. 46–50.
33. Tai, Y.; Li, P. An analytical model for thermoelastic damping in microresonators based on entropy generation. *J. Vib. Acoust. Trans. ASME* **2014**, *136*, 031012. [CrossRef]
34. Metcalf, T.; Pate, B.; Photiadis, D.; Houston, B. Thermoelastic damping in micromechanical resonators. *Appl. Phys. Lett.* **2009**, *95*, 165–178. [CrossRef]
35. Cagnoli, G.; Lorenzini, M.; Cesarini, E.; Piergiovanni, F.; Granata, M.; Heinert, J.; Martelli, F.; Nawrodt, R.; Amato, A.; Cassar, Q.; et al. Mode-dependent mechanical losses in disc resonators. *Phys. Lett. A* **2018**, *382*, 2165–2173. [CrossRef]
36. Huang, Y.; Sturt, R.; Willford, M. A damping model for nonlinear dynamic analysis providing uniform damping over a frequency range. *Comput. Struct.* **2019**, *212*, 101–109. [CrossRef]
37. Amabili, M. Nonlinear damping in large-amplitude vibrations: Modelling and experiments. *Nonlinear Dyn.* **2018**, *93*, 5–18. [CrossRef]
38. Serra, E.; Bonaldi, M. A finite element formulation for thermoelastic damping analysis. *Int. J. Numer. Methods Eng.* **2009**, *78*, 671–691. [CrossRef]
39. Landau, L.; Lifshitz, E. Theory of elasticity. In *Course of Theoretical Physics*, 2nd ed.; Pergamon Press: Oxford, UK, 1986.
40. Logg, A.; Mardal, K.A.; Wells, G. *Automated Solution of Differential Equations by the Finite Element Method: The FEniCS Book*; Springer Science & Business Media: Berlin/Heidelberg, Germany, 2012; Volume 84.
41. Adams, V.; Askenazi, A. *Building Better Products with Finite Element Analysis*, 1st ed.; OnWord Press: Santa Fe, NW, USA, 1999.
42. Cremer, L.; Heckl, M.; Petersson, B. *Structure-Borne Sound: Structural Vibrations and Sound Radiation at Audio Frequencies*, 3rd ed.; Springer: Berlin, Germany, 2010.

# Natural Frequencies Optimization of Thin-Walled Circular Cylindrical Shells Using Axially Functionally Graded Materials

Nabeel Taiseer Alshabatat

Mechanical Engineering Department, Tafila Technical University, Tafila 66110, Jordan; nabeel@ttu.edu.jo

**Abstract:** One method to avoid vibration resonance is shifting natural frequencies far away from excitation frequencies. This study investigates optimizing the natural frequencies of circular cylindrical shells using axially functionally graded materials. The constituents of functionally graded materials (FGMs) vary continuously in the longitudinal direction based on a trigonometric law or using interpolation of volume fractions at control points. The spatial change of material properties alters structural stiffness and mass, which then affects the structure's natural frequencies. The local material properties at any place in the structure are obtained using Voigt model. First-order shear deformation theory and finite element method are used for estimating natural frequencies, and a genetic algorithm is used for optimizing material volume fractions. To demonstrate the proposed method, two optimization problems are presented. The goal of the first one is to maximize the fundamental frequency of an FGM cylindrical shell by optimizing the material volume fractions. In the second problem, we attempt to find the optimal material distribution that maximizes the distance between two adjoining natural frequencies. The optimization examples show that building cylindrical shells using axially FGM is a useful technique for optimizing their natural frequencies.

**Keywords:** cylindrical shell; functionally graded material; natural frequency; optimization

## 1. Introduction

Cylindrical shells are extensively employed in various applications such as aerospace, automobile, marine, and construction industries. They are usually exposed to dynamic excitations, which cause vibrations and make noise. Reducing vibrations and noise radiation from these structures are important issues in the early stages of structural design process. A common technique to decrease vibrations is shifting the natural frequencies of a structure away from the excitation frequencies to prevent resonance. Different passive methods are employed for optimizing the natural frequencies of cylindrical shells, and these include adjusting the thickness, adding stiffeners, and tailoring material. For example, Alzahabi [1] shifted the natural frequencies of a submarine hull by controlling the local thickness. The length of the hull was divided into equal segments, and then the thickness of each segment was optimized to achieve the sought natural frequency. Nasrekani et al. [2] studied the effect of changing the thickness, continuously along a cylindrical shell, on its first axisymmetric natural frequency. Bagheri et al. [3] investigated a multi-objective problem for simultaneously maximizing the fundamental frequency and minimizing the weight of cylindrical shell using ring stiffeners. Mehrabami et al. [4] maximized the fundamental frequency-to-weight ratio of cylindrical shell using rings and strings. Akl et al. [5] minimized the vibration and/or the sound radiation from under water cylindrical shells using stiffeners. In addition to optimizing the natural frequencies, the literature about the optimization of circular cylindrical shells with other objective functions is very wide. For example, Biglar et al. [6] optimized the locations and orientations of piezoelectric patches on cylindrical shells to reduce the vibration. Sadeghifar et al. [7] studied maximizing the critical buckling load and minimizing the weight of orthogonally stiffened cylindrical shells. Belradi et al. [8] utilized anisogrid composite lattice structures in building circular cylindrical shells to optimize the critical buckling load with strength and stiffness constraints.

Laminated composite materials are used in building circular cylindrical shells due to their high stiffness-to-weight ratio. Various studies have dealt with the vibrations of laminated composite cylindrical shells [9–17]. The optimization of the stacking sequence for natural frequency optimization is a common technique. For example, Hu and Tsi [18] investigated the maximization of the fundamental frequency for cylindrical shells with and without cutout. The optimization problem was solved using Golden section method. Koide and Luersen [19] used ant colony optimization to maximize the fundamental frequency. Trias et al. [20] investigated the maximization of the fundamental frequency using bound formulation. Miller and Ziemiański [21,22] maximized the fundamental frequency and the distance between two adjoining natural frequencies using the genetic algorithm (GA) in combination with neural networks. Recently, Jing [23] used sequential permutation search algorithm to maximize the fundamental frequency. The fundamental frequency of laminated composite cylindrical shells reinforced with shape memory alloy fibers was investigated by Nekouei et al. [24]. They showed that a small amount of shape memory alloy fibers increased the fundamental frequency significantly.

Due to sudden change in the mechanical properties between laminars, laminated composites are very prone to interlamination damages. Thus, using functionally graded materials (FGMs) overcomes the disadvantages of using multilayer composites, such as delamination and stress concentration [25]. FGMs are composites typically built from two materials with continuous variation of the materials' composition through the dimensions of structures. These continuous variations of compositions results in a smooth variation of material properties. Owing to their distinct characteristics, FGMs are used in the design optimization of structures [26–28]. Several researchers investigated the free vibration of thin-walled FGMs cylindrical shells. Loy et al. [29] investigated the free vibration of simply-supported FGM cylindrical shell utilizing Love's shell theory and Rayleigh–Ritz method. Using a similar approach, Arshad et al. [30,31] studied FGM cylinders with various boundary conditions. Using first-order shear deformation theory (FSDT) and Rayleigh–Ritz method, Sue et al. [32] investigated a unified solution technique to estimate the natural frequencies of cylindrical and conical shells. Jin et al. [33] employed FSDT and Haar wavelet discretization. Punera and Kant [34] used several higher-order theories and Navier's method to investigate the free vibration of open FGM cylinders. Ni et al. [35] adopted Reissner shell theory and the Hamiltonian method to find the natural frequencies of FGM cylinders set in elastic mediums. Liu et al. [36] used FSDT and wave function expansions to find the natural frequencies. Alshabatat and Zannon [37] studied the free vibrations of FGM cylindrical shell by employing third-order shear deformation theory and Carrera's unified formulation. All previous studies on materials' volume fraction variations in the radial direction of cylindrical shells (i.e., through thickness) based on simple power law [29–31,34,35,37], exponential law [30,31], trigonometric law [30,31], and four-parameter power laws [32,33,36].

To the best of our knowledge, the optimization of natural frequencies of thin-walled cylindrical shells using axial grading material has not been attempted yet. Hence, in this work, analyses and optimization of FGM cylindrical shells are presented. First, the effect of material gradient through the axial direction on the natural frequencies of thin cylindrical shells is investigated. Then, the optimization of the natural frequencies is conducted. The material volume fractions are graded through the length of the cylinders according to control points method and a trigonometric law. FSDT and finite element method (FEM) are employed to find the natural frequencies. To validate the present method, numerical results are compared with those available in the literature. A genetic algorithm (GA) is utilized to look for the ideal material distributions that optimize the natural frequencies. Two optimization problems are considered to show the efficiency of the presented method. The methods for the manufacturing and processing of FGMs are not considered here and can be found in [38–40].

## 2. Materials and Methods

### 2.1. Functionally Graded Material

Consider a circular cylindrical shell made from FGM, which consists of two materials mixture. The constituents' volume fractions are graded in the axial direction of the cylinder and follow a simple power law distribution or a trigonometric distribution. The volume fraction of the first material is given by

$$V_1(x) = \left(\frac{x}{L}\right)^\gamma \tag{1}$$

$$\text{or} \quad V_1(x) = (\alpha + \alpha \cos(\eta x + \phi))^\gamma \tag{2}$$

where $L$ is the length of the cylinder, and $x$ is the axial coordinate ($0 \leq x \leq L$). $\gamma, \alpha, \eta$, and $\phi$ are the design variables used to control the volume fraction profile. These parameters are chosen such that $0 \leq V_1(x) \leq 1$. Note that $V_2(x) = 1 - V_1(x)$. In addition to these laws, this work suggests using control points method for optimization problems. In this method, the volume fraction profile is controlled by the volume fractions at some control points that are distributed evenly through the length of the cylinder. The location of $i$th control point, along the cylinder length, is $x_i = (i-1)L/(N-1)$ in which $N$ is the number of control points. The volume fraction distribution in the cylinder, $V_1(x)$, can then be estimated by the piecewise cubic hermit interpolating polynomial (PCHIP) [41,42]. The design variables in this case are $V_{1,1}, V_{1,2}, \ldots, V_{1,N}$.

The local material properties of FGM are estimated using Voigt model (or the rule of mixture) as

$$E(x) = (E_1 - E_2)V_1(x) + E_2, \tag{3}$$

$$\rho(x) = (\rho_1 - \rho_2)V_1(x) + \rho_2, \tag{4}$$

$$\nu(x) = (\nu_1 - \nu_2)V_1(x) + \nu_2 \tag{5}$$

where $E_i$ is the Young's modulus, $\rho_i$ is the density, $\nu_i$ is the Poisson's ratio, and the subscript $i$ denotes material 1 or 2.

### 2.2. Kinematic Relations

A circular cylindrical FGM shell is considered with the length $L$, thickness $h$, and radius $R$ as shown in Figure 1. The shell is assumed thin (i.e., $h/R \leq 0.05$). According to Reddy et al. [43,44], the solution of Love–Kirchhoff theory is not accurate for composite structures. Thus, FSDT is used in this study. Based on FSDT, the displacements of any point in the cylinder can be written as

$$u(x,y,z,t) \simeq u_o(x,y,t) + z\theta_x(x,y,t) \tag{6}$$

$$v(x,y,z,t) \simeq v_o(x,y,t) + z\theta_y(x,y,t) \tag{7}$$

$$w(x,y,z,t) \simeq w_o(x,y,t) \tag{8}$$

where $u, v$, and $w$ are the displacements in the axial ($x$), circumferential ($y$), and radial ($z$) directions, respectively. $u_o, v_o$, and $w_o$ are the middle surface displacements. $\theta_x$ and $\theta_y$ are the middle surface rotations, and $t$ is the time.

The strain–displacement relations for thin shells are given as [45]

$$\{\varepsilon\} = \begin{Bmatrix} \varepsilon_{xx} \\ \varepsilon_{yy} \\ \gamma_{xy} \\ \gamma_{xz} \\ \gamma_{yz} \end{Bmatrix} = \begin{Bmatrix} \frac{\partial u_o}{\partial x} + z\frac{\partial \theta_x}{\partial x} \\ \frac{\partial v_o}{\partial y} + z\frac{\partial \theta_y}{\partial y} + \frac{w_o}{R} \\ \frac{\partial v_o}{\partial x} + \frac{\partial u_o}{\partial y} + z\left(\frac{\partial \theta_y}{\partial x} + \frac{\partial \theta_x}{\partial y}\right) \\ \frac{\partial w_o}{\partial x} + \theta_x \\ \frac{\partial w_o}{\partial y} - \frac{v_o}{R} + \theta_y \end{Bmatrix}. \tag{9}$$

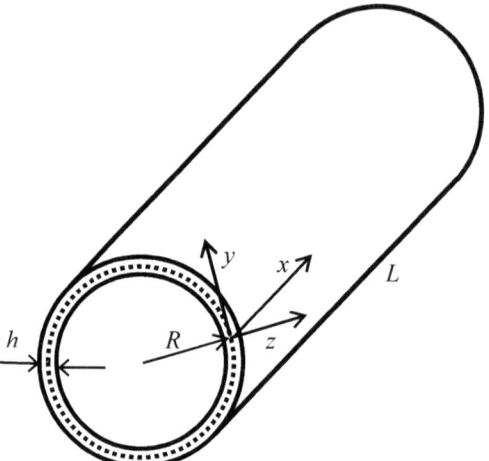

**Figure 1.** The geometry of the circular cylindrical shell.

The stress $\sigma_{zz}$ is negligible for a thin shell. The stress–strain relations of FGM cylindrical shell are given as

$$\begin{Bmatrix} \sigma_{xx} \\ \sigma_{yy} \\ \tau_{xy} \\ \tau_{xz} \\ \tau_{yz} \end{Bmatrix} = \frac{E(x)}{1-\nu^2(x)} \begin{bmatrix} 1 & \nu(x) & 0 & 0 & 0 \\ \nu(x) & 1 & 0 & 0 & 0 \\ 0 & 0 & \frac{1-\nu(x)}{2} & 0 & 0 \\ 0 & 0 & 0 & \frac{1-\nu(x)}{2} & 0 \\ 0 & 0 & 0 & 0 & \frac{1-\nu(x)}{2} \end{bmatrix} \begin{Bmatrix} \varepsilon_{xx} \\ \varepsilon_{yy} \\ \gamma_{xy} \\ \gamma_{xz} \\ \gamma_{yz} \end{Bmatrix} \quad (10)$$

or $\quad \{\sigma\} = [Q]\{\varepsilon\}. \quad (11)$

The elements of matrix $Q$ are functions of the longitudinal coordinate ($x$). The governing equations are obtained by applying Hamilton's principle as

$$\int_{t_1}^{t_2} \delta(T-U)dt = 0 \quad (12)$$

where $T$ and $U$ are the kinetic and strain energies of the vibrating cylindrical shell, respectively. The kinetic energy can be given by

$$T = \frac{1}{2} \iiint \rho(x) \left[ \left(\frac{\partial u}{\partial t}\right)^2 + \left(\frac{\partial v}{\partial t}\right)^2 + \left(\frac{\partial w}{\partial t}\right)^2 \right] dV, \quad (13)$$

and the strain energy can be given by

$$U = \frac{1}{2} \iiint \sigma \varepsilon dV. \quad (14)$$

The equations of motion are acquired by substituting Equations (13) and (14) into Equation (12). Owing to the complication of the present case, FEM is adopted to solve the equations and find the natural frequencies. A quadrilateral shell element with four nodes is employed to discretize the FGM cylinders. Each node has five degrees of freedom. The material properties of each finite element are apportioned by Equations (3)–(5), in which $V_1(x)$ represents the volume fraction of the first material at the center of the element. The degrees of freedom include three translational ($u_o^i$, $v_o^i$, and $w_o^i$) and two rotational ($\theta_x^i$ and $\theta_y^i$). The element kinetic and strain energies are given by [46]

$$T_e = \frac{1}{2}\{\dot{\delta}\}_e^T [m]_e \{\dot{\delta}\}_e \tag{15}$$

$$\text{and} \quad U_e = \frac{1}{2}\{\delta\}_e^T [k]_e \{\delta\}_e, \tag{16}$$

where $\{\delta\}_e^T = \{\delta_1^T, \delta_2^T, \delta_3^T, \delta_4^T\}$, $\delta_i^T = \{u_o^i, v_o^i, w_o^i, \theta_x^i, \theta_y^i\}$, $[m]_e$ is the element mass matrix, and $[k]_e$ is the element stiffness matrix. The element mass matrix is given by

$$[m]_e = \iiint \rho N^T N |J| dV_e, \tag{17}$$

and the element stiffness matrix is

$$[k]_e = \iiint B^T \overline{Q} B |J| dV_e, \tag{18}$$

where $N$ is the shape functions matrix, $J$ is the Jacobian matrix, $B$ is the strain-displacement matrix, and $\overline{Q}$ is the stress-strain matrix with respect to global coordinates. The shape functions are quadratic. Equations (17) and (18) can be evaluated using Gause–Legendre numerical integration. Equation (17) can be evaluated using $3 \times 3$ integration points in the in-plane coordinates [46]. For thin shells, Equation (18) can be evaluated using $2 \times 2$ integration points in the in-plane coordinates [47]. The details of the finite element formulations are given in [46]. The governing equations for the free vibration of FGM cylinders are given as

$$[M]\{\ddot{\delta}\} + [K]\{\delta\} = \{0\}, \tag{19}$$

where $[M]$ and $[K]$ are the global mass and stiffness matrices, respectively, and $\{\delta\}$ and $\{\ddot{\delta}\}$ are the nodal displacement and acceleration vectors, respectively. The natural frequencies ($\omega_n$) and mode shapes can be found by solving the following eigenvalue problem:

$$\left[K - \omega_n^2 M\right]\{\delta\} = \{0\}. \tag{20}$$

Note that the fundamental frequency is the smallest value of $\omega_n$.

## 3. Results and Discussion

### 3.1. Validation and Convergence

In this study, FEM is employed for free vibration analysis of cylindrical shells. The eigenvalue problem (Equation (20)) is solved using Block Lanczos iteration method [48], which is performed using in-house program developed with MATLAB software. To evaluate the accuracy of the algorithm in the present method, validation and convergence study is performed for clamped-free (CF) isotropic circular cylindrical shells with Naeem and Sharma [49] and Arshad et al. [31]. In this study, the dimensions of the cylinders are $R = 0.2423$ m, $h = 0.000648$ m, and $L = 0.6255$ m. The material properties are $E = 68.95$ GPa and $\rho = 2714.5$ kg/m$^3$. The numerical results, which are summarized in Table 1, are close to those available in the literature.

**Table 1.** Comparison of the lowest natural frequencies of clamped-free isotropic cylindrical shells.

|  |  | Frequency (Hz) | | | | |
|---|---|---|---|---|---|---|
|  | Elements | 1 | 2 | 3 | 4 | 5 |
| Present | 20 × 10 | 94.08 | 108.09 | 108.6 | 138.6 | 167.3 |
|  | 24 × 10 | 93.26 | 106.25 | 108.49 | 135.29 | 167.78 |
|  | 36 × 20 | 92.58 | 105.28 | 108.15 | 133.95 | 168.28 |
| Ref. [49] |  | 92.55 | 105.05 | 108.47 | 133.60 | 169.12 |
| Ref. [31] |  | 95.38 | 106.33 | 114.39 | 134.23 | 171.84 |

Here, we are interested in FGM circular cylindrical shell. Thus, convergence study is performed for FGM cylinders composed of aluminum (Al) and zirconia (ZrO$_2$) with

CF boundary conditions. The material volume fractions are graded over the cylinder length according to a power Law (Equation (1)). The materials properties are listed in Table 2. In this study, the dimensions of the cylinders are $R = 1$ m, $h/R = 0.05$, and $L/R = 10$. Table 3 shows the convergence study of the fundamental frequencies with different power-law exponents ($\gamma$) and number of elements. The fundamental frequency can be considered as convergent at $36 \times 40$ elements (i.e., 36 elements in the circumferential direction and 40 elements in the longitudinal direction). In the following investigations, the $36 \times 40$ element mesh is used.

**Table 2.** Material properties of aluminum and zirconia.

| Property | Al | $ZrO_2$ |
|---|---|---|
| $E$ (GPa) | 20 | 205 |
| $\rho$ (kg/m$^3$) | 2700 | 6050 |
| $v$ | 0.3 | 0.31 |

**Table 3.** Convergence of the fundamental frequency (Hz) of clamped-free FGM cylinders with number of finite elements.

| Elements | $\gamma$ | | |
|---|---|---|---|
| | 0.5 | 1 | 10 |
| 12 × 5 | 23.67 | 24.88 | 21.84 |
| 24 × 5 | 24.26 | 25.47 | 22.38 |
| 36 × 5 | 24.57 | 25.80 | 22.67 |
| 36 × 10 | 25.17 | 26.40 | 23.19 |
| 36 × 20 | 25.32 | 26.53 | 23.35 |
| 36 × 40 | 25.34 | 26.54 | 23.37 |

*3.2. Parametric Study*

In this study, the possibility of shifting the fundamental frequency of thin cylinders using FGM graded in the axial direction is investigated. The effect of the power-law exponent ($\gamma$) on the fundamental frequency of CF thin cylinder with different dimensions is presented. The FGM compositions are aluminum (AL) and zirconia ($ZrO_2$). The properties of aluminum and zirconia are listed in Table 2. The material constituents vary in axial direction as per of the power law (Equation (1)), where $V_1(x)$ represents the volume fraction of aluminum. The FGM cylinders have zirconia at $x = 0$ and aluminum at $x = L$, as shown in Figure 2.

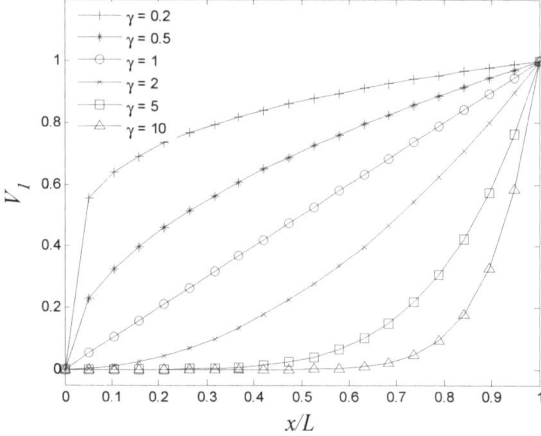

**Figure 2.** Variations of volume fractions based on simple power law (Equation (1)).

Table 4 shows the effect of varying the power law exponent ($\gamma$) on the fundamental frequency for different length-to-radius ($L/R$) ratios (assuming $h/R = 0.05$). Table 5 shows the effect of varying the power law exponent ($\gamma$) on the fundamental frequency for different thickness-to-radius ($h/R$) ratios (assuming $L/R = 10$). Similar to isotropic cylindrical shells, the fundamental frequencies of long cylinders (i.e., high $L/R$ ratios) are smaller than those of short cylinders (i.e., increasing $L/R$ ratio decreases the bending stiffness-to-mass ratio, which decreases the fundamental frequency). The fundamental frequency also increases by increasing the thickness (i.e., increasing $h/R$ ratio increases the bending stiffness-to-mass ratio, which increases the fundamental frequency). Tables 4 and 5 also show that increasing the exponent ($\gamma$) only increases the cylinder's fundamental frequency to a certain value. Any further increase of the exponent ($\gamma$) does not increase the cylinder's fundamental frequency, as an increase in the cylinder mass leads to the reduction in the cylinder's fundamental frequency. This trend differs from that of FGM cylinders when material constituents vary in thickness ($z$). In the case of material variations in thickness, the natural frequencies of the FGM thin cylindrical shells are limited between the natural frequencies of the first- and second-base materials [31,37]. The natural frequencies depend on material distributions through the axial direction of FGM circular cylindrical shells.

**Table 4.** Variations of the fundamental frequencies (Hz) with the power-law exponent and length to radius ratio ($L/R$).

| | | $\gamma$ | | | | | | |
|---|---|---|---|---|---|---|---|---|
| $L/R$ | Al | 0.5 | 1 | 2 | 5 | 10 | 100 | $ZrO_2$ |
| 0.2 | 1347.0 | 1689.1 | 1758.1 | 1754.9 | 1678.2 | 1622.3 | 1549.0 | 1539.9 |
| 0.5 | 469.0 | 559.8 | 581.0 | 586.6 | 572.9 | 559.1 | 538.77 | 536.21 |
| 1 | 231.0 | 282.6 | 293.2 | 296.1 | 286.1 | 277.4 | 265.54 | 264.09 |
| 2 | 119.3 | 136.1 | 140.6 | 142.3 | 140.7 | 139.1 | 136.7 | 136.4 |
| 5 | 41.45 | 47.71 | 49.31 | 49.76 | 49.05 | 48.40 | 47.50 | 47.39 |
| 10 | 18.84 | 25.35 | 26.54 | 26.33 | 24.63 | 23.38 | 21.75 | 21.54 |
| 15 | 8.68 | 11.75 | 12.31 | 12.20 | 11.38 | 10.79 | 10.02 | 9.92 |
| 20 | 4.94 | 6.71 | 7.03 | 6.97 | 6.50 | 6.16 | 5.71 | 5.65 |

**Table 5.** Variations of the fundamental frequencies (Hz) with the power-law exponent and thickness to radius ratio ($h/R$).

| | | $\gamma$ | | | | | | |
|---|---|---|---|---|---|---|---|---|
| $h/R$ | Al | 0.5 | 1 | 2 | 5 | 10 | 100 | $ZrO_2$ |
| 0.001 | 3.52 | 4.38 | 4.50 | 4.57 | 4.43 | 4.27 | 4.05 | 4.02 |
| 0.005 | 7.09 | 9.17 | 9.57 | 9.52 | 9.01 | 8.64 | 8.16 | 8.10 |
| 0.01 | 9.13 | 10.92 | 11.33 | 11.37 | 11.04 | 10.80 | 10.48 | 10.44 |
| 0.02 | 14.68 | 16.14 | 16.59 | 16.83 | 16.87 | 16.85 | 16.80 | 16.79 |
| 0.03 | 18.82 | 22.26 | 22.79 | 23.23 | 23.57 | 23.36 | 21.73 | 21.52 |
| 0.04 | 18.83 | 25.34 | 26.53 | 26.32 | 24.62 | 23.37 | 21.74 | 21.53 |
| 0.05 | 18.84 | 25.34 | 26.54 | 26.33 | 24.63 | 23.38 | 21.75 | 21.54 |

### 3.3. Optimization Examples

The efficiency of using FGMs for optimizing the natural frequencies of circular cylindrical shells is demonstrated in two design examples. The first one involves maximizing the fundamental frequency of a circular cylindrical shill. In the second optimization example, the natural frequencies are shifted out of a frequency band. Practically, these design examples can be applied for cases in which we cannot control the excitation frequency (or frequencies) that happen to coincide with the cylindrical shell natural frequencies. In the optimization problems, we look for the optimal distribution of the materials' constituents through the length of the cylindrical shells that optimize the objective function. In both examples, the cylindrical shell under consideration is a CF circular cylindrical shell ($R = 1$ m,

$L/R = 10$, and $h/R = 0.05$). It is composed of aluminum as the first material and zirconia as the second material. The first two natural frequencies of aluminum and zirconia cylinders are $f_{1,Al} = 18.84$ Hz, $f_{2,Al} = 33.75$ Hz, $f_{1,ZrO_2} = 21.54$ Hz, and $f_{2,ZrO_2} = 38.59$ Hz. The mode shapes are shown in Figure 3. The optimization process couples FEM and a GA to find the optimal design parameters that provide the optimal material distribution through the length of the cylinders (i.e., $V_1(x)$ and $V_2(x)$). For more information about the GA used, see Alshabatat et al. [27].

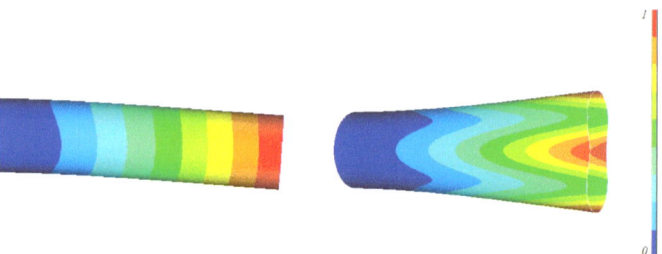

**Figure 3.** The first two mode shapes of thin clamped-free cylindrical shell.

3.3.1. Maximizing the Fundamental Frequency of FGM Cylindrical Shell

In most structures, the excessive vibrations occur when excitation frequencies are near the fundamental frequency. One method to solve this problem is maximizing the fundamental frequency of a structure. In this optimization problem, we look for the optimal distribution of the materials' constituents that maximizes the fundamental frequency of a CF cylinder. The optimization of material distribution is attained by finding the optimal volume fractions at the control points. The constraint optimization problem is defined as follows:

$$\begin{aligned} \text{Maximize} \quad & f_1 \\ \text{Design variables} \quad & \{V_{1,1}, V_{1,2}, \ldots, V_{1,N}\} \\ \text{Subject to} \quad & 0 \leq V_1(x) \leq 1 \end{aligned} \quad (21)$$

In this problem, we use different number of control points $N = 5, 7,$ and 11. The volume fractions of the first material, $V_1(x)$, are estimated by the piecewise cubic hermit interpolating polynomial (PCHIP) [41,42]. A GA is employed to obtain the optimal design parameters $\{V_{1,1}, V_{1,2}, \ldots, V_{1,11}\}$. In each generation of optimization process, the number of the population is 150. The optimization process stops when the change in the frequency is less than $10^{-2}$. The optimal volume fractions at the control points are summarized in Table 6, and the corresponding aluminum volume fraction distribution is shown in Figure 4. The maximum fundamental frequency is resulted by using 11 control points. The maximum fundamental frequency of the axially FGM cylinder is 29.39 Hz, which is more than the fundamental frequencies of aluminum and zirconia cylinders by 56.0% and 36.4%, respectively. To reduce the computational efforts that resulted from the high number of design variables and the interpolation method, a trigonometric law (Equation (2)) is proposed to describe the volume fraction distribution in the optimization process. In this law, there are four design variables only (i.e., $\alpha, \eta, \phi,$ and $\gamma$). The design variables must be carefully selected to ensure that $0 \leq V_1(x) \leq 1$. The constraints are assumed as $0 \leq \alpha \leq 0.5, -\pi \leq \eta, \phi \leq \pi,$ and $0 \leq \gamma \leq 10$. In this case, the number of the population in each generation is assumed to be 40. The optimal design parameters in Equation (2) are listed in Table 6, and the corresponding aluminum volume fraction distributions are shown in Figure 5. The maximum fundamental frequency is 28.60 Hz, which is greater than the fundamental frequency of aluminum and zirconia cylinders by 51.8% and 32.8%, respectively. Figures 4 and 5 show that assigning the stiffer zirconia near the support (i.e., high bending moment region) and the lighter aluminum near the area of high modal displacement increases the fundamental frequency.

Table 6. The optimal frequencies and design variables of the first design problem.

| Material Type | $f_1$ (Hz) | Optimal Design Variables |
|---|---|---|
| Aluminum | 18.84 | - |
| Zirconia | 21.54 | - |
| FGM (5 Control Points) | 28.93 | $\{0, 0, 0.39, 1, 1\}$ |
| FGM (7 Control Points) | 29.25 | $\{0, 0, 0.01, 0.42, 0.99, 1, 1\}$ |
| FGM (11 Control Points) | 29.39 | $\{0, 0, 0, 0.019, 0.090, 0.450, 0.931, 0.982, 1, 1, 1\}$ |
| FGM (Equation (2)) | 28.60 | $\alpha = 0.5, \eta = -0.397, \phi = -2.813, \gamma = 1.351$ |

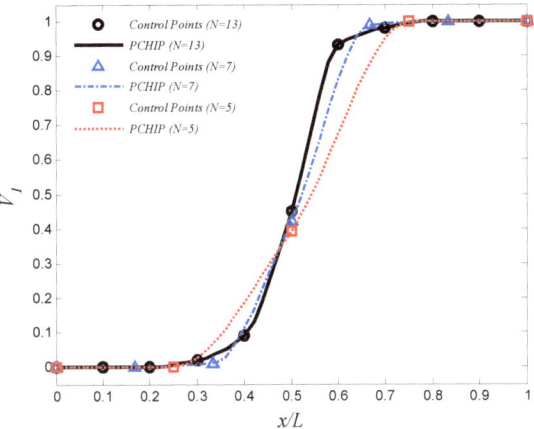

Figure 4. The optimal aluminum volume fraction profile which maximizes the fundamental frequency (control points approach).

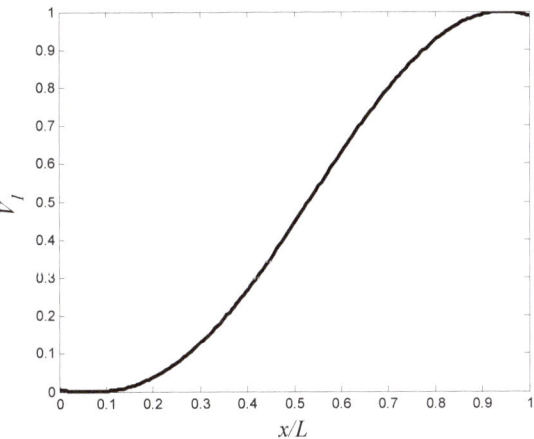

Figure 5. The optimal aluminum volume fraction profile which maximizes the fundamental frequency (trigonometric law, Equation (2)).

3.3.2. Maximizing the Gap between Two Adjoining Natural Frequencies in FGM Cylindrical Shell

A common method to decrease vibration is by maximizing the gap between two adjoining natural frequencies to avoid the coincidence of these natural frequencies with the excitation frequencies that lie between them [50]. In this optimization problem, we look for the volume fractions of the materials' constituents to maximize the distance between the first and second natural frequencies of the CF cylinder. The cylinder under discussion is similar in size and base materials to one in the previous example. We assume that the cylinder is exposed to a force with excitation frequencies in the range of 15–35 Hz. As mentioned previously, the first and second natural frequencies of the aluminum cylinder are $f_{1,Al}$ = 18.84 Hz and $f_{2,Al}$ = 33.75 Hz, and the first natural frequency of the zirconia cylinder is $f_{1,ZrO_2}$ = 21.54 Hz. These natural frequencies are located in the frequency band of the force. Thus, our target is to shift the natural frequencies out the range of 15–35 Hz. This constraint optimization problem can be stated as follows:

$$\begin{aligned}&\text{Maximize} & & f_2 - f_1 \\ &\text{Design variables} & & \{V_{1,1}, V_{1,2}, \ldots, V_{1,N}\} \\ &\text{Subject to} & & 0 \leq V_1(x) \leq 1 \\ & & & f_1 < 15 \text{ Hz} \\ & & & f_2 > 35 \text{ Hz}\end{aligned} \qquad (22)$$

The GA parameters are similar to those in the previous example. The optimal volume fractions at the control points are summarized in Table 7, and the corresponding aluminum volume fraction distributions that maximizes the gap between the natural frequencies are shown in Figure 6. The optimal frequencies are summarized in Table 7. The maximum gab between the first and second natural frequencies is resulted by using 11 control points. The gap between the first and second natural frequencies of the optimal FGM cylinder is 24.64 Hz, which is greater than the gap between the first and second natural frequencies of aluminum and zirconia cylinders by 65.3% and 44.5%, respectively. By using the trigonometric law (Equation (2)), with similar constraints, number of the population, and stopping criterion to the previous example, the optimal gap between the first and second natural frequencies of the FGM cylinder is 23.84 Hz, which is more than the gap between the first and second natural frequencies of aluminum and zirconia cylinders by 59.9% and 39.8%, respectively. the optimal design variables are listed in Table 7, and the corresponding aluminum volume fraction distribution is shown in Figure 7.

Table 7. The optimal frequencies and design variables of the second design problem.

| Material Type | $f_1$(Hz) | $f_2$(Hz) | $f_2 - f_1$(Hz) | Optimal Design Variables |
|---|---|---|---|---|
| Aluminum | 18.84 | 33.75 | 14.91 | - |
| Zirconia | 21.54 | 38.59 | 17.05 | - |
| FGM (5 Control Points) | 13.87 | 37.94 | 24.07 | $\{1, 1, 0.39, 0, 0\}$ |
| FGM (7 Control Points) | 13.57 | 37.95 | 24.38 | $\{1, 1, 1, 0.43, 0.01, 0, 0\}$ |
| FGM (11 Control Points) | 13.32 | 37.96 | 24.64 | $\{1, 1, 1, 0.995, 0.994, 0.469, 0.007, 0.004, 0, 0, 0\}$ |
| FGM (Equation (2)) | 14.13 | 37.97 | 23.84 | $\alpha = 0.5, \eta = -0.406, \phi = 0.473, \gamma = 1.381$ |

Figures 6 and 7 show that assigning the low stiffness aluminum near the support (i.e., high bending moment region in the first mode) decreases the first natural frequency. In addition, assigning the high stiffness zirconia at the area of high modal displacement in the second mode outweighs the disadvantage of increasing masshus. The second natural frequencies of the optimized cylinders are close to natural frequency of zirconia cylinder.

Using the control points method in material distribution optimization is efficient for finding the optimal natural frequencies, but it has a large number of design variables, making it computationally expensive. Moreover, the high number of control points may cause a very high gradient of volume fraction. This high gradient may lead to a sudden change in mechanical properties, which causes cracks and delamination. On the contrary, the trigonometric law can achieve good optimization results with acceptable computational efforts.

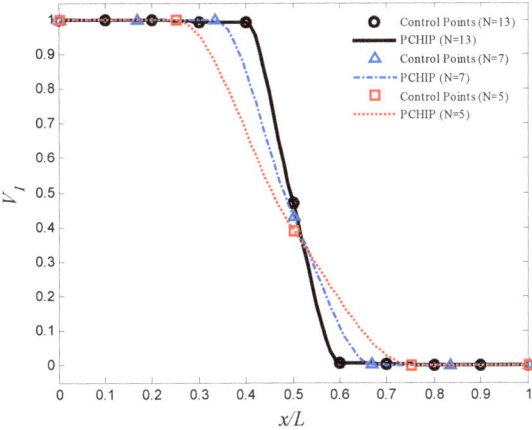

**Figure 6.** The optimal aluminum volume fraction profile which maximizes the gap between $f_1$ and $f_2$ (control points approach).

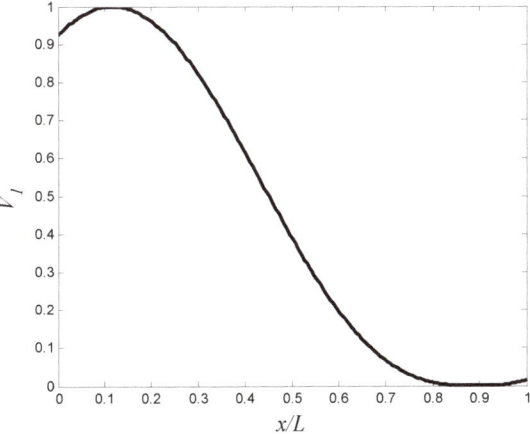

**Figure 7.** The optimal aluminum volume fraction profile which maximizes the gap between $f_1$ and $f_2$ (trigonometric law, Equation (2)).

## 4. Conclusions

In this article, a technique to optimize the natural frequencies of circular cylindrical shells using FGM was presented. The constituents of the FGM were graded in the axial direction. Material volume fraction distribution was described by simple power law, four-parameter trigonometric law, and piecewise cubic interpolation of volume fractions at control points. The effective material properties were estimated by applying Voigt model. FSDT and FEM were employed for free vibrations analysis. To validate the FE program, numerical results were compared with the results of other authors. Two optimization problems were presented. The first problem focused on maximizing the fundamental frequency of a clamped-free FGM circular cylindrical shell by optimizing the distribution of aluminum and zirconia through the axial direction. The maximum fundamental frequency, based on the control points method, was 56.0% and 36.4% more than the fundamental frequencies of aluminum and zirconia cylinders, respectively. In the second example, the gap between the first and second natural frequencies of a FGM circular cylindrical shell was maximized. Based on the control points method, the maximum gap between these frequencies was

65.3% and 44.5% more than those of aluminum and zirconia cylinders, respectively. It was found that using light material near the area of high modal displacement and using stiff material near the area of high bending moment increased the corresponding natural frequency. The control points method is efficient in optimizing the natural frequencies. Increasing the number of control points gave better results. However, the high number of control points may cause a very high gradient of volume fraction which may cause cracks and delamination in real life cases. In addition, the high number of control points makes the design method computationally expensive. Thus, a good alternative is to use the suggested trigonometric law. (Equation (2)) to reduce the drawbacks of control points method. This law provides a flexible description of material volume fraction profile with small number of design variables, which gives engineers a powerful tool for design optimization with accepted computational efforts.

The presented examples show the capabilities of axially FGM in optimizing the natural frequencies of cylindrical shells. A further study of the proposed method is recommended to take in consideration other boundary conditions. Future extension of this work may include minimizing the vibration and sound radiation from such structures.

**Funding:** This research received no external funding.

**Institutional Review Board Statement:** Not applicable.

**Informed Consent Statement:** Not applicable.

**Data Availability Statement:** The data presented in this study are available on request from the author.

**Conflicts of Interest:** The author declares no conflict of interest.

## References

1. Alzahabi, B. Modal Vibration Control of Submarine hulls. *WIT Trans. Built Environ.* **2004**, *76*, 515–525.
2. Nasrekani, F.; Kumar, S.; Narayan, S. Structural Dynamic Modification of Cylindrical Shells with Variable Thickness. In Proceedings of the Pressure Vessels and Piping Conference, Virtual, Online, 3 August 2020.
3. Bagheri, M.; Jafari, A.; Sadeghifar, M. Multi-objective optimization of ring stiffened cylindrical shells using a genetic algorithm. *J. Sound Vib.* **2011**, *330*, 374–384. [CrossRef]
4. Mehrabani, M.; Jafari, A.; Azadi, M. Multidisciplinary optimization of a stiffened shell by genetic algorithm. *J. Mech. Sci. Technol.* **2012**, *26*, 517–530. [CrossRef]
5. Akl, W.; Ruzzene, M.; Baz, A. Optimal design of underwater stiffened shells. *Struc. Multidiscip. Optim.* **2002**, *23*, 297–310. [CrossRef]
6. Biglar, M.; Mirdamadi, H.R.; Danesh, M. Optimal locations and orientations of piezoelectric transducers on cylindrical shell based on gramians of contributed and undesired Rayleigh-Ritz modes using genetic algorithm. *J. Sound Vib.* **2014**, *333*, 1224–1244. [CrossRef]
7. Sadeghifar, M.; Bagheri, M.; Jafari, A.A. Multiobjective optimization of orthogonally stiffened cylindrical shells for minimum weight and maximum axial buckling load. *Thin-Walled Struct.* **2010**, *48*, 979–988. [CrossRef]
8. Belardi, V.G.; Fanelli, P.; Vivio, F. Structural analysis and optimization of anisogrid composite lattice cylindrical shells. *Compos. B Eng.* **2018**, *139*, 203–215. [CrossRef]
9. Hui, L.; Haiyu, L.; Jianfei, G.; Jian, X.; Qingkai, H.; Jinguo, L.; Zhaoye, Q. Nonlinear vibration characteristics of fiber reinforced composite cylindrical shells in thermal environment. *Mech. Syst.* **2021**, *156*, 107665.
10. Yang, J.S.; Xiong, J.; Ma, L.; Feng, L.N.; Wang, S.Y.; Wu, L.Z. Modal response of all-composite corrugated sandwich cylindrical shells. *Compos. Sci. Technol.* **2015**, *115*, 9–20. [CrossRef]
11. Matsunaga, H. Vibration and buckling of cross-ply laminated composite circular cylindrical shells according to a global higher-order theory. *Int. J. Mech. Sci.* **2007**, *49*, 1060–1075. [CrossRef]
12. Yadav, D.; Verma, N. Free vibration of composite circular cylindrical shells with random material properties. Part II: Applications. *Compos. Struct.* **2001**, *51*, 371–380. [CrossRef]
13. Hufenbach, W.; Holste, C.; Kroll, L. Vibration and damping behaviour of multi-layered composite cylindrical shells. *Compos. Struct.* **2002**, *58*, 165–174. [CrossRef]
14. Lee, Y.; Lee, K. On the dynamic response of laminated circular cylindrical shells under impulse loads. *Compos. Struct.* **1997**, *63*, 149–157. [CrossRef]
15. Jafari, A.; Khalili, S.; Azarafza, R. Transient dynamic response of composite circular cylindrical shells under radial impulse load and axial compressive loads. *Thin. Walled. Struct.* **2005**, *43*, 1763–1786. [CrossRef]

16. Jin, G.; Ye, T.; Chen, Y. An exact solution for the free vibration analysis of laminated composite cylindrical shells with general elastic boundary conditions. *Compos. Struct.* **2013**, *106*, 114–127. [CrossRef]
17. Qu, Y.; Hua, H.; Meng, G. A domain decomposition approach for vibration analysis of isotropic and composite cylindrical shells with arbitrary boundaries. *Compos. Struct.* **2013**, *95*, 307–321. [CrossRef]
18. Hu, H.; Tsai, J. Maximization of the fundamental frequencies of laminated cylindrical shells with respect to fiber orientations. *J. Sound Vib.* **1999**, *225*, 723–740. [CrossRef]
19. Koide, R.M.; Luersen, M.A. Maximization of fundamental frequency of laminated composite cylindrical shells by ant colony algorithm. *J. Aerosp. Technol. Manag.* **2013**, *5*, 75–82. [CrossRef]
20. Trias, D.; Maimí, P.; Blanco, N. Maximization of the fundamental frequency of plates and cylinders. *Compos. Struct.* **2016**, *156*, 375–384. [CrossRef]
21. Miller, B.; Ziemiański, L. Maximization of eigenfrequency gaps in a composite cylindrical shell using genetic algorithms and neutral networks. *Appl. Sci.* **2019**, *9*, 2754. [CrossRef]
22. Miller, B.; Ziemiański, L. Optimization of dynamic behavior of thin-walled laminated cylindrical shells by genetic algorithms and deep neural networks supported by modal shape identification. *Adv. Eng. Softw.* **2020**, *147*, 102830. [CrossRef]
23. Jing, Z. Optimal design of laminated composite cylindrical shells for maximum fundamental frequency using sequential permutation search with mode identification. *Compos. Struct.* **2022**, *279*, 114736. [CrossRef]
24. Nekouei, M.; Raghebi, M.; Mohammadi, M. Free vibration analysis of hybrid laminated composite cylindrical shells reinforced with shape memory alloy fibers. *J. Vib. Control.* **2020**, *26*, 610–626. [CrossRef]
25. Liu, Y.; Qin, Z.; Chu, F. Nonlinear forced vibrations of functionally graded piezoelectric cylindrical shells under electric-thermo-mechanical loads. *Int. J. Mech. Sci.* **2021**, *201*, 106474. [CrossRef]
26. Nikbakht, S.; Kamarian, S.; Shakeri, M. A review on optimization of composite structures Part II: Functionally graded materials. *Compos. Struct.* **2019**, *214*, 83–102. [CrossRef]
27. Alshabatat, N.; Myers, K.; Naghshineh, K. Design of in-plane functionally graded material plates for optimal vibration performance. *Noise Control Eng. J.* **2016**, *64*, 268–278. [CrossRef]
28. Alshabatat, N.; Naghshineh, K. Minimizing the radiated sound power from vibrating plates by using in-plane functionally graded materials. *J. Vibroengineering* **2021**, *23*, 744–758. [CrossRef]
29. Loy, C.T.; Lam, K.Y.; Reddy, J.N. Vibration of functionally graded cylindrical shells. *Int. J. Mech. Sci.* **1999**, *41*, 309–324. [CrossRef]
30. Arshad, S.H.; Naeem, M.N.; Sultana, N. Frequency analysis of functionally graded material cylindrical shells with various volume fraction laws. *Proc. Inst. Mech. Eng. Part C J. Mech. Eng. Sci.* **2007**, *221*, 1483–1495. [CrossRef]
31. Arshad, S.H.; Naeem, M.N.; Sultana, N.; Iqbal, Z.; Shah, A.G. Effects of exponential volume fraction law on the natural frequencies of FGM cylindrical shells under various boundary conditions. *Arch. Appl. Mech.* **2011**, *81*, 999–1016. [CrossRef]
32. Zhu, S.; Guoyong, J.; Shuangxia, S.; Tiangui, Y.; Xingzhao, J. A unified solution for vibration analysis of functionally graded cylindrical, conical shells and annular plates with general boundary conditions. *Int. J. Mech. Sci.* **2014**, *80*, 62–80.
33. Guoyong, J.; Xiang, X.; Zhigang, L. The Haar wavelet method for free vibration analysis of functionally graded cylindrical shells based on the shear deformation theory. *Compos. Struct.* **2014**, *108*, 435–448.
34. Punera, D.; Kant, T. Free vibration of functionally graded open cylindrical shells based on several refined higher order displacement models. *Thin-Walled Struct.* **2017**, *119*, 707–726. [CrossRef]
35. Yiwen Ni, Y.; Tong, Z.; Rong, D.; Zhou, Z.; Xu, X. A new Hamiltonian-based approach for free vibration of a functionally graded orthotropic circular cylindrical shell embedded in an elastic medium. *Thin-Walled Struct.* **2017**, *120*, 236–248.
36. Liu, T.; Wang, A.; Wang, Q.; Qin, B. Wave based method for free vibration characteristics of functionally graded cylindrical shells with arbitrary boundary conditions. *Thin-Walled Struct.* **2020**, *148*, 106580. [CrossRef]
37. Alshabatat, N.; Zannon, M. Natural frequencies analysis of functionally graded circular cylindrical shells. *J. Appl. Comput. Mech.* **2021**, *15*, 1–18. [CrossRef]
38. Kieback, B.; Neubrand, A.; Riedel, H. Processing techniques for functionally graded materials. *Mater. Sci. Eng. A* **2003**, *362*, 81–106. [CrossRef]
39. Jin, G.Q.; Li, W.D. Adaptive rapid prototyping/manufacturing for functionally graded material-based biomedical models. *Int. J. Adv. Manuf. Technol.* **2013**, *65*, 97–113. [CrossRef]
40. Gupta, A.; Talha, M. Recent development in modeling and analysis of functionally graded materials and structures. *Prog. Aerosp. Sci.* **2015**, *79*, 1–14. [CrossRef]
41. Esfandiari, R.S. *Numerical Methods for Engineers and Scientists Using MATLAB*, 2nd ed.; CRC Press: Boca Raton, FL, USA, 2017.
42. Fritsch, F.N.; Carlson, R.E. Monotone Piecewise Cubic Interpolation. *SIAM J. Numer. Anal.* **1980**, *17*, 238–246. [CrossRef]
43. Reddy, J.N.; Liu, C.F. A higher-order shear deformation theory of laminated elastic shells. *Int. J. Eng. Sci.* **1985**, *23*, 319–330. [CrossRef]
44. Khdeir, A.A.; Reddy, J.N.; Frederick, D. A study of bending, vibration and buckling of cross-ply circular cylindrical shells with various shell theories. *Int. J. Eng. Sci.* **1989**, *27*, 1337–1351. [CrossRef]
45. Kraus, H. *Thin Elastic Shells*, 1st ed.; John Wiley & Sons: New York, NY, USA, 1999.
46. Petyt, M. *Introduction to Finite Element Vibration Analysis*, 2nd ed.; Cambridge University Press: Cambridge, UK, 2015.
47. Zienkiewicz, O.; Too, J.; Taylor, R. Reduced integration technique in general analysis of plates and shells. *Int. J. Num. Meth. Eng.* **1971**, *3*, 275–290. [CrossRef]

48. Chen, P.; Sun, S.; Zhao, Q.; Gong, Y.; Yuan, M. Advances in solution of classical generalized eigenvalue problem. *Interact. Multiscale Mech.* **2008**, *1*, 211–230. [CrossRef]
49. Naeem, M.N.; Sharma, C.B. Prediction of natural frequencies for thin circular cylindrical shells. *Proc. Inst. Mech. Eng. Part C J. Mech. Eng. Sci.* **2000**, *214*, 1313–1328. [CrossRef]
50. Bendsøe, M.P.; Olhoff, N. A method of design against vibration resonance of beams and shafts. *Optim. Control Appl. Methods* **1985**, *6*, 191–200. [CrossRef]

Article

# Effect of Conditioning on PU Foam Matrix Materials Properties

Lubomír Lapčík [1,2,*], Martin Vašina [3,*], Barbora Lapčíková [1,2] and Yousef Murtaja [2]

[1] Faculty of Technology, Tomas Bata University in Zlin, Nam. TGM 275, 760 01 Zlin, Czech Republic; lapcikova@utb.cz
[2] Department of Physical Chemistry, Faculty of Science, Palacky University Olomouc, 17. Listopadu 12, 771 46 Olomouc, Czech Republic; yousef.murtaja01@upol.cz
[3] Department of Hydromechanics and Hydraulic Equipment, Faculty of Mechanical Engineering, VSB-Technical University of Ostrava, 17. listopadu 15/2172, Poruba, 708 33 Ostrava, Czech Republic
* Correspondence: lapcikl@seznam.cz (L.L.); martin.vasina@vsb.cz (M.V.)

**Abstract:** This article deals with the characterization of the thermal-induced aging of soft polyurethane (PU) foams. There are studied thermal and mechanical properties by means of thermal analysis, tensile, compression and dynamic mechanical vibration testing. It was found in this study, that the increasing relative humidity of the surrounding atmosphere leads to the initiation of the degradation processes. This is reflected in the observed decreased mechanical stiffness. It is attributed to the plasticization of the PU foams wall material. It is in agreement with the observed increase of the permanent deformation accompanied simultaneously with the decrease of Young's modulus of elasticity. The latter phenomenon is studied by the novel non-destructive forced oscillations vibration-damping testing, which is confirmed by observed lower mechanical stiffness thus indicating the loss of the elasticity induced by samples conditioning. In parallel, observed decreasing of the matrix hardness is confirming the loss of elastic mechanical performance as well. The effect of conditioning leads to the significant loss of the PU foam's thermal stability.

**Keywords:** thermal analysis; polyurethanes; open-cell foams; mechanical vibrations; tensile testing; compression testing

**Citation:** Lapčík, L.; Vašina, M.; Lapčíková, B.; Murtaja, Y. Effect of Conditioning on PU Foam Matrix Materials Properties. *Materials* **2022**, *15*, 195. https://doi.org/10.3390/ma15010195

Academic Editors: Andrea Sorrentino and Halina Kaczmarek

Received: 10 November 2021
Accepted: 24 December 2021
Published: 28 December 2021

**Publisher's Note:** MDPI stays neutral with regard to jurisdictional claims in published maps and institutional affiliations.

**Copyright:** © 2021 by the authors. Licensee MDPI, Basel, Switzerland. This article is an open access article distributed under the terms and conditions of the Creative Commons Attribution (CC BY) license (https://creativecommons.org/licenses/by/4.0/).

## 1. Introduction

Polyurethanes are a group of polymers formed by the reaction of multifunctional isocyanates with polyalcohols. Polyurethanes (polycarbamates, or PU) are characterized by the (-NH-CO-O-) group [1]. In constitution and properties, they lie between polyureas (-NH-CO-NH-) and polycarbonates (-O-CO-O-). Polyurethane products are very different in nature. The diversity is attributable to the urethane bond's qualities (-NH-C-O-O) and the existence of other groups in the chain, which enable the application of intermolecular forces, cross-linking, and crystallization orientation, flexibility, and chain stiffening [2]. While PU can be synthesized in various methods, it is always based on diisocyanates in practice. Due to the toxicity of isocyanate raw materials and their synthesis method utilizing phosgene, new synthetic routes for polyurethanes without using isocyanates have been developed at present [2].

The study of polymer thermal aging and degradation at specific humidity conditions reached a relatively high importance in technological praxis due to the wide applications of porous polyurethane based materials in automotive, aerospace [3], food packaging [4,5], and consumer products industries [6,7]. In general, polymers are thermally stable until reaching the decomposition threshold [8]. Two main types of polymer thermal decomposition processes are usually recognized for polymer main macromolecular chain: depolymerization and random decomposition. These two processes can occur concurrently or independently. On the other hand, individual polymers are heated in accordance with the specified time-temperature regimes and often experience melting first, followed by various stages of the thermal degradation process [8]. Temperature changes can stimulate a variety of physical and chemical processes in polymer systems. Important examples of these processes include thermal degradation, cross-linking, crystallization, glass transition, hydrolysis [9] etc.

Authors [10] found that the polyurethane biodegradation is controlled by diffusion of oxygen into the polymer. In PUs, the urethane or carbamate linkage is also susceptible to hydrolysis, as it is an ester linkage of a substituted carbamic acid. However, hydrolytic degradation occurs less readily than with the carboxylic ester linkage. The ultimate strain of the elastomers was decreasing with increasing degradation time.

Changes of the PUs fluorescence mapping signals induced by the weathering indicated degradation process proceeding in the PU matrix, as reflected in the increase in fluorescence intensity over time in the mapped fluorescence signals excited at 470 nm and detected in a wavelength range exceeding 500 nm. These signals revealed as response to the ongoing progress of polymer ageing [11].

Most polyurethanes are derived from three basic building blocks: a polyol (a multifunctional alcohol, usually polymeric, e.g., polybutadiene with hydroxyl-terminated hydroxyls), a diisocyanate, a chain extender (e.g., low molecular weight or polymeric diol or diamine), and a cross-linker (e.g., triol). Two basic procedures are used to synthesize polyurethanes: one-step and two-step, i.e., either a one-step mixing of all components of the reaction mixture or a prepolymer preparation procedure. One-step synthesis of polyurethanes is carried out by simultaneous mixing of polyol, diisocyanate (MDI, TDI), extenders (diols, diamines), and cross-linkers (multifunctional alcohols or amines) [12]. Cross-linking is often carried out with sufficient speed at room temperature and is completed within about a few hours. The reaction time can be shortened as required by adding accelerators (dibutyltindilaurate, stannous octoate, etc.), which are added just before application. The two-step synthesis produces the final polyurethane in two temporally separated reaction steps. In the first stage, the polyol reacts with the diisocyanate to form an intermediate with NCO-terminal groups, the so-called prepolymer. In the second step, the prepolymer reacts with extenders and cross-linkers to form the final product. In general, the prepolymer can be prepared with any excess isocyanate (isocyanate index rNCO/OH > 2). The viscosity of the prepolymer increases strongly (up to 5 times for the original hydroxyl-terminated polybutadiene); the excess liquid isocyanate can prevent or mitigate this increase (dilution effect). The hydroxyl-terminated polybutadiene may be partially replaced by a proportion of polyester diol, polyether diol, or polyol in general. From a homogeneity standpoint, it is frequently preferable to convert these components of the mixture independently into prepolymers before combining. The two-step method improves the characteristics of crosslinked matrices [6].

Presently, the research and development of recyclable and sustainable PU foams are focused on advanced bio-based systems in packaging application, e.g., for refrigerated and frozen foodstuff storage and sale. For this reason, the modern trend in the field is moving towards fully bio-based and compostable foams. However, the direction via partially bio-based PU foams remains the most consistent performer [13]. For example, the application of the algae and algal cellulose added during PU-composite foam production led to the algae base cellulose integration into the PU foam structure and resulted in more open cells and softer foam structure. The PU-composite showed greater shock absorbent capacity in comparison with the usual PU foam suitable for the application in the food packaging industry [14]. Another current trend in PU material development is the non-isocyanate PU foam synthesis [15,16].

According to the structure, soft and rigid PU foams with different specific gravity, porous structure, and mechanical properties are distinguished. The basic components of soft foams are a mixture of 2,4- and 2,6-toluenedioisocyanate and polyols (polyesters-linear or slightly branched, polyethers). The main blowing agent is gaseous carbon dioxide, formed by the reaction of water with the isocyanate group. This reaction produces highly polar groups, forming H-bridges and affecting physical properties. Soft PU foams have a foam structure with open, interconnected pores. Rigid PU foams are networks with denser interchain bonds or cross-linking sites, and a higher content of rigid urethane sections. The closed pore structure predominates in the foam because the pore walls are more rigid. The mechanical strength of the foam increases with specific gravity. Soft PU rubbers, cross-linked elastomers and soft PU foams have approximately the same chain stiffness. High chain stiffness and intermolecular forces characterize rigid PU foams.

Results of the vibrational dynamical mechanical analysis combined with the FTIR analysis allowed detailed understanding of the induced chemical changes leading to both the crosslinking, as well as degradation processes, in the complex PU matrix [17,18]. The final changes of the mechanical properties were dependent on the type of PU matrix, where poly(ester) based PUs exhibited higher stiffness in comparison with the poly(ether) [19].

The combination of the novel, non-destructive dynamic-mechanical vibrational analysis with the conventional tensile, hardness, and permanent deformation measurements was applied in this study. The latter experimental techniques were combined with the thermal analysis. Studied samples were subjected to varying conditioning times at different temperatures and relative humidity.

## 2. Materials and Methods

### 2.1. Materials

Open cell soft PU foams of $(35.4 \pm 0.3)$ g/cm$^3$ density were purchased in the local construction hobby market and studied. PU samples were cut into the required shapes and dimensions of the specimen, e.g., dog bone for tensile, cubic for mechanical vibration analysis, permanent deformation, and hardness testing. A photo of the studied foam is shown in Figure 1. Small bits of the tested PU foams were used for thermal analysis. All materials under study were subjected to the conditioning at two different temperatures of 45 and 80 °C and relative humidity kept at 45 and 80% for different time intervals ranging from 0 to 300 h in the climate chamber Discovery 105 (Angelantoni Test Technologies, Massa Martana, Italy).

**Figure 1.** Photo of the studied PU foam material.

### 2.2. Methods

Uniaxial tensile testing was performed on Autograph AGS-X instrument (Shimadzu, Kyoto, Japan) equipped with the Compact Thermostatic Chamber TCE Series. Measurements were performed according to the ČSN EN ISO 527-1, 527-2 standards at 100 mm/min deformation rate [20]. Each measurement was repeated ten times and mean values were calculated.

Thermogravimetry (TG) and differential thermal analysis (DTA) experiments were measured on simultaneous DSC/TGA, SDT 650 Discovery with TRIOS software for thermal analysis (TA Instrument, New Castle, DE, USA). The apparatus was calibrated using indium as a standard. Prior to each measurement, samples were grated into an aluminum pan. Throughout the experiment, the sample temperature and weight-heat flow changes were continuously monitored. The measurements were performed at a heat flow rate of 10 °C/min in a static air atmosphere at the temperature range of 30 to 300 °C.

The material's ability to damp harmonically excited mechanical vibration of single-degree-of-freedom (SDOF) systems is characterized by the displacement transmissibility $T_d$, which is expressed by the Equation (1) [21]:

$$T_d = \frac{y_2}{y_1} = \frac{a_2}{a_1} = \sqrt{\frac{k^2 + (c \cdot \omega)^2}{(k - m \cdot \omega^2)^2 + (c \cdot \omega)^2}} \qquad (1)$$

where $y_1$ is the displacement amplitude on the input side of the tested sample, $y_2$ is the displacement amplitude on the output side of the tested sample, $a_1$ is the acceleration amplitude on the input side of the tested sample, $a_2$ is the acceleration amplitude on the output side of the tested sample, $k$ is the stiffness, $c$ is the damping coefficient, $m$ is the mass, and $\omega$ is the angular frequency [22,23].

The mechanical vibration damping testing of the investigated PU foams was performed by the forced oscillation method [24]. The displacement transmissibility $T_d$ was experimentally measured using the BK 4810 vibrator combined with a BK 3560-B-030 signal pulse multi-analyzer, and a BK 2706 power amplifier at the frequency range from 2 to 1000 Hz [25]. The acceleration amplitudes $a_1$ and $a_2$ on the input and output sides of the investigated specimens were evaluated by means of BK 4393 accelerometers (Brüel & Kjær, Nærum, Denmark) [26]. The tested specimen dimensions were (60 × 60 × 50) mm (length × width × thickness). Each measurement was repeated three times at an ambient temperature of 25 °C and mean values of the displacement transmissibility were calculated.

Foam hardness was measured according to ČSN EN ISO 2439 (645440) standard method A "Flexible cellular polymeric materials-Determination of hardness (indentation technique)". During the hardness measurements, the specimens were compressed to 60% of their original height, and the force value (N) was read after 30 s of the deformation. The tested specimen dimensions were (50 × 50 × 50) mm (length × width × thickness). Each measurement was repeated five times at selected conditioning temperatures of 45 and 80 °C and relative humidity of 40 and 80%.

Permanent deformation was measured according to ČSN EN ISO 1856 standard. The tested block specimen dimensions were (50 × 50 × 50) mm (length × width × thickness). Each measurement was repeated five times. Tested samples after conditioning for 300 h at the same temperatures and relative humidity as given in the section above were inserted between the plates of the pressing device compressed to 50% of the original dimensions for 30 min. At the end of the period of compression time, the specimens were released and left lying freely on the table for 30 min. The thickness was then measured again and the permanent deformation expressed as a percentage relative to the original samples dimension was calculated.

## 3. Results

Results of the tensile testing of the studied PU foam materials are shown in Figure 2. Observed data indicated an increase in mechanical strength and plasticity of the cellular materials after 300 h conditioning as reflected in the increased magnitudes of the stress at break by 19% from 155 kPa to 184 kPa (for sample treated at 45 °C and 80% relative humidity). Simultaneously, the increase in ductility from 87% (at the conditioning temperature of 45 °C and the relative humidity of 45%) to 114% (at the conditioning temperature of 80 °C and the relative humidity of 45%) was observed. For the conditioning temperature of 80 °C, this dependence remained relatively constant. For the relative humidity of 80% and the conditioning temperature of 45 °C, the mechanical strength after 24 h increased, followed by the sharp decline observed after 48 h of conditioning. However, it increased again after 300 h of conditioning. Latter mentioned behavior indicated that at 80% relative humidity, significant structural changes ascribed to the cross-linking were obtained after 48 h of conditioning. At the temperature of 80 °C and a relative humidity of 45%, the stress at break did not change significantly. However, it increased about 10% at the higher relative humidity conditions.

Figure 2 shows that the PU matrix continued to plasticize, as seen by the rise in ductility of the material. We assume that this may be a combination of a change in the elasticity of the individual cell walls of the foam matrix. However, we also have to consider the resulting air pressure increase induced by the change in individual cell volumes during the deformation of the foam structure. The latter change of volume is also associated with a change in the geometry of the individual cells. Simultaneously, the change in the viscoelastic and viscoplastic properties of the PU matrix should be assumed due to the PU matrix cross-linking [27], and partial hydrolysis [28]. As a complex material response, the varying degrees of swelling of the PU cell walls material were proposed. We assume that the viscous friction and the wall stiffness are the dominant mechanisms of the transferred mechanical energy dissipation [29].

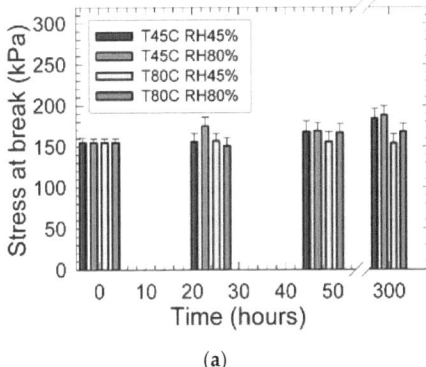 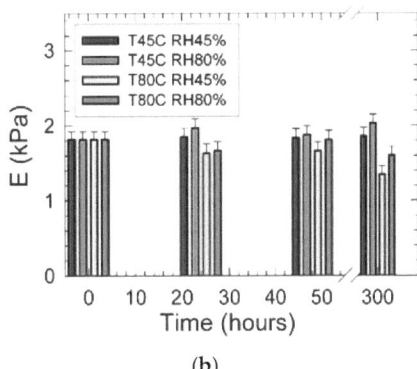

**Figure 2.** Results of the uniaxial tensile testing of the studied PU materials: (**a**) Stress at break vs. conditioning time at different temperatures and relative humidity; (**b**) Young's modulus of elasticity (*E*) vs. conditioning time at different temperatures and relative humidity. Inset: T—temperature (°C), RH—relative humidity (%).

The observed increase of the stress at the break for high relative humidity conditioning (80%) at 45 °C indicated an increase of the PU matrix's stiffness, accompanied by the increase of Young's modulus of elasticity resulting in a simultaneous increase in matrix brittleness. However, signs of matrix hydrolysis were identified at 80 °C conditioning for both relative humidity conditions resulting in the decrease of the *E* from 1816 Pa (original sample without conditioning) to 1350 Pa (conditioned at 80 °C and 45% RH) and to 1600 Pa (conditioned at 80 °C and 80% RH).

Results of the mechanical vibration measurements are summarized in Figures 2 and 3. It was found that the displacement transmissibility gradually increased with the increasing conditioning temperature and high relative humidity of 80% (Figure 3a), indicating higher mechanical stiffness of the PU matrix in the frequency range of 250 to 400 Hz. A similar effect was also found for the 40% relative humidity conditioning. We assume that the chemical cross-linking process of the residual non-reacted PU matrix components proceeded in the time scale of the first 48 h of the conditioning treatment. This hypothesis was also supported by our previous tensile testing measurements discussed above. Simultaneously, the increasing trend of the PU matrix mechanical stiffness was also observed. This was reflected in the increased magnitude of $T_d$ with the increasing relative humidity, as shown in Figure 3b [30]. Results shown in Figure 4a indicated increased elasticity with the increased conditioning time of the studied foams as reflected in the appearance of the two maxima of $T_d$ (at the frequencies of 280 and 430 Hz and magnitudes of 0.014 and 0.017). Observed stiffening appeared in the relatively wide frequency range of 200 to 600 Hz. After prolonged conditioning, plasticization patterns in mechanical behavior were found, as indicated in Figure 4b. After 24 h of conditioning at 80 °C and 80% RH, the foam elasticity was decreased in the frequency range of 200 to 600 Hz in comparison with the virgin PU sample indicating the probable start of the hydrolysis process of the wall material decomposition [31,32].

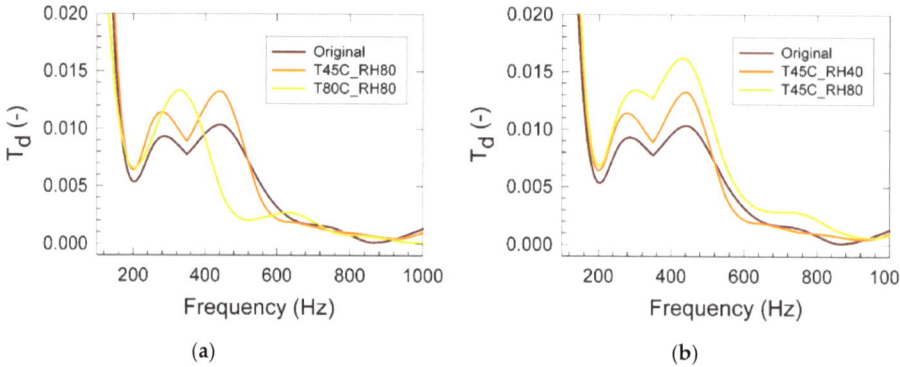

**Figure 3.** Displacement transmissibility vs. frequency as obtained from the vibration damping measurements of the studied PU materials after 300 h conditioning: (**a**) Effect of the conditioning temperature at 80% RH; (**b**) Effect of the conditioning relative humidity at 45 °C. Inset: T—temperature (°C), RH—relative humidity (%), Original—non-conditioned sample.

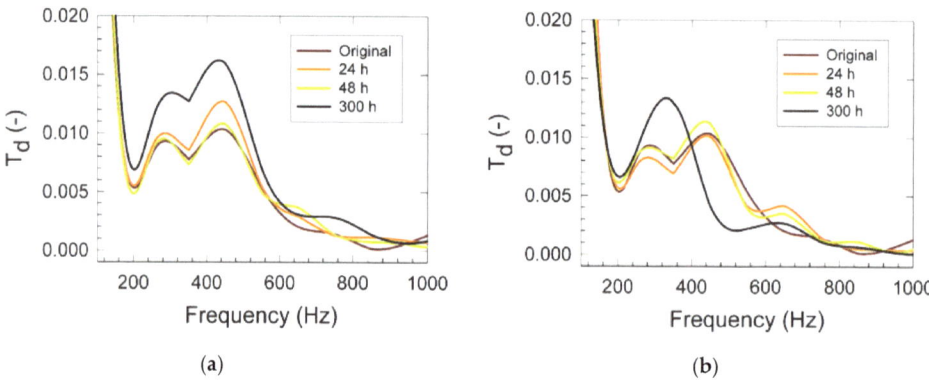

**Figure 4.** Displacement transmissibility vs. frequency as obtained from the vibration damping measurements of the studied PU materials: (**a**) Effect of the conditioning time at 45 °C and 80% RH; (**b**) Effect of the conditioning time at 80 °C and 80% RH. Inset: Conditioning time (hours), Original—non-conditioned sample.

Results of the effect of the sample conditioning on PU foam hardness are shown in Figure 5. It was found that with increasing conditioning time, the hardness decreased for all tested materials except conditioning at 45 °C. Most probably, this fact was attributed to the increased PU foams wall matrix crosslinking density as obtained at latter mentioned conditioning parameters. After 300 h conditioning in the temperature range of 45 to 80 °C and at the relative humidity of 45 and 80%, the hardness decreased from 10 to 18%. However, at the 45 °C temperature and the relative humidity of 45%, the hardness slightly increased from 18 N to 19 N. We assumed the onset of the PU molecules degradation due to increased temperature and relative humidity. It was found by Weise et al. [19] that crosslinks occur to a greater extent for the PUs containing more poly(ester) because the ester groups can more easily undergo hydrolysis than the PUs containing more poly(ether). It was confirmed by FTIR analysis [19], that the hydrolysis of ester groups yields carboxylic acids, which provide new and rigid ionic crosslinks. In the case of PUs containing only the ether macrodiols, they became slightly more mechanically stiff at the short weathering time less than 1000 h followed by their weakening to the point of fracture after prolonged weathering. This indicated that although shorter and more ridged crosslinks were formed during weathering, the sum of the degradation ultimately weakened the mechanical properties of poly(ether) type PUs [19].

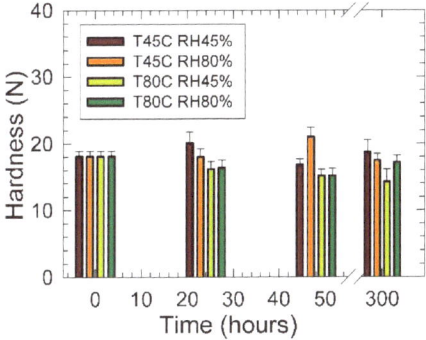

**Figure 5.** Hardness vs. conditioning time dependencies of studied PU foams at different temperatures and relative humidity. Inset: T—temperature (°C), RH—relative humidity (%).

Results of the permanent deformation measurements are shown in Figure 6. It was found that conditioning at the lower temperature of 45 °C triggered higher elasticity of the PU foams independently of the surrounding atmosphere relative humidity. These results were in excellent agreement with the observed increased Young´s modulus of elasticity, as shown in Figure 2b. With increased conditioning temperature, the permanent deformation vigorously increased from 18% (for the original sample) to 46% after 300 h of conditioning time. The latter permanent deformations were independent of the applied relative humidity. Again, these results agreed with the obtained decrease of the $E$ reflecting loss of elasticity and increased plastic mechanical behavior.

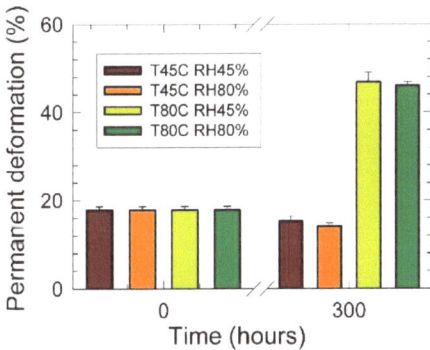

**Figure 6.** Permanent deformation vs. conditioning time dependencies of studied PU foams at different temperatures and relative humidity. Inset: T—temperature (°C), RH—relative humidity (%).

It was found in the literature [19], that the single component PUs containing either a poly(ester) or poly(ether) macrodiols were mechanically more unstable than the two component PUs containing both macrodiols. This led to a synergistic effect where their damping abilities increased and their mechanical stabilities after weathering increased as well. The increase in the damping abilities of the blended PUs was attributed to a decrease in the packing efficiency of the macrodiols chains by combining the two different functional groups. Furthermore, the increase in the mechanical stabilities after weathering was observed as a result of the competing degradation processes and simultaneous protection of the urethane group by the ester.

A typical degradation pattern of the studied PU foams is shown in Figure 7. The degradation process was started at the temperature of 260 °C with the observed weight loss of about 86%, accompanied by the exothermic heat of fusion. Observed degradation pattern was ascribed to the decomposition of the PU material into the starting polyol and diisocyanate. It was followed by the thermal decomposition of the diisocyanate. Benzonitrile, methylbenzonitrile, phenyl isocyanate, toluene, and benzene were detected as the main decomposition products [33,34]. The latter decomposition process reflected the decomposition of diisocyanates and polyols in the PU matrix accompanied by the evolution of gaseous carbon oxide and hydrogen cyanide [7].

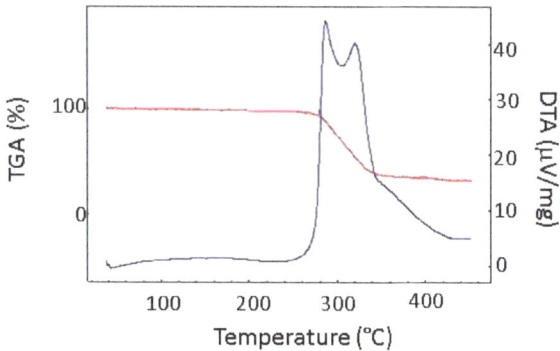

**Figure 7.** Typical TG DTA pattern of the studied PU foam material: red line—TGA, blue line—DTA.

In Figure 8 results of the TGA analysis are summarized. Observed data shows a minor increase in the weight loss with the increasing conditioning time for both conditioning temperatures and relative humidity. It was ascribed to the change of the chemical composition of the PU foam, most probably due to the addition of the free isocyanate groups. It can be concluded that conditioning led to the loss of the PU foam's thermal stability.

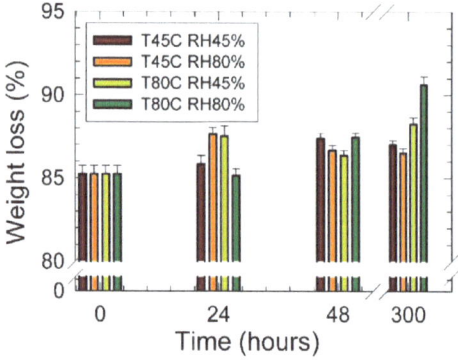

**Figure 8.** Weight loss vs. conditioning time of the studied PU foams. Inset: T—temperature (°C), RH—relative humidity (%).

## 4. Conclusions

This study was aimed on the physicochemical and mechanical characterization of the thermal-induced degradation of the commercial soft polyurethane foams. It was found that the increased conditioning temperature and the relative humidity of the surrounding atmosphere led to the initiation of the degradation patterns of the studied PU foams. There was an observed decrease in mechanical stiffness due to the plasticization of the

PU foams wall material, as confirmed by the simultaneous increase of the permanent deformation accompanied by the decrease of the Young's modulus of elasticity. This phenomenon was also confirmed by the nondestructive dynamical-mechanical vibration testing, which confirmed samples´ higher vibration damping, resulting in the loss of elasticity. The last changes of the mechanical behavior agreed with the observed decrease of the matrix hardness, again confirming the loss of elastic mechanical performance. It has been also found that the effect of conditioning led to the significant loss of the PU foam's thermal stability.

**Author Contributions:** Conceptualization, L.L. and M.V.; methodology, B.L., L.L. and M.V.; investigation, B.L., L.L., Y.M. and M.V.; resources, M.V.; data curation, B.L., L.L., Y.M. and M.V.; writing—original draft preparation, L.L. and M.V.; writing—review and editing, B.L., L.L. and M.V.; visualization, B.L., L.L., Y.M. and M.V.; supervision, L.L.; project administration, M.V.; funding acquisition, M.V. All authors have read and agreed to the published version of the manuscript.

**Funding:** This study was funded by the European Regional Development Fund in the Research Centre of Advanced Mechatronic Systems project, project number CZ.02.1.01/0.0/0.0/16_019/0000867 within the Operational Programme Research, Development and Education and by Internal Grant of Palacky University in Olomouc, grant number IGA_PrF_2021_031.

**Institutional Review Board Statement:** Not applicable.

**Informed Consent Statement:** Not applicable.

**Data Availability Statement:** Not applicable.

**Conflicts of Interest:** The authors declare no conflict of interest.

# References

1. Krol, P. Polyurethanes—A Review of 60 Years of their Syntheses and Applications. *Polimery* **2009**, *54*, 489–500. [CrossRef]
2. Stachak, P.; Lukaszewska, I.; Hebda, E.; Pielichowski, K. Recent Advances in Fabrication of Non-Isocyanate Polyurethane-Based Composite Materials. *Materials* **2021**, *14*, 3497. [CrossRef] [PubMed]
3. Lapcik, L.; Cetkovsky, V.; Lapcikova, B.; Vasut, S. Materials for noise and vibration attenuation. *Chem. Listy* **2000**, *94*, 117–122.
4. Shaw, S.D.; Harris, J.H.; Berger, M.L.; Subedi, B.; Kannan, K. Brominated Flame Retardants and Their Replacements in Food Packaging and Household Products: Uses, Human Exposure, and Health Effects. In *Toxicants in Food Packaging and Household Plastics: Exposure and Health Risks to Consumers*; Springer: London, UK, 2014; pp. 61–93. [CrossRef]
5. Song, H.Y.; Cheng, X.X.; Chu, L. Effect of Density and Ambient Temperature on Coefficient of Thermal Conductivity of Heat-Insulated EPS and PU Materials for Food Packaging. *Res. Food Packag. Technol.* **2014**, *469*, 152–155. [CrossRef]
6. Volcik, V.; Lapcikova, B.; Lapcik, L.; Asuquo, R. Uses of polyurethane matrixes in the environmental field. *Plasty Kauc.* **2002**, *39*, 164–169.
7. Tomin, M.; Kmetty, A. Polymer foams as advanced energy absorbing materials for sports applications-A review. *J. Appl. Polym. Sci.* **2022**, *139*, 51714. [CrossRef]
8. Lapcikova, B.; Lapcik, L., Jr. TG and DTG Study of Decomposition of Commercial PUR Cellular Materials. *J. Polym. Mater.* **2011**, *28*, 353–366.
9. Scholz, P.; Wachtendorf, V.; Panne, U.; Weidner, S.M. Degradation of MDI-based polyether and polyester-polyurethanes in various environments—Effects on molecular mass and crosslinking. *Polym. Test.* **2019**, *77*, 105881. [CrossRef]
10. Oprea, S.; Oprea, V. Mechanical behavior during different weathering tests of the polyurethane elastomers films. *Eur. Polym. J.* **2002**, *38*, 1205–1210. [CrossRef]
11. Scholz, P.; Wachtendorf, V.; Elert, A.-M.; Falkenhagen, J.; Becker, R.; Hoffmann, K.; Resch-Genger, U.; Tschiche, H.; Reinsch, S.; Weidner, S. Analytical toolset to characterize polyurethanes after exposure to artificial weathering under systematically varied moisture conditions. *Polym. Test.* **2019**, *78*, 105996. Available online: https://www.scopus.com/inward/record.uri?eid=2-s2.0-85069503515&doi=10.1016%2fj.polymertesting.2019.105996&partnerID=40&md5=5d3e6a62259aeed6cdeb3912a2ec95d6 (accessed on 14 December 2021). [CrossRef]
12. Kuranska, M.; Pinto, J.A.; Salach, K.; Barreiro, M.F.; Prociak, A. Synthesis of thermal insulating polyurethane foams from lignin and rapeseed based polyols: A comparative study. *Ind. Crops Prod.* **2020**, *143*, 111882. [CrossRef]
13. Mort, R.; Vorst, K.; Curtzwiler, G.; Jiang, S. Biobased foams for thermal insulation: Material selection, processing, modelling, and performance. *RSC Adv.* **2021**, *11*, 4375–4394. [CrossRef]
14. Jonjaroen, V.; Ummartyotin, S.; Chittapun, S. Algal cellulose as a reinforcement in rigid polyurethane foam. *Algal Res. Biomass Biofuels Bioprod.* **2020**, *51*, 102057. [CrossRef]
15. Cornille, A.; Auvergne, R.; Figovsky, O.; Boutevin, B.; Caillol, S. A perspective approach to sustainable routes for non-isocyanate polyurethanes. *Eur. Polym. J.* **2017**, *87*, 535–552. [CrossRef]

16. Rodrigues, J.D.O.; Andrade, C.K.Z.; Quirino, R.L.; Sales, M.J.A. Non-isocyanate poly(acyl-urethane) obtained from urea and castor (*Ricinus communis* L.) oil. *Prog. Org. Coat.* **2022**, *162*, 106557. [CrossRef]
17. Wilhelm, C.; Rivaton, A.; Gardette, J.-L. Infrared analysis of the photochemical behaviour of segmented polyurethanes: 3. Aromatic diisocyanate based polymers. *Polymer* **1998**, *39*, 1223–1232. Available online: https://www.scopus.com/inward/record.uri?eid=2-s2.0-0032028577&doi=10.1016%2fS0032-3861%2897%2900353-4&partnerID=40&md5=4ffb65e5dd9477d91e7672f4dff2de01 (accessed on 14 December 2021). [CrossRef]
18. Wilhelm, C.; Gardette, J.-L. Infrared analysis of the photochemical behaviour of segmented polyurethanes: Aliphatic poly(ether-urethane)s. *Polymer* **1998**, *39*, 5973–5980. Available online: https://www.scopus.com/inward/record.uri?eid=2-s2.0-0032210353&doi=10.1016%2fS0032-3861%2897%2910065-9&partnerID=40&md5=71595238af20bfe5abd9a439287d94e8 (accessed on 14 December 2021). [CrossRef]
19. Weise, N.K.; Bertocchi, M.J.; Wynne, J.H.; Long, I.; Mera, A.E. High performance vibrational damping poly(urethane) coatings: Blending 'soft' macrodiols for improved mechanical stability under weathering. *Prog. Org. Coat.* **2019**, *136*, 105240. Available online: https://www.scopus.com/inward/record.uri?eid=2-s2.0-85071912699&doi=10.1016%2fj.porgcoat.2019.105240&partnerID=40&md5=aeb9c93665d568497d6faf5157645215 (accessed on 14 December 2021). [CrossRef]
20. Lapcik, L.; Manas, D.; Lapcikova, B.; Vasina, M.; Stanek, M.; Cepe, K.; Vlcek, J.; Waters, K.E.; Greenwood, R.W.; Rowson, N.A. Effect of filler particle shape on plastic-elastic mechanical behavior of high density poly(ethylene)/mica and poly(ethylene)/wollastonite composites. *Compos. Part B Eng.* **2018**, *141*, 92–99. [CrossRef]
21. Rao, S.S. *Mechanical Vibrations*, 5th ed.; Prentice Hall: Upper Saddle River, NJ, USA, 2010; p. 1105.
22. Liu, K.; Liu, J. The damped dynamic vibration absorbers: Revisited and new result. *J. Sound Vib.* **2005**, *284*, 1181–1189. [CrossRef]
23. Hadas, Z.; Ondrusek, C. Nonlinear spring-less electromagnetic vibration energy harvesting system. *Eur. Phys. J.-Spec. Top.* **2015**, *224*, 2881–2896. [CrossRef]
24. Carrella, A.; Brennan, M.J.; Waters, T.P.; Lopes, V., Jr. Force and displacement transmissibility of a nonlinear isolator with high-static-low-dynamic-stiffness. *Int. J. Mech. Sci.* **2012**, *55*, 22–29. [CrossRef]
25. Dupuis, R.; Duboeuf, O.; Kirtz, B.; Aubry, E. Characterization of Vibrational Mechanical Properties of Polyurethane Foam. *Dyn. Behav. Mater.* **2016**, *1*, 123–128. [CrossRef]
26. Lapcik, L.; Vasina, M.; Lapcikova, B.; Stanek, M.; Ovsik, M.; Murtaja, Y. Study of the material engineering properties of high-density poly(ethylene)/perlite nanocomposite materials. *Nanotechnol. Rev.* **2020**, *9*, 1491–1499. [CrossRef]
27. Platonova, E.; Chechenov, I.; Pavlov, A.; Solodilov, V.; Afanasyev, E.; Shapagin, A.; Polezhaev, A. Thermally Remendable Polyurethane Network Cross-Linked via Reversible Diels-Alder Reaction. *Polymers* **2021**, *13*, 1935. [CrossRef] [PubMed]
28. Nemade, A.M.; Mishra, S.; Zope, V.S. Kinetics and Thermodynamics of Neutral Hydrolytic Depolymerization of Polyurethane Foam Waste Using Different Catalysts at Higher Temperature and Autogenious Pressures. *Polym. Plast. Technol. Eng.* **2010**, *49*, 83–89. [CrossRef]
29. Casati, F.; Herrington, R.; Broos, R.; Miyazaki, Y. Tailoring the performance of molded flexible polyurethane foams for car seats (Reprinted from Polyurethanes World Congress '97, 29 September–1 October 1997). *J. Cell. Plast.* **1998**, *34*, 430–466. [CrossRef]
30. Suh, K.; Park, C.; Maurer, M.; Tusim, M.; De Genova, R.; Broos, R.; Sophiea, D. Lightweight cellular plastics. *Adv. Mater.* **2000**, *12*, 1779–1789. [CrossRef]
31. Vakil, A.U.; Petryk, N.M.; Shepherd, E.; Beaman, H.T.; Ganesh, P.S.; Dong, K.S.; Monroe, M.B.B. Shape Memory Polymer Foams with Tunable Degradation Profiles. *ACS Appl. Bio. Mater.* **2021**, *4*, 6769–6779. [CrossRef]
32. Zahedifar, P.; Pazdur, L.; Vande Velde, C.M.L.; Billen, P. Multistage Chemical Recycling of Polyurethanes and Dicarbamates: A Glycolysis-Hydrolysis Demonstration. *Sustainability* **2021**, *13*, 3583. [CrossRef]
33. Gaboriaud, F.; Vantelon, J.P. Thermal-Degradation of Polyurethane Based on Mdi and Propoxylated Trimethylol Propane. *J. Polym. Sci. Part A Polym. Chem.* **1981**, *19*, 139–150. [CrossRef]
34. Ballistreri, A.; Foti, S.; Maravigna, P.; Montaudo, G.; Scamporrino, E. Mechanism of Thermal-Degradation of Polyurethanes Investigated by Direct Pyrolysis in the Mass-Spectrometer. *J. Polym. Sci. Part A Polym. Chem.* **1980**, *18*, 1923–1931. [CrossRef]

Article

# Study of the Mechanical, Sound Absorption and Thermal Properties of Cellular Rubber Composites Filled with a Silica Nanofiller

Marek Pöschl [1] and Martin Vašina [2,3,*]

[1] Centre of Polymer Systems, Tomas Bata University in Zlin, Třída Tomáše Bati 5678, 760 01 Zlin, Czech Republic; poschl@utb.cz
[2] Faculty of Technology, Tomas Bata University in Zlin, Nám. T.G. Masaryka 275, 760 01 Zlin, Czech Republic
[3] Faculty of Mechanical Engineering, VŠB-Technical University of Ostrava, 17. listopadu 15/2172, Poruba, 708 00 Ostrava, Czech Republic
* Correspondence: vasina@utb.cz; Tel.: +420-57-603-5112

**Citation:** Pöschl, M.; Vašina, M. Study of the Mechanical, Sound Absorption and Thermal Properties of Cellular Rubber Composites Filled with a Silica Nanofiller. *Materials* **2021**, *14*, 7450. https://doi.org/10.3390/ma14237450

Academic Editor: Klaus Werner Stöckelhuber

Received: 15 November 2021
Accepted: 2 December 2021
Published: 4 December 2021

**Publisher's Note:** MDPI stays neutral with regard to jurisdictional claims in published maps and institutional affiliations.

**Copyright:** © 2021 by the authors. Licensee MDPI, Basel, Switzerland. This article is an open access article distributed under the terms and conditions of the Creative Commons Attribution (CC BY) license (https://creativecommons.org/licenses/by/4.0/).

**Abstract:** This paper deals with the study of cellular rubbers, which were filled with silica nanofiller in order to optimize the rubber properties for given purposes. The rubber composites were produced with different concentrations of silica nanofiller at the same blowing agent concentration. The mechanical, sound absorption and thermal properties of the investigated rubber composites were evaluated. It was found that the concentration of silica filler had a significant effect on the above-mentioned properties. It was detected that a higher concentration of silica nanofiller generally led to an increase in mechanical stiffness and thermal conductivity. Conversely, sound absorption and thermal degradation of the investigated rubber composites decreased with an increase in the filler concentration. It can be also concluded that the rubber composites containing higher concentrations of silica filler showed a higher stiffness to weight ratio, which is one of the great advantages of these materials. Based on the experimental data, it was possible to find a correlation between mechanical stiffness of the tested rubber specimens evaluated using conventional and vibroacoustic measurement techniques. In addition, this paper presents a new methodology to optimize the blowing and vulcanization processes of rubber samples during their production.

**Keywords:** cellular rubber; mechanical stiffness; sound absorption; vibration damping; thermal behavior; silica nanofiller; excitation frequency

## 1. Introduction

Natural rubber (NR) comes from a tree called "*Hevea brasiliensis*" that is grown in many countries around the world, e.g., in Malaysia, Indonesia, India, Vietnam and Thailand. This tree produces a milky liquid called natural rubber latex [1]. Natural rubber latex is a dispersion of rubber particles in aqueous medium containing approximately 33% cis-1,4 polyisoprene rubber particles dispersed in water. This latex is obtained from the tree by making a cut in its bark in tropical countries of its origin [2]. The rubber is prepared by coagulation of the latex with acids. The obtained precipitate is subsequently dried and pressed into packaging. Natural rubber is very important in the rubber industry due to its elasticity, tensile strength and self-reinforcing character, which is said to originate from its strain-induced crystallization (SIC). SIC is caused by its linear chains, which are able to orient under stress. Rubbers with a macromolecular structure like polychloroprene rubber (CR) and polyisoprene rubber (IR) similar to NR can crystallize under stress. Due to the SIC behavior, these rubbers have excellent mechanical properties and good resistance to crack growth [3–5]. The advantages of NR also include low heat build-up [6].

However, unvulcanized rubbers without ingredients are not often used because they do not have the required properties. Additionally, additives (fillers, vulcanizing agents, UV stabilizers, etc.) are added to these rubbers to achieve the required properties. NR with additives is referred to as a rubber compound that is plastic and malleable. The composition of rubber compounds is given in units of phr (i.e., parts per hundred rubber). Crosslinking process occurs during high-temperature heating, when the rubber compound turns into an elastic material, i.e., vulcanizate [7,8]. Sulphur is one of the most commonly used curing agents [7]. Peroxides usually for ethylene propylene diene methylene (EPDM) and oxides of divalent metals for CR rubbers are also used. The main components of rubber compounds are fillers because they reduce the cost of rubber products and modify physical and mechanical properties of vulcanizates, such as tensile strength and abrasion [8]. The stiffening effect of rubber compounds generally depends on filler particle sizes. The highest stiffness is usually achieved with rubber compounds containing filler particle sizes from 10 to 100 nm that are characterized by a large specific surface area. In practice, carbon black and silica nanofillers are most commonly used as reinforcing fillers [9]. Silica is obtained from silica rocks by natural mining of minerals in various parts of the world and by the subsequent grinding or crushing of these raw materials. Such silica is called ground silica and is characterized by a low reinforcing effect due to the large primary particles. By chemical treatment of the mined material, it is possible to obtain silica with small particle sizes, which have a stiffening effect comparable to that of carbon black filler. The smallest primary particle sizes (i.e., from 1 to 3 nm) are obtained by pyrolysis process. Such silica is called fumed silica [10]. These primary particles form aggregates of size from 100 to 250 nm. Silica is characterized by the best stiffening effect due to its large specific surface area (i.e., from 50 to 400 $m^2 \cdot g^{-1}$) [11]. Larger particle sizes (>40 nm) are obtained by precipitation of sodium silicate ($Na_2SiO_3$) with acids. Such silica is called precipitated silica [10,12].

Other additives are also added to rubber compounds. Nucleating agents are used to improve the cell morphology of rubber compounds (i.e., increasing cell density, decreasing cell size and narrowing cell-size distribution) by providing heterogeneous nucleation sites [13]. Blowing agents are added in order to produce foam rubber products. Along with urethane and polyvinyl chloride rubber materials, the foam rubber products are competitive with cushioning or padding plastic foams such as cloth, polyester fiberfill, hair, jute and the like. Foam rubber products are very important in many applications due to their unique structural properties such as their low weight, performance, impact damping, thermal and acoustic insulation properties, moderate energy absorption and low price [14–16]. In general, foam rubbers can be divided into sponge, cellular and microcellular rubbers. Sponge rubber has a similar structure to a real sea sponge. It consists of open and interconnected cavities. The channels thus formed allow a high absorption of liquids and release the absorbed liquid after their compression. Cellular rubber is characterized by a higher specific weight compared to sponge rubber. Its structure, characterized by cell size, cell density and expansion ratio [17], consists of smaller cavities that are not interconnected. The cellular rubber is thus more compressible, non-absorbent and achieves excellent damping properties [14]. Microcellular rubbers are characterized by cell densities greater than 109 cells/$cm^3$ and cells smaller than 10 μm [18]. These rubber types are produced by using physical or chemistry blowing agents. Physical blowing is performed with blowing liquids (e.g., pentane and hexane) and is based on the liquid–vapor transition. The essence of chemical blowing is the chemical decomposition of the blowing agent or the reaction to release gases. In the rubber blowing process, it is important that the blowing process takes place before vulcanization. Once the cellular structure is formed, vulcanization should occur to fix this cellular structure [14,19,20]. Another type of foam rubber is latex foam, which is obtained from natural rubber latex by mechanical whipping and subsequent vulcanization. New applications of the NR latex foam rubbers allow their use as absorbent materials for oils and carbon dioxide ($CO_2$). NR latex foam rubber filled with silica can serve as a $CO_2$ absorbing material [21].

Many studies have already investigated various properties of silica-filled rubber composites. The aim of this research is to investigate the mechanical, sound absorption and thermal properties of cellular rubber composites that were produced with various concentrations of silica nanofiller and the same blowing agent concentration. Mechanical properties of rubber materials can be determined by conventional measurement techniques such as tensile, shear, hardness, permanent deformation, rebound resilience and viscoelasticity tests. In this research, the mechanical stiffness of the investigated silica-filled rubber composites determined from permanent deformation and rebound resilience measurements is verified by non-conventional methods for determining the stiffness of these materials, namely by vibration damping and sound absorption methods. The latter two methods are relatively simple, inexpensive, fast and non-destructive compared to the conventional methods used to determine the mechanical stiffness of solids. Therefore, the vibration damping and sound absorption methods can also be easily applied to compare the mechanical stiffness of different materials. In addition, the rubber molding process is optimized using rubber processing analysis, where the rubber blowing phase should generally take place before the vulcanization phase for a given rubber recipe.

## 2. Materials and Methods

*2.1. Materials*

Natural rubber SMR (Standard Malaysian Rubber)–20 was used as the base for rubber compounds. Silica Perkasil KS–408 from WR Grace & Co. (Columbia, MD, USA) with the specific surface area of 175 $m^2 \cdot g^{-1}$ and bulk density of 0.17 $g \cdot cm^{-3}$ was used as a nanofiller in rubber compounds. The additives, including activator ZnO from SlovZink a.s. (Košeca, Slovakia), activator stearic acid from Setuza a.s. (Ústí nad Labem, Czech Republic), antiozonant N-(1,3-dimethylbutyl)-N′-phenyl-p-phenylenediamine (Vulkanox 4020) from Lanxess AG (Brussel, Belgium), plasticizer (paraffinic oil Nyflex 228) from Nynas AB (Stockholm, Sweden), blowing agent azodicarbonamide $C_2H_4N_4O_2$ (Porofor ADC) from Lanxess AG (Brussel, Belgium), accelerator N-tert-butyl-benzothiazole sulfonamide (TBBS) from Duslo a.s. (Šaľa, Slovakia) and curing agent sulphur type Crystex OT33 from Eastman Chemical company (Kingsport, TN, USA), were also compounded. Five rubber compounds with different silica concentrations were prepared. Their recipe is detailed in Table 1.

Table 1. Recipe of the investigated rubber compounds.

| Ingredients | Rubber Designation | | | | |
|---|---|---|---|---|---|
| | S0 | S10 | S20 | S30 | S45 |
| | Loading (phr) | | | | |
| SMR–20 | 100 | 100 | 100 | 100 | 100 |
| Silika KS–408 | 0 | 10 | 20 | 30 | 45 |
| ZnO | 2 | 2 | 2 | 2 | 2 |
| Vulkanox 4020 | 1 | 1 | 1 | 1 | 1 |
| Porofor ADC | 5 | 5 | 5 | 5 | 5 |
| Nyflex 228 | 10 | 10 | 10 | 10 | 10 |
| Stearic acid | 1 | 1 | 1 | 1 | 1 |
| TBBS | 1 | 1 | 1 | 1 | 1 |
| Sulphur OT33 | 2 | 2 | 2 | 2 | 2 |

The optimization of mechanical properties of rubber compounds is highly dependent on the silica concentration. During the mixing process of rubber compounds, the viscosity of these compounds generally increases with increasing the silica concentration. High concentrations of silica filler (i.e., over 30 phr) in rubber compounds lead to a difficult mixing process, a lower dispersion of additives and an increase in curing time [22]. For this reason, rubber samples produced with the highest silica concentration of 45 phr are characterized by an imperfect blowing process, which results in a reduction in damping properties of cellular rubbers.

In the first stage of the production of rubber composites, the basic rubber compounds (i.e., SMR–20, ZnO, Vulkanox 4020, stearic acid, silica and Nyflex 228) were mixed in a two-roll mill (150 mm × 330 mm, Farrel, Milan, Italy) at a rotation speed ratio of 15/20 at a temperature of 50 °C. After mixing these ingredients, a three-minute homogenization took place on the two-roll mill and subsequently the compound was withdrawn from this mill. The mixing process in the second stage, in which the remaining ingredients (i.e., Porofor ADC, TBBS and sulphur OT33) were mixed, followed after a 20-min pause on the two-roll mill under the same conditions as in the first stage. After this mixing, a five-minute homogenization was performed, and the obtained rubber compounds were subsequently withdrawn from the mill.

The rubber compounds were compression-molded at a temperature of 170 °C using a hydraulic hot press during the time $t_m$, which is given by the equation:

$$t_m = t_{90} + c_m \cdot t \tag{1}$$

where $c_m$ = 0.5 min/mm is the molding constant, $t_{90}$ (min) is the optimum cure time (see Chapter 3.1) and $t$ (mm) is the sample thickness.

Finally, rubber specimens of two different diameters of 20 and 100 mm and three different thicknesses of 4, 8 and 12 mm were produced.

*2.2. Measurement Methodology*

2.2.1. Curing Characteristics

Curing characteristics were measured according to ASTM D 5289 standard on the Premier MDR moving die rheometer from Alpha Technology (Hudson, OH, USA) at a constant temperature of 170 °C. Minimum ($M_L$) and maximum ($M_H$) torques, scorch ($t_{s1}$) and optimum cure ($t_{90}$) times were subsequently evaluated.

2.2.2. Rubber Processing Analysis

The rubber processing (RP) analysis, which is based on cone-cone principle, was performed on the Premier RPA rubber process analyzer from Alpha Technology (Hudson, OH, USA). As a result of RP analysis, time dependencies of pressure curves during isothermal vulcanization were determined. The rate of blowing was subsequently obtained from the first time derivative of the pressure curve.

2.2.3. Optical Microscopy

Pore sizes on cut surfaces of the investigated rubber specimens were evaluated using an optical microscope "Carl Zeiss Stemi 2000–C" from Carl Zeiss Microimaging GmbH (Jena, Germany) [23]. The magnification of the microscope objective was from 0.65× to 5× and the stereo angle was set to 11°.

2.2.4. Rebound Resilience

The rebound resilience was measured according to ISO 4662 standard. Rubber specimens measuring 12 mm in thickness were examined. Average values of the rebound resilience and their standard deviations were subsequently determined from six measured values. The measurements were carried out at an ambient temperature of 22 °C.

2.2.5. Compression Testing

Experimental measurements of permanent deformation of cellular rubber composites were performed according to ČSN EN ISO 1856 standard based on compression tests. These measurements were carried out on rubber samples measuring 12 mm in thickness and 20 mm in diameter. The samples were compressed to 50% initial thickness ($t_1$) for 24 h at three different temperatures (i.e., 24, 40 and 65 °C). After compression, the samples were released for 30 min, and their thicknesses ($t_2$) were measured again. Finally, the compression set $CS$ (%) of the rubber samples was determined from the equation [24]:

$$CS = \frac{t_1 - t_2}{t_1} \times 100 \tag{2}$$

Average values of the compression set, including their standard deviations, were determined from five measured values.

2.2.6. Mechanical Vibration Damping Testing

The ability of a material to dampen harmonically excited mechanical vibration is expressed by the displacement transmissibility $T_d$ (–) that is defined for a linear single degree of freedom as follows [25,26]:

$$T_d = \frac{y_2}{y_1} = \frac{a_2}{a_1} = \sqrt{\frac{k^2 + (c\omega)^2}{(k - m\omega^2)^2 + (c\omega)^2}} = \sqrt{\frac{1 + (2\zeta r)^2}{(1 - r^2)^2 + (2\zeta r)^2}} \quad (3)$$

where $y$ is the displacement amplitude on the input (1) and output (2) sides of the harmonically loaded sample, $a$ is the acceleration amplitude on input (1) and output (2) sides of the harmonically loaded sample, $k$ is the stiffness (spring constant), $c$ is the viscous damping coefficient, $m$ is the mass, $\omega$ is the frequency of oscillation, $\zeta$ is the damping ratio and $r$ is the frequency ratio. The damping and frequency ratios are expressed by the equations [27–29]:

$$\zeta = \frac{c}{2 \cdot \sqrt{k \cdot m}} \quad (4)$$

$$r = \frac{\omega}{\omega_n} = \frac{\omega \cdot \sqrt{m}}{\sqrt{k}} \quad (5)$$

where $\omega_n$ is the natural frequency. There are three different types of mechanical vibration depending on the displacement transmissibility value, namely damped ($T_d < 1$), undamped ($T_d = 1$) and resonance ($T_d > 1$) vibration. Under the condition $dT_d/dr = 0$ in Equation (3), it is possible to determine the frequency ratio $r_0$, at which the displacement transmissibility reaches a maximum value [25]:

$$r_0 = \frac{\sqrt{\sqrt{1 + 8\zeta^2} - 1}}{2\zeta} \quad (6)$$

It is apparent from Equation (6) that the local extreme of the displacement transmissibility ($T_{dmax}$) is generally shifted to lower values of the frequency ratio $r$ with an increase in the damping ratio $\zeta$.

The mechanical vibration damping testing of the investigated rubber composites was performed by the forced oscillation method. The displacement transmissibility $T_d$ was experimentally measured using a BK 4810 shaker in combination with a BK 3560-B-030 signal pulse multi-analyzer and a BK 2706 power amplifier (Brüel & Kjær, Nærum, Denmark) in the frequency range of 2–1500 Hz. Harmonic sine waves were generated by the shaker. The displacement transmissibility was determined from Equation (3) based on the acceleration amplitudes $a_1$ and $a_2$ on the input and output sides of the investigated samples by means of BK 4393 accelerometers (Brüel & Kjær, Nærum, Denmark). Measurements of the displacement transmissibility were performed for different inertial masses $m$ (i.e., 0, 90 and 500 g), which were located on the upper side of the harmonically loaded investigated samples. Furthermore, vibration damping testing of the investigated rubber samples with a ground plane dimension of 60 mm × 60 mm was performed for three different sample thicknesses (i.e., 4, 8 and 12 mm). Each measurement was repeated five times at an ambient temperature of 23 °C.

### 2.2.7. Sound Absorption Measurements

Sound absorption properties of materials are characterized by the sound absorption coefficient α (–), which is defined by the ratio [30]:

$$\alpha = \frac{P_d}{P_i} \tag{7}$$

where $P_d$ is the dissipated acoustic power and $P_i$ is the incident acoustic power. The ability of a material to absorb sound is generally affected by different factors, namely by excitation frequency of acoustic waves, material thickness, structure, density, temperature, humidity and so on. Sound absorption properties of materials are also characterized by the noise reduction coefficient NRC (–), which considers the excitation frequency influence on the sound absorption coefficient. This coefficient is defined as the average value of the sound absorption coefficients of a given material at the excitation frequencies of 250, 500, 1000 and 2000 Hz [31]:

$$NRC = \frac{\alpha_{250} + \alpha_{500} + \alpha_{1000} + \alpha_{2000}}{4} \tag{8}$$

Sound absorption measurements can also be used to determine the longitudinal elastic coefficient K of a powder bed. This coefficient is similar to Young´s modulus of elasticity and is defined by the formula [32,33]:

$$K = c^2 \cdot \rho_b = (4 \cdot h \cdot f_{p1})^2 \cdot \rho_b \tag{9}$$

where $c$ is the speed of sound of elastic wave propagated through a powder bed, $\rho_b$ is the powder bulk density, $h$ is the powder bed height and $f_{p1}$ is the primary absorption peak frequency.

Frequency dependencies of the normal incidence sound absorption coefficient α of the tested rubber samples and the loose silica powder were determined by the transfer function method ISO 10534-2 [34] using a BK 4206 two-microphone impedance tube in combination with a BK 3560-B-030 signal pulse multi-analyzer and a BK 2706 power amplifier (Brüel & Kjær, Nærum, Denmark) in the frequency range of 150–6400 Hz. The transfer function method is based on the partial standing wave principle. In this case, the normal incidence sound absorption coefficient α is expressed by the equation [35]:

$$\alpha = 1 - \left| \frac{H_{12} - e^{-k_0 \cdot (x_1 - x_2) \cdot j}}{e^{k_0 \cdot (x_1 - x_2) \cdot j} - H_{12}} \cdot e^{2k_0 \cdot x_1 \cdot j} \right|^2 \tag{10}$$

where $H_{12}$ is the complex acoustic transfer function, $k_0$ is the wave number, $x_1$ and $x_2$ are the distances of two microphone positions from the reference plane ($x = 0$) and $j$ is the imaginary unit. The complex acoustic transfer function is defined as follows:

$$H_{12} = \frac{p_2}{p_1} = \frac{e^{k_0 \cdot x_2 \cdot j} + r \cdot e^{-k_0 \cdot x_2 \cdot j}}{e^{k_0 \cdot x_1 \cdot j} + r \cdot e^{-k_0 \cdot x_1 \cdot j}} \tag{11}$$

where $p_1$ and $p_2$ are the complex acoustic pressures at two microphone positions. Sound damping properties of the tested rubber samples and the silica powder bed were investigated in the impedance tube having inside diameter of 30 mm for different sample thicknesses (i.e., 4, 8 and 12 mm) of the studied rubber composites and various silica powder bed heights (ranging from 10 to 40 mm). Experimental measurements of the sound absorption coefficient were realized at an ambient temperature of 22 °C.

### 2.2.8. Thermal Conductivity

Experimental measurements of the thermal conductivity over the range of 0.01–10 W·m$^{-1}$·K$^{-1}$ of the rubber composites were carried out according to ČSN 72 1105 standard using a C-Therm TCi thermal conductivity analyzer (C-Therm Technologies,

Frederiction, NB, Canada). Average values of the thermal conductivity and their standard deviations were subsequently determined from five measured values.

2.2.9. Thermogravimetric Analysis

Thermogravimetric properties of the investigated cellular rubber samples were realized on a thermal analyzer DTG/60 (Shimadzu, Japan). Experimental measurements were performed at a heating rate of 10 °C/min under nitrogen atmosphere (50 mL/min) in the temperature range from 30 to 500 °C. Thermogravimetric analysis, which consisted in measurements of the weight loss $\Delta m$ of the tested rubber samples, was evaluated using ta60 Version 1.40 (Shimadzu, Japan) software. The measured results were performed in triplicate.

## 3. Results and Discussion
### 3.1. Curing Characteristics

Curing curves of the tested rubber composites are demonstrated in Figure 1. Curing parameters of these materials are shown in Table 2. It is visible (see Figure 1) that the minimum ($M_L$) and maximum ($M_H$) torques increase with an increase in the silica concentration due to a higher mechanical stiffness of the rubber composites at higher silica concentrations. Similarly, the optimum cure time $t_{90}$ increases with increasing the silica concentration. Table 2 also shows the scorch time $t_{s1}$, which is characterized by the fact that the rubber composite was still plastic during this time. Therefore, the blowing process of the tested rubbers should also be carried out during the time $t_{s1}$.

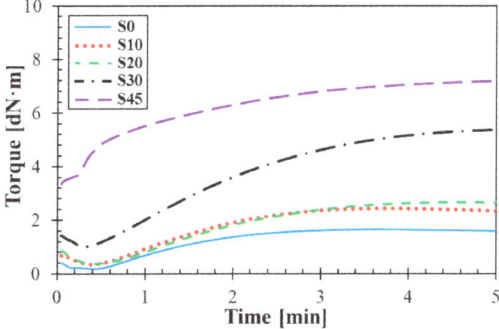

**Figure 1.** Time dependencies of vulcanization characteristics.

**Table 2.** Curing parameters of the studied rubber composites.

| Rubber Type | $M_H$ [dN·m] | $M_L$ [dN·m] | $t_{s1}$ [min] | $t_{90}$ [min] |
|---|---|---|---|---|
| S0  | 1.67 | 0.20 | 0.54 | 1.76 |
| S10 | 2.45 | 0.37 | 0.57 | 1.73 |
| S20 | 2.67 | 0.41 | 1.09 | 2.31 |
| S30 | 5.28 | 1.03 | 0.87 | 3.36 |
| S45 | 7.14 | 3.37 | 0.31 | 3.54 |

### 3.2. Rubber Processing Analysis

Time dependencies of pressure curves and the rate of blowing of the investigated rubber composites are shown in Figures 2 and 3. It is evident from these figures that the blowing process of the rubbers took place in three phases. In the first phase (i.e., preheating), the rubber composite is heated, its viscosity is reduced to a minimum and the blowing agent was not decomposed. In addition, a skin is formed in order to prevent profile collapse.

In the second stage (i.e., expansion), the blowing agent was decomposed with increasing the vulcanization rate. A relative balance between pressure gas and vulcanizate formation created a cell structure. The degree of vulcanization reached its maxima in the third stage (i.e., curing). If the decomposition of the blowing agent starts late, the production of released gas from the decomposition of the blowing agent could be limited during the curing process, which could lead to imperfect cellular structures of rubbers. It is evident from Figure 2 that a rapid increase in pressure (i.e., blowing) took place in the second phase, when the decomposition of the blowing agent occurs only with the increasing rate of vulcanization. As shown in Figure 3, the fastest decomposition of the blowing agent occurred at the maximum rate of blowing. The vulcanization and blowing processes of the rubber composite containing 45 phr of silica were more difficult due to higher filler content. These processes were slower because the silica filler bound the vulcanization system due to the content of polar groups. Therefore, the rubber composite containing 45 phr of silica became more viscous during the blowing process, and thus the blowing of this rubber composite was imperfect.

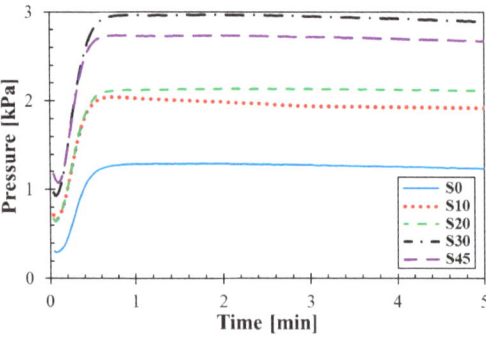

**Figure 2.** Time dependencies of pressure curves.

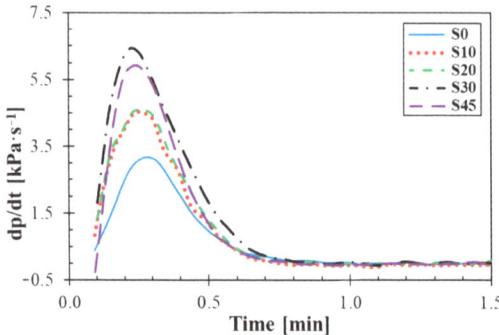

**Figure 3.** Time dependencies of rate of blowing.

*3.3. Optical Microscopy*

As can be seen from the optical microscopy images (see Figure 4), it is possible to observe distinct cavities caused by evaporation of the blowing agent. It is evident that the most distinct cavities are observed for the rubber composite containing 0 phr of silica. Smaller cavities are observable in the rubber samples containing 10, 20 and 30 phr of silica. The size of these cavities is still sufficient to ensure good thermal insulation properties of these rubbers. It is also obvious from Figure 4 that there was practically no cellular structure of the rubber composite containing 45 phr of silica. This is due to the fact that silica is reinforcing filler, which increases the viscosity of rubber composites at high silica concentrations. Therefore, sufficiently large cavities were not created during the blowing process in the rubber composite with 45 phr of silica. This fact also corresponds with the time dependence of the rate of blowing (see Figure 3), where the maximum rate of blowing for the rubber composite containing 45 phr of silica was lower compared to the maximum rate of blowing of the rubber composite with 30 phr of silica. For this reason, the rubber composite containing 45 phr of silica was not suitable in terms of thermal insulation properties.

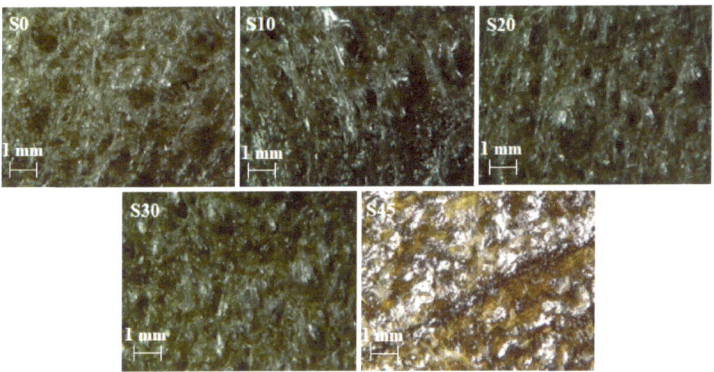

**Figure 4.** Optical microscopy images of the studied rubber samples.

*3.4. Mechanical Properties*

3.4.1. Rebound Resilience

The results of the average values of the rebound resilience and their standard deviations are shown in Table 3. The rebound resilience decreased from 51% to 35% for S0 and S45 rubber composites, respectively. It can be concluded that higher silica concentrations in rubber composites led to a higher ability to transform mechanical energy into heat, and thus to a decrease in the rebound resilience.

**Table 3.** Values of rebound resilience of the tested rubber composites.

| Rubber Type | Resilience [%] |
|---|---|
| S0 | 51 ± 1 |
| S10 | 48 ± 1 |
| S20 | 44 ± 1 |
| S30 | 41 ± 1 |
| S45 | 35 ± 1 |

3.4.2. Compression Testing

Dependencies of permanent deformation vs. silica concentration for three different temperatures (i.e., 25, 40 and 65 °C) are demonstrated in Figure 5. It can be observed that the compression set generally increased with increasing temperature. It is also evident that the permanent deformation decreased at higher silica concentrations. This was due to the higher silica content in rubber composites and the formation of smaller cavities that were able to return the rubber composites to their original state [24]. The highest compression set values (i.e., 52.2 and 48.7%) at the given compression temperatures (i.e., 65 and 40 °C) were obtained for the rubber composites containing 10 phr of silica. At the compression temperature of 25 °C, the highest compression set of 36.2% was found for the rubber composite containing 20 phr of silica.

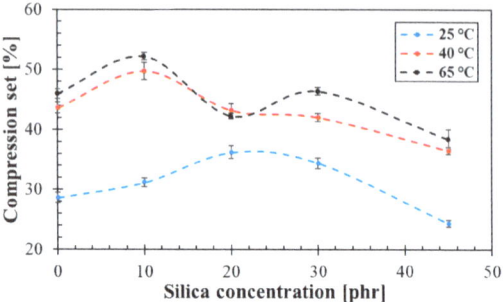

**Figure 5.** Effect of silica concentration on compression set of the investigated rubber composites under different compression temperatures.

### 3.4.3. Mechanical Vibration Damping Testing

Frequency dependencies of the displacement transmissibility of the tested rubber composites measuring $t$ = 4 mm in thickness and containing different concentrations of silica are shown in Figure 6. In addition, the rubber samples were loaded with an inertial mass of 90 g. It is evident from this comparison that the concentration of silica had a significant influence on vibration damping properties of the tested composite materials. The ability of a material to damp mechanical vibration is generally decreasing with increasing the concentration of silica, which was in accordance with measurements of the rebound resilience and the permanent deformation. It was again due to higher stiffness $k$ (or lower damping ratio $\zeta$) of the rubber composites, which were produced with higher concentrations of silica. These facts resulted in a lower transformation of input mechanical energy into heat during forced oscillations [36], and in an increase in the values of the damped and undamped natural frequencies (see Equation (6)) [37]. Therefore, the first resonance frequency ($f_{R1}$) value was shifted to the right (see Figure 6) with increasing concentration of silica, namely from 218 Hz (i.e., sample S0) to 721 Hz (i.e., sample S45), as indicated in Table 4.

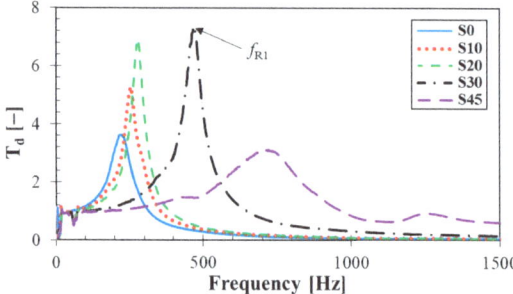

**Figure 6.** Effect of silica concentration on frequency dependencies of the displacement transmissibility of the tested rubber composites measuring $t$ = 4 mm in thickness and loaded with inertial mass $m$ = 90 g.

The vibration damping properties of the investigated harmonically loaded rubber samples were also influenced by their thickness $t$, the inertial mass $m$ and the excitation frequency $f$.

**Table 4.** First resonance frequency ($f_{R1}$) in Hz of the studied rubber composites as induced by harmonic force vibration depending on rubber thickness ($t$) and inertial mass ($m$).

| Rubber Type | $t$ [mm] | $m$ [g] | | |
|---|---|---|---|---|
| | | 0 | 90 | 500 |
| S0 | 4 | 399 ± 13 | 218 ± 10 | 100 ± 5 |
| | 8 | 192 ± 8 | 115 ± 5 | 62 ± 3 |
| | 12 | 176 ± 6 | 80 ± 3 | 55 ± 2 |
| S10 | 4 | 417 ± 15 | 251 ± 11 | 106 ± 4 |
| | 8 | 235 ± 10 | 122 ± 4 | 70 ± 3 |
| | 12 | 220 ± 9 | 103 ± 4 | 59 ± 2 |
| S20 | 4 | 434 ± 14 | 279 ± 12 | 120 ± 5 |
| | 8 | 294 ± 12 | 178 ± 6 | 77 ± 3 |
| | 12 | 257 ± 11 | 146 ± 4 | 73 ± 3 |
| S30 | 4 | 525 ± 15 | 468 ± 18 | 196 ± 6 |
| | 8 | 436 ± 16 | 343 ± 11 | 140 ± 4 |
| | 12 | 396 ± 13 | 278 ± 9 | 119 ± 4 |
| S45 | 4 | 1239 ± 25 | 721 ± 20 | 284 ± 9 |
| | 8 | 804 ± 19 | 548 ± 15 | 201 ± 7 |
| | 12 | 545 ± 14 | 457 ± 12 | 154 ± 5 |

The effect of the inertial mass on vibration damping properties for the S20 rubber type measuring $t$ = 4 mm in thickness is shown in Figure 7a. It is visible that better vibration damping properties were generally observed for higher inertial masses $m$. For this reason, the inertial mass had a positive influence on mechanical vibration damping, which is reflected in a shift of the first resonance frequency peak position to lower excitation frequencies, i.e., in the decrease of the frequency $f_{R1}$ (see Table 4) from 434 Hz ($m$ = 0 g) to 120 Hz ($m$ = 500 g). This finding is consistent with Equations (4) and (6). It should be noted that the magnitude of inertial mass is limited only to the range of elastic deformations of the harmonically loaded rubber specimens.

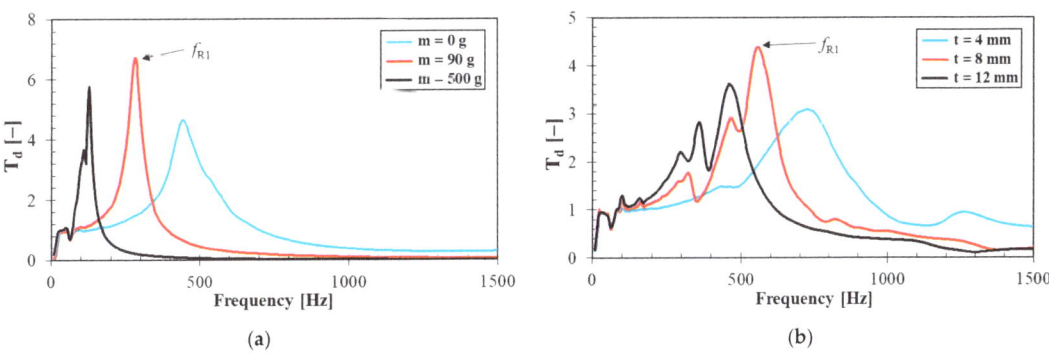

**Figure 7.** Frequency dependencies of the displacement transmissibility: (**a**) Rubber samples containing 20 phr of silica measuring $t$ = 4 mm in thickness and loaded with different inertial masses; (**b**) rubber samples of various thicknesses containing 45 phr of silica and loaded with inertial mass $m$ = 90 g.

The effect of the sample thickness on vibration damping properties is shown in Figure 7b for the S45 rubber type loaded by an inertial mass of 90 g. It is visible that a higher material thickness led to lower values of the frequency $f_{R1}$, i.e., from 721 Hz ($t$ = 4 mm) to 457 Hz ($t$ = 12 mm), as indicated in Table 4. For this reason, the rubber thickness generally has a positive effect on vibration damping and leads to a higher transformation of input mechanical energy into heat during dynamic loading.

It is also visible from Figures 6 and 7 that the vibration damping properties were also significantly influenced by the excitation frequency $f$. It is evident from these frequency dependencies that the resonant mechanical vibration (i.e., $T_d > 1$) was achieved at low excitation frequencies depending on the rubber composite type, its thickness $t$ and the inertial mass $m$. For example, for the S45 rubber type measuring 4 mm in thickness and without inertial mass (i.e., $m$ = 0 g), the resonant mechanical vibration was observed over the entire frequency range. In the case of the S0 rubber type measuring 12 mm in thickness and loaded with the highest inertial mass (i.e., $m$ = 500 g), the resonant mechanical vibration was observed at significantly lower excitation frequencies (namely at $f$ < 86 Hz). On the contrary, the damped mechanical vibration (i.e., $T_d < 1$) was generally obtained at higher excitation frequencies (see Figures 6 and 7).

### 3.5. Sound Absorption Properties

Frequency dependencies of the sound absorption coefficient of the loose silica powder depending on its height are shown in Figure 8. The obtained results of acoustical and mechanical quantities from sound absorption measurements are shown in Table 5. It is clear from the calculated values of the noise reduction coefficient NRC (see Equation (8)) that sound damping properties of the tested loose silica filler generally increased with increasing the powder bed height. As is shown in Figure 8, sound absorption properties of the silica powder were similar at higher excitation frequencies (i.e., at $f$ > 2 kHz) independently of the powder bed height. It is also visible that the maximum value of the sound absorption coefficient, which is proportional to the primary absorption peak frequency $f_{p1}$ (i.e., $\alpha_{max} = \alpha_{fp1}$), generally shifted towards lower excitation frequencies with increasing the powder bed height (see Figure 8). The speed of sound $c$ and the longitudinal elastic coefficient $K$ were subsequently determined from Equation (9). It was found that a higher powder bed height led to a higher speed of sound $c$, and thus, to higher values of the longitudinal elastic coefficient $K$. For this reason, the mechanical stiffness of the studied silica filler generally increased with increasing the silica bed height $h$.

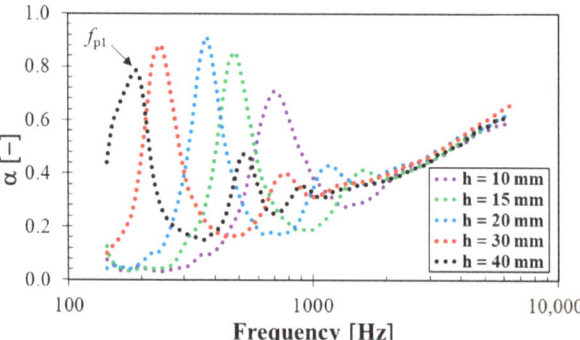

**Figure 8.** Frequency dependencies of the sound absorption coefficient of the silica powder for different loose powder bed heights.

**Table 5.** Results of measured and calculated acoustical and mechanical quantities for different loose powder bed heights of the applied silica filler.

| Quantity | $h$ [mm] | | | | |
|---|---|---|---|---|---|
| | 10 | 15 | 20 | 30 | 40 |
| NRC [–] | 0.259 | 0.359 | 0.302 | 0.458 | 0.384 |
| $f_{p1}$ [Hz] | 696 | 480 | 364 | 244 | 184 |
| $\alpha_{fp1}$ [–] | 0.709 | 0.854 | 0.911 | 0.879 | 0.784 |
| $c$ [m·s$^{-1}$] | 27.8 | 28.8 | 29.1 | 29.3 | 29.4 |
| $K$ [MPa] | 0.132 | 0.141 | 0.144 | 0.146 | 0.147 |

Frequency dependencies of the sound absorption coefficient of the tested rubber composites measuring $t = 12$ mm in thickness with different concentrations of silica are shown in Figure 9a. It is evident that sound absorption properties generally increased with decreasing silica concentrations. This fact was influenced by the rubber stiffness, which increases with increasing the silica concentration, as well as the longitudinal elastic coefficient of the loose powder beds (see Table 5). For this reason, better sound damping properties were obtained for rubber samples containing lower silica concentrations, at which the rubber samples exhibit lower stiffness and their structure is more spongy-like. This finding is consistent with the fact that porous, spongy and fiber materials belong to suitable materials in order to absorb sound [38]. The above-mentioned phenomena were also in good agreement with the vibration damping measurements, when a better ability of damp mechanical vibrations was generally obtained in rubber composites containing lower concentrations of silica.

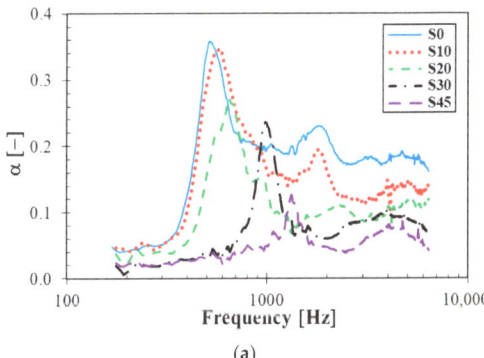
(a)

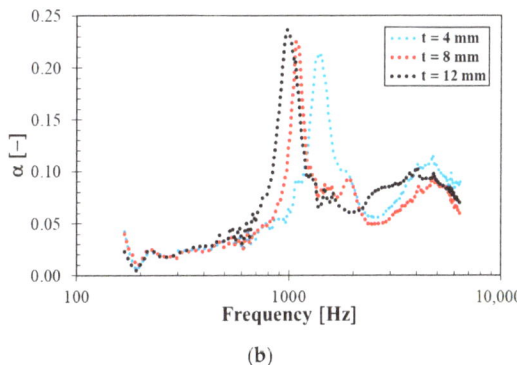
(b)

**Figure 9.** Frequency dependencies of the sound absorption coefficient: (a) Rubber samples measuring $t = 12$ mm in thickness at different concentrations of silica filler; (b) rubber samples of various thicknesses with 30 phr of silica filler.

It was also found that sound damping properties increase with increasing the sample thickness (see Figure 9b), especially at lower excitation frequencies. This finding was confirmed on the basis of the noise reduction coefficient. As indicated in Table 6, the noise reduction coefficient generally increased with an increase in rubber thickness and the decreasing concentration of silica. Its value ranges from 0.028 (S45, $t = 4$ mm) to 0.182 (S0, $t = 12$ mm). For these reasons, sound absorption properties of the investigated rubber composites were relatively low.

**Table 6.** Values of noise reduction coefficient (NRC) of the studied rubber composites.

| Rubber Type | t [mm] | | |
|---|---|---|---|
| | 4 | 8 | 12 |
| S0 | 0.086 | 0.113 | 0.182 |
| S10 | 0.066 | 0.107 | 0.158 |
| S20 | 0.065 | 0.098 | 0.110 |
| S30 | 0.047 | 0.067 | 0.084 |
| S45 | 0.028 | 0.030 | 0.037 |

*3.6. Thermal Properties*

3.6.1. Thermal Conductivity

The effect of silica concentration on thermal conductivity of the studied rubber composites is shown in Figure 10. Observed data show an exponential increase in thermal conductivity with the increasing silica concentration. Low silica content led to larger pore sizes in rubber composites, which is accompanied by lower values of thermal conductivity. Therefore, rubber composites containing pores of larger sizes can be used as thermal insulators. As shown in Figure 10, a significant increase in thermal conductivity was achieved at higher concentrations (>30 phr) of silica. In these cases, the rubber composite is too stiff and viscous, which prevented the formation of larger pores. The thermal conductivity of 0.508 W·m$^{-1}$·K$^{-1}$ for the rubber composite containing 45 phr of silica was approximately double compared to the thermal conductivity of the rubber composite with 30 phr of silica. For this reason, a porous structure of the rubber composite containing 45 phr of silica was practically not formed, which led to a significant decrease in thermal insulation properties of rubber composites containing high concentrations (i.e., over 45 phr) of silica.

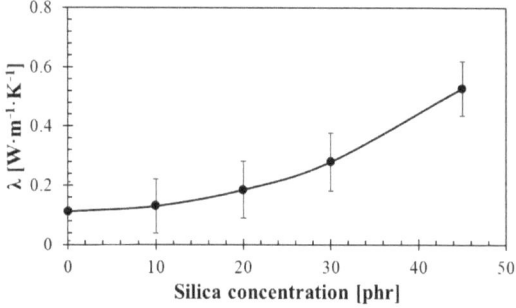

**Figure 10.** Effect of silica concentration on thermal conductivity of the tested rubber composites.

3.6.2. Thermogravimetric Analysis

As illustrated in Figure 11, the rubber composites were characterized by a two-step symmetric degradation of similar curve shapes, regardless of filler amount. The first declination on thermogravimetric curves started above 150 °C and continued until 300 °C with the corresponding weight loss of about 10% ($w/w$). Exceeding this temperature limit (i.e., about 300 °C), a rapid degradation occurred, and was accompanied by a noticeable decrease in the sample weight. At the temperatures from 420 to 440 °C, the degradation process stopped, while the degradation extent was apparently dependent on the filler concentration.

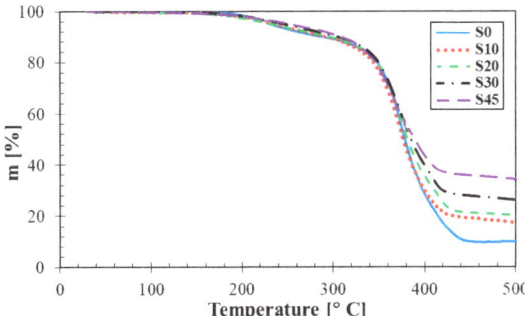

**Figure 11.** Thermogravimetric analysis results of the studied silica/rubber composites.

As is obvious from Table 7, higher reduction in the weight loss $\Delta m$ was generally observed at lower filler concentrations. It was found that the rubber sample without any silica filler (i.e., 0 phr) exhibited the weight loss of 90.42% ($w/w$) in the entire temperature range. On the other hand, the rubber sample containing the highest filler concentration (i.e., 45 phr) showed the highest thermal stability at the weight loss $\Delta m = 66.63\%$ ($w/w$). Therefore, higher concentrations of the silica filler generally led to a higher thermal stability of the investigated rubber composites.

**Table 7.** Values of weight loss ($\Delta m$) of the investigated rubber composites containing different silica concentrations in the temperature range of 30 to 500 °C as the results of thermogravimetric analysis.

| Rubber Type | $\Delta m$ [% $w/w$] |
|---|---|
| S0 | 90.42 ± 0.40 |
| S10 | 83.29 ± 0.53 |
| S20 | 79.34 ± 0.43 |
| S30 | 73.33 ± 0.58 |
| S45 | 66.63 ± 0.83 |

## 4. Conclusions

Rubber composites are widely applied in many areas at the present time. It is often necessary to optimize rubber properties for given purposes. This research was focused on the study of mechanical, sound absorption and thermal properties of cellular rubber composites that were filled with different concentrations of silica nanofiller. Nevertheless, the tested rubber composites were produced with the same blowing agent concentration. In addition, the production of the rubber samples was carried out according to a new methodology in order to optimize the blowing and vulcanization processes of these samples. Based on this research, it can be concluded that the concentration of silica filler in the rubber composites had a significant influence on the mechanical stiffness of the rubbers, and thus on their sound absorption and thermal properties.

The mechanical properties examined included rebound resilience, compression testing and mechanical vibration damping properties of the investigated rubber composites. It was found that the mechanical stiffness of the rubber composites generally increased with increasing concentration of silica filler. The rebound resilience of the rubber composite containing 45 phr of silica was approximately 32% lower compared to the rubber composite containing 0 phr of silica. Similarly, the compression set of the rubber composites with 45 phr of silica was approximately 16% lower compared to those of the rubber composites with 0 phr of silica. These findings were in good agreement with vibration damping measurements that were performed using the non-destructive forced oscillation method based on the displacement transmissibility. It can be stated that the first resonance frequency peak position increased with an increase in mechanical stiffness leading to lower vibration damping properties of the rubber composites. Therefore, a higher concentration of silica filler in rubber composites generally led to a lower transformation of input mechanical energy into heat under dynamic loading of these rubbers. The mechanical properties of the investigated rubber composites were also in agreement with their sound absorption properties. It was found based on the sound damping measurements that higher concentrations of the loose silica powder led to higher values of the longitudinal elastic coefficient, and thus to a higher mechanical stiffness of the silica filler. Furthermore, sound absorption properties of the rubber composites generally decreased with increasing silica concentration. For this reason, a higher ability to damp sound was found for the rubber composites of more spongy-like structure, which is typical for rubbers containing lower concentrations of silica filler.

It was also found that higher concentrations of silica nanofiller generally led to a decrease in thermal degradation and an increase in thermal conductivity of the investigated rubber composites. Therefore, in practice, rubber composites containing lower concentrations of silica filler can be used as thermal insulators.

From the experimental data, it can be concluded that the mechanical stiffness of the tested silica-filled rubber composites can be also evaluated by the non-conventional methods of mechanical vibration damping and sound absorption. These methods are relatively simple, inexpensive, fast and non-destructive, which are their undeniable advantages compared to conventional methods used to determine the mechanical stiffness.

The investigated silica-filled rubber composites can be used in applications where higher shape recovery, stiffness-to-weight ratio and thermal conductivity are required, e.g., in the manufacture of rubber components in automotive, aerospace and footwear industries.

**Author Contributions:** Conceptualization, M.V. and M.P.; methodology, M.V. and M.P.; software, M.V. and M.P.; validation, M.V. and M.P.; formal analysis, M.V. and M.P.; investigation, M.V. and M.P.; resources, M.V. and M.P.; data curation, M.V. and M.P.; writing—original draft preparation, M.V. and M.P.; writing—review and editing, M.V. and M.P.; visualization, M.P.; supervision, M.V.; project administration, M.V. and M.P.; funding acquisition, M.V. and M.P. All authors have read and agreed to the published version of the manuscript.

**Funding:** This work was supported by the Ministry of Education, Youth and Sports of the Czech Republic—DKRVO (RP/CPS/2020/004); as well as with direct support of the European Regional Development Fund in the Research Centre of Advanced Mechatronic Systems project, project number CZ.02.1.01/0.0/0.0/16_019/0000867 within the Operational Programme Research, Development and Education.

**Institutional Review Board Statement:** Not applicable.

**Informed Consent Statement:** Not applicable.

**Data Availability Statement:** Not applicable.

**Acknowledgments:** The authors would like to thank the Ministry of Education, Youth and Sports of the Czech Republic for its direct support of this research through the project DKRVO (RP/CPS/2020/004), as well as the European Regional Development Fund in the Research Centre of Advanced Mechatronic Systems for its financial support of this research through the project number CZ.02.1.01/0.0/0.0/16_019/0000867 within the Operational Programme of Research, Development, and Education. Authors would like to express their special thanks to Ing. T. Valenta, and D. Měřínská, (both from Faculty of Technology, Tomas Bata University in Zlin) for thermal measurements.

**Conflicts of Interest:** The authors declare that they have no conflict of interest.

## References

1. Shizok, K. *Natural Rubber: From the Odyssey of the Hevea Tree to the Age of Transport*, 1st ed.; A Smithers Group Company: Shawbury, UK, 2015; pp. 3–4.
2. Gent, A.N.; Kawahara, S.; Zhao, J. Crystalization and strength of natural rubber and synthetic cis-1,4 polyisoprene. *Rubber Chem. Technol.* **1998**, *71*, 668–678. [CrossRef]
3. Parrella, F.W.; Gaspari, A.A. Natural rubber latex protein reduction with an emphasis on enzyme treatment. *Methods* **2002**, *27*, 77–86. [CrossRef]
4. Nie, Y.; Gu, Z.; Wei, Y.; Hao, T.; Zhou, Z. Features of strain-induced crystallization of natural rubber revealed by experiments and simulations. *Polym. J.* **2017**, *49*, 309–317. [CrossRef]
5. Chenal, J.-M.; Gauthier, C.; Chazeaur, L.; Guy, L.; Bomaly, Y. Parameters governing strain induced crystallization in filled natural rubber. *Polymer* **2007**, *48*, 6893–6901. [CrossRef]
6. Sarkawi, S.; Dierkes, W.K.; Noordermeer, J.W.M. Reinforcement of natural rubber by precipitated silica: The influence of processing temperature. *Rubber Chem. Tech.* **2014**, *87*, 103–119. [CrossRef]
7. Morton, M. *Rubber Technology*, 3rd ed.; Springer Science+Business Media: Dordrech, OH, USA, 1999; pp. 30–47.
8. Alan, N.G. *Engineering with Rubber—How to Design Rubber Components*, 3rd ed.; Hanscher Publischers: Ohio, OH, USA, 2012; pp. 18–20.
9. Thomas, S.; Stephen, R. *Rubber Nanocomposites Preparation, Properties and Applications*, 1st ed.; John Wiley & Sons Pte Ltd.: Singapore, 2010; pp. 21–39.
10. Hewitt, N. *Compositeing Precipitated Silica in Elastomers*, 1st ed.; William Andrew Publishing/Plastics Design Library: New York, NY, USA, 2007; pp. 1–9.
11. Cassagnau, P. Melt rheology of organoclay and fumed silica nanocomposites. *Polymer* **2008**, *49*, 2183–2196. [CrossRef]
12. Wilgis, T.A.; Heinrich, G.K. *Reinforcement of Polymer Nano-Composites—Theory, Experiments and Applications*, 1st ed.; Cambridge University Press: Cambridge, UK, 2009; pp. 96–100.
13. Leung, S.N.; Wong, A.; Park, C.B.; Zhong, J.H. Ideal surface geometries of nucleating agents to enhance cell nucleation in polymeric foaming processes. *J. Appl. Polym. Sci.* **2008**, *108*, 3997–4003. [CrossRef]
14. Arthur, H.L. *Handbook of Plastics Foams*, 1st ed.; Elsevier Books: Cambridge, UK, 1995; pp. 246–250.
15. Nijib, N.N.; Ariff, Z.M.; Manan, N.A.; Bakar, A.A.; Sipaut, C.S. Effect of blowing agent concentration on cell morphology and impact properties of natural rubber foam. *J. Phys. Sci.* **2009**, *20*, 13–25.
16. Gayathri, R.; Vasanthakumari, R.; Padmanabhan, C. Sound absorption, thermal and mechanical behavior of polyurethane foam modified with nano silica, nano clay and crumb rubber fillers. *Int. J. Sci. Eng. Res.* **2013**, *4*, 301–308.
17. Pang, Y.; Cao, Y.; Zheng, W.; Park, C.B. A comprehensive review of cell structure variation and general rules for polymer microcellular foams. *Chem. Eng. J.* **2021**, *430*, 132662. [CrossRef]
18. Park, C.B.; Baldwin, D.F.; Suh, N.P. Effect of the pressure drop rate on cell nucleation in continuous processing of microcellular polymers. *Polym. Eng. Sci.* **1995**, *35*, 432–440. [CrossRef]
19. Bayat, H.; Fasihi, M.; Zare, Y.; Rhee, K.Y. An experimental study on one-step and two-step foaming of natural rubber/silica nanocomposites. *Nanotechnol. Rev.* **2020**, *9*, 427–435. [CrossRef]
20. Sun, X.; Zhang, G.; Zhang, G.; Shi, Q.; Tang, B.; Wu, Z. Study on foaming water-swellable EPDM rubber. *J. App. Polym. Sci.* **2002**, *84*, 3712–3717. [CrossRef]
21. Panploo, K.; Chalermsinsuwan, B.; Poompradus, S. Natural rubber latex foam with particulate fillers for carbon dioxide adsorption and regeneration. *RSC Adv.* **2019**, *9*, 28916–28923. [CrossRef]
22. Sombatsompop, N.; Thongsang, S.; Markpin, T.; Wimolmala, E. Fly ash particles and precipitated silica as fillers in rubbers. I. Untreated fillers in natural rubber and styrene–butadiene rubber compounds. *J. Appl. Polym. Sci.* **2004**, *93*, 2119–2130. [CrossRef]
23. Rodriguez-Vear, R.; Genovese, K.; Rayas, J.A.; Mendoza-Santoyo, F. Vibration Analysis at Microscale by Talbot Fringe Projection Method. *Strain* **2009**, *45*, 249–258. [CrossRef]
24. Prasopdee, T.; Smitthipong, W. Effect of fillers on the recovery of rubber foam: From theory to applications. *Polymer* **2020**, *12*, 2745. [CrossRef]
25. Rao, S.S. *Mechanical Vibrations*, 5th ed.; Pearson Education, Inc.: Upper Saddle River, NJ, USA, 2011; pp. 281–287.
26. Carrella, A.; Brennan, M.J.; Waters, T.P.; Lopes, V., Jr. Force and displacement transmissibility of a nonlinear isolator with high-static-low-dynamic-stiffness. *Int. J. Mech. Sci.* **2012**, *55*, 22–29. [CrossRef]
27. Liu, K.; Liu, J. The damped dynamic vibration absorbers: Revisited and next result. *J. Sound Vib.* **2005**, *284*, 1181–1189. [CrossRef]
28. Stephen, N. On energy harvesting from ambient vibration. *J. Sound Vib.* **2006**, *293*, 409–425. [CrossRef]
29. Vette, A.H.; Wu, N.; Masani, K.; Popovic, M.R. Low-intensity functional electrical stimulation can increase multidirectional trunk stiffness in able-bodied individuals during sitting. *Med. Eng. Phys.* **2015**, *37*, 777–782. [CrossRef] [PubMed]
30. Sgard, F.; Castel, F.; Atalla, N. Use of a hybrid adaptive finite element/modal approach to assess the sound absorption of porous materials with meso-heterogeneities. *Appl. Acoust.* **2011**, *72*, 157–168. [CrossRef]
31. Tiwari, W.; Shukla, A.; Bose, A. Acoustic properties of cenosphere reinforced cement and asphalt concrete. *Appl. Acoust.* **2004**, *65*, 263–275. [CrossRef]
32. Okudaira, Y.; Kurihara, Y.; Ando, H.; Satoh, M.; Miyanami, K. Sound absorption measurements for evaluating dynamic physical properties of a powder bad. *Powder Tech.* **1993**, *77*, 39–48. [CrossRef]

33. Yanagida, T.; Matchett, A.J.; Coulthard, J.M.; Asmar, B.N.; Langston, P.A.; Walters, J.K. Dynamic measurements for the stiffness of loosely packed powder beds. *AIChE J.* **2002**, *48*, 2510–2517. [CrossRef]
34. International Organization for Standardization. *ISO 10534-2, Acoustics-Determination of Sound Absorption Coefficient and Impedance in Impedance Tubes-Part 2: Transfer-Function Method*; ISO/TC 43/SC2 Building Acoustics; CEN, European Committee for Standardization: Brussels, Belgium, 1998; pp. 10534–10542.
35. Han, F.S.; Seiffert, G.; Zhao, Y.Y.; Gibbs, B. Acoustic absorption behaviour of an open-celled alluminium foam. *J. Phys. D Appl. Phys.* **2003**, *36*, 294–302. [CrossRef]
36. Júnior, J.H.S.A.; Júnior, H.L.O.; Amico, S.C.; Amado, F.D.R. Study of hybrid intralaminate curaua/glass composites. *Mater. Des.* **2012**, *42*, 111–117. [CrossRef]
37. Rajoria, H.; Jalili, N. Passive Vibration Damping Enhancement using Carbon Nanotube-epoxy Reinforced Composites. *Compos. Sci. Tech.* **2005**, *65*, 2079–2093. [CrossRef]
38. Kalauni, K.; Pawar, S.J. A review on the taxonomy, factors associated with sound absorption and theoretical modeling of porous sound absorbing materials. *J. Porous Mater.* **2019**, *26*, 1795–1819. [CrossRef]

MDPI AG
Grosspeteranlage 5
4052 Basel
Switzerland
Tel.: +41 61 683 77 34

*Materials* Editorial Office
E-mail: materials@mdpi.com
www.mdpi.com/journal/materials

Disclaimer/Publisher's Note: The title and front matter of this reprint are at the discretion of the Guest Editor. The publisher is not responsible for their content or any associated concerns. The statements, opinions and data contained in all individual articles are solely those of the individual Editor and contributors and not of MDPI. MDPI disclaims responsibility for any injury to people or property resulting from any ideas, methods, instructions or products referred to in the content.

www.ingramcontent.com/pod-product-compliance
Lightning Source LLC
LaVergne TN
LVHW072331090526
838202LV00019B/2400